Quantum Inspired Meta-heuristics for Image Analysis

Quantum Inspired Meta-heuristics for Image Analysis

Sandip Dey
Global Institute of Management and Technology
Krishnanagar, Nadia, West Bengal
India

Siddhartha Bhattacharyya
RCC Institute of Information Technology
Kolkata
India

Ujjwal Maulik
Jadavpur University
Kolkata
India

The right of Sandip Dey, Siddhartha Bhattacharyya, and Ujjwal Maulik to be identified as the authors of this work has been asserted in accordance with law.

Registered Offices
John Wiley & Sons, Inc., 111 River Street, Hoboken, NJ 07030, USA
John Wiley & Sons Ltd, The Atrium, Southern Gate, Chichester, West Sussex, PO19 8SQ, UK

Editorial Office
The Atrium, Southern Gate, Chichester, West Sussex, PO19 8SQ, UK

For details of our global editorial offices, customer services, and more information about Wiley products visit us at www.wiley.com.

Wiley also publishes its books in a variety of electronic formats and by print-on-demand. Some content that appears in standard print versions of this book may not be available in other formats.

Library of Congress Cataloging-in-Publication Data

Names: Dey, Sandip, 1977- author. | Bhattacharyya, Siddhartha, 1975- author.
  | Maulik, Ujjwal, author.
Title: Quantum Inspired Meta-heuristics for Image Analysis / Dr. Sandip Dey
  (Global Institute of Management and Technology Krishnanagar, Nadia, West Bengal, India),
  Professor Siddhartha Bhattacharyya (RCC Institute of Information Technology, Kolkata, India),
  Professor Ujjwal Maulik (Jadavpur University, Kolkata, India).
Description: Hoboken, NJ : Wiley, 2019. | Includes bibliographical references
  and index. |
Identifiers: LCCN 2019001402 (print) | LCCN 2019004054 (ebook) | ISBN
  9781119488774 (Adobe PDF) | ISBN 9781119488781 (ePub) | ISBN 9781119488750
  (hardcover)
Subjects: LCSH: Image segmentation. | Image analysis. | Metaheuristics. |
  Heuristic algorithms.
Classification: LCC TA1638.4 (ebook) | LCC TA1638.4 .D49 2019 (print) | DDC
  006.4/2015181–dc23
LC record available at https://lccn.loc.gov/2019001402

Cover Design: Wiley
Cover Image: © issaro prakalung/Shutterstock

Set in 10/12pt WarnockPro by SPi Global, Chennai, India
Printed and bound in Singapore by Markono Print Media Pte Ltd

10 9 8 7 6 5 4 3 2 1

Sandip Dey would like to dedicate this book to the loving memory of his father, the late Dhananjoy Dey, his mother Smt. Gita Dey, his wife Swagata Dey Sarkar, his children Sunishka and Shriaan, his siblings Kakali, Tanusree, Sanjoy and his nephews Shreyash and Adrishan.

Siddhartha Bhattacharyya would like to dedicate this book to his father, the late Ajit Kumar Bhattacharyya, his mother, the late Hashi Bhattacharyya, his beloved wife Rashni Bhattacharyya, and Padmashree Professor Ajoy Kumar Ray, Honorable Chairman, RCC Institute of Information Technology, Kolkata, India.

Ujjwal Maulik would like to dedicate this book to his son Utsav, his wife Sanghamitra and all his students and friends.

Contents

Preface

In the present information era, the processing and retrieval of useful image information and multimedia-based data, for the purpose of faithful and realistic analysis, are supposed to be of the highest importance. One significant image processing chore is to separate objects or other important information in digital images through thresholding of the image under consideration. Efficient techniques are required in order to develop an appropriate analysis of noisy and noise-free image data to obtain suitable object-specific information.

The *soft computing approaches* have certain tools and techniques among various other approaches, which integrate intelligent thinking and principles. Fuzzy logic, Neural networks, Fuzzy sets, and Evolutionary Computation are used as the computing framework, which successfully combines these intelligent principles.

This book attempts to address the problem of image thresholding using classical algorithms. Attempts have also been made to take out the intrinsic limitations in the present soft computing methods as initiated by theoretical investigations. New versions of quantum inspired meta-heuristics algorithms have also been introduced, taking into cognizance the time and space complexity of present approaches.

The introductory chapter of the book presents a brief summary on image analysis, quantum computing, and optimization. The chapter highlights quantum solutions of NP-complete problems in brief. This introductory chapter also presents a related literature survey using different approaches, such as *quantum-based approaches, meta-heuristic-based approaches* and *multi-objective-based approaches*. The chapter also discusses the scope and organization of the book.

Chapter 2 focuses on the review of image analysis. This chapter discusses the mathematical formalism of image segmentation technique. It also highlights different digital image analysis approaches. It also throws light on popular image thresholding techniques in the binary, multi-level and gray-scale domains. A short summary of the applications of image analysis is also presented. Finally, the chapter ends with a relevant conclusion; a chapter summary and a set of exercise questions are provided.

Chapter 3 focuses on the overview of some popular meta-heuristics. The fundamentals of each of them are briefly discussed in this chapter. Pseudo-code of the corresponding meta-heuristic is also presented after each section. The summary of the chapter and a set of exercise questions are provided at the end of this chapter.

Chapter 4 addresses the intrinsic limitations of classical algorithms to deal with image data for binary thresholding. The chapter develops two different quantum-based standard conventional algorithms for binary image thresholding. The basic quantum computing principles have been recognized to develop the proposed approaches. The features of quantum computing have been properly linked with the framework of popular classical algorithms for the formation of the proposed algorithms. Experiments have been conducted with different combinations of the parameter settings. The proposed algorithms are compared with several other algorithms. The implementation results are presented for the proposed algorithms and other comparable algorithms.

In line with the objectives of Chapter 4, several novel versions of quantum inspired classical algorithms are proposed in Chapter 5. This chapter concentrates on the functional modification of the quantum inspired meta-heuristic algorithms as an attempt to extend them to multi-level and gray-scale domain. Application of the proposed algorithms is demonstrated on a set of synthetic/real-life gray-scale images. As a sequel to these algorithms, experiments were conducted with several other algorithms for comparative purposes. The experiments were conducted with different parameter values. Implementation results are reported for all the participating algorithms.

Parallel extensions to these quantum inspired classical algorithms are presented in Chapter 6. This approach introduces thresholding of color image information. The parallel operation of the proposed framework reduces the time complexity of the color image thresholding. Application of the proposed versions of quantum inspired algorithms is exhibited using thresholding of multi-level and color images. As a result of the comparative study, the proposed algorithms are compared with other popular algorithms. Implementation results with several parameter adjustments are reported.

Chapter 7 introduces several quantum inspired multi-objective algorithms using different approaches. First, an NSGA-II-based quantum inspired algorithm is proposed in a multi-objective framework. Later, several quantum inspired classical algorithms are developed in a multi-objective flavor for bi-level, multi-level and gray-scale image thresholding.

Different parameters settings are used for the clarification of proposed algorithms. Application of these algorithms is demonstrated on several real-life grayscale images. A number of other popular algorithms are used for comparative purposes. The test results are reported for all of the participating algorithms.

Finally, the concluding chapter ends the book. This chapter presents an outlook of future directions of research in this area.

Sodepur

2 November 2018

Sandip Dey
Siddhartha Bhattacharyya
Ujjwal Maulik

Acronyms

ACO	Ant Colony Optimization
AI	Artificial Intelligence
BSA	Backtracking Search Optimization Algorithm
CNOT	Controlled NOT Gate
CoDE	Composite DE
DE	Differential Evolution
EA	Evolutionary Algorithm
EC	Evolutionary Computation
EP	Evolutionary Programming
ES	Evolutionary Strategies
GA	Genetic Algorithm
GP	Genetic Programming
HDQ	Heuristic Search Designed Quantum Algorithm
HV	Hypervolume
IGD	Inverted Generational Distance
MA	Metropolis Algorithm
MADS	Mesh Adaptive Direct Search
MBF	Modified Bacterial Foraging
MDS	Multidirectional Search
MODE	Multi-Objective Differential Evolutionary Algorithm
MOEA	Multi-Objective Evolutionary Algorithm
MOGA	Multi-Objective Genetic Algorithm
MOO	Multi-Objective Optimization
MOSA	Multi-Objective Simulated Annealing
MRI	Magnetic Resonance Imaging
MTT	Maximum Tsallis entropy Thresholding
NM	Nelder-Mead
NPGA	Niched-Pareto Genetic Algorithm
NSGA-II	Non-dominated Sorting Genetic Algorithm II
NSGA	Non-dominated Sorting Genetic Algorithm
OCEC	Organizational Coevolutionary algorithm for Classification
PAES	Pareto Archived Evolution Strategy
PESA	Pareto Envelope-based Selection Algorithm
PET	Positron Emission Tomography
PO	Pareto Optimal

PSNR	Peak Signal-to-Noise Ratio
PSO	Particle Swarm Optimization
QC	Quantum Computer
QEA	Quantum Evolutionary Algorithm
QFT	Quantum Fourier Transform
QIACO	Quantum Inspired Ant Colony Optimization
QIACOMLTCI	Quantum Inspired Ant Colony Optimization for Color Image Thresholding
QIDE	Quantum Inspired Deferential Evolution
QIDEMLTCI	Quantum Inspired Deferential Evolution for Color Image Thresholding
QIEA	Quantum Inspired Evolutionary Algorithms
QIGA	Quantum Inspired Genetic Algorithm
QIGAMLTCI	Quantum Inspired Genetic Algorithm for Color Image Thresholding
QIMOACO	Quantum Inspired Multi-objective Ant Colony Optimization
QIMOPSO	Quantum Inspired Multi-objective Particle Swarm Optimization
QINSGA-II	Quantum Inspired NSGA-II
QIPSO	Quantum Inspired Particle Swarm Optimization
QIPSOMLTCI	Quantum Inspired Particle Swarm Optimization for Color Image Thresholding
QISA	Quantum Inspired Simulated Annealing
QISAMLTCI	Quantum Inspired Simulated Annealing for Color Image Thresholding
QISAMO	Quantum Inspired Simulated Annealing for Multi-objective algorithms
QITS	Quantum Inspired Tabu Search
QITSMLTCI	Quantum Inspired Tabu Search for Color Image Thresholding
RMSE	Root mean-squared error
SA	Simulated Annealing
SAGA	Simulated Annealing and Genetic Algorithm
SC	Soft Computing
SOO	Single-Objective Optimization
SPEA2	Strength Pareto Evolutionary Algorithm 2
SPEA	Strength Pareto Evolutionary Algorithm
SVM	Support Vector Machines
TS	Tabu Search
TSMO	Two-Stage Multithreshold Otsu
TSP	Travelling Salesman Problems
VEGA	Vector-Evaluated Genetic Algorithm
X-ray	X-radiation

1

Introduction

A quantum computer, as the name suggests, fundamentally works on several quantum physical characteristics. It is also considered as the field of study, primarily focused on evolving computer technology using the features of quantum theory, which expounds the nature of energy and substance and its behavior on the quantum level, i.e., at the atomic and subatomic level. Developing a quantum computer would mark an advance in computing competency far superior to any current computers. Thus, the use of quantum computers could be an immense improvement on current computers because they have enormous processing capability, even exponentially, compared to classical computers. The supremacy of processing is gained through the capacity of handling multiple states at once, and performing tasks exploiting all the promising permutations in chorus. The term *quantum computing* is fundamentally a synergistic combination of thoughts from quantum physics, classical information theory, and computer science.

Soft computing (SC), introduced by Lotfi A. Zadeh [282], manages the soft meaning of thoughts. SC, comprising a variety of thoughts and practices, is fundamentally used to solve the difficulties stumbled upon in real-life problems. This can be used to exploit the uncertainty problem almost with zero difficulty. This can also handle real-world state of affairs and afford lower solution costs [29]. The advantageous features of SC can best be described as leniency of approximation, vagueness, robustness, and partial truth [103, 215]. This is a comparatively novel computing paradigm which involves a synergistic amalgamation of essentially several additional computing paradigms, which may include fuzzy logic, evolutionary computation, neural networks, machine learning, support vector machines, and also probabilistic reasoning. SC can combine the afore-mentioned computing paradigms to offer a framework for designing many information processing applications that can function in the real world. This synergism was called computational intelligence by Bezdek [24]. These SC components are different from each other in more than one way. These can be used to operate either autonomously or conjointly, depending on the application domain.

Evolutionary computation (EC) is a search and optimization procedure which uses biological evolution inspired by Darwinian principles [14, 83, 136]. It is stochastic and delivers robust search and optimization methods. It starts with a pool of trial solutions in its search space, which is called the *population*. Numerous in-built operators are generally applied to each individual of the population, which may cause population diversity and also leads to better solutions. A metric, called the fitness function (objective function), is employed to determine the suitability of an individual in the population at any

Quantum Inspired Meta-heuristics for Image Analysis, First Edition.
Sandip Dey, Siddhartha Bhattacharyya, and Ujjwal Maulik.
© 2019 John Wiley & Sons Ltd. Published 2019 by John Wiley & Sons Ltd.

particular generation. As soon as the fitness of the existing individuals in the population is computed, the operators are successively applied to produce a new population for the successive generations. Distinct examples of EC may include the Differential Evolution [242], Genetic Algorithms [127, 210], Particle Swarm Optimization [144], and Ant Colony Optimization [196], to name but a few. Simulated annealing [147] is another popular example of meta-heuristic and optimization techniques in this regard. This technique exploits the features of statistical mechanics concerning the behavior of atoms at very low temperature to find minimal cost solutions of any given optimization problem. EC techniques are also useful when dealing with several conflicting objectives, called the multi-objective evolutionary techniques. These search procedures provide a set of solutions, called optimal solutions. Some typical examples of these techniques may include the multi-objective differential evolutionary algorithm (MODE) [275], the multi-objective genetic algorithm (MOGA) [172, 183], and multi-objective simulated annealing (MOSA) [237], to name but a few.

Fuzzy logic tenders more elegant alternatives to conventional (Boolean) logic. Fuzzy logic is able to handle the notion of partial truth competently [139, 141, 215, 282, 283]. A neural network is a computing framework comprising huge numbers of simple, exceedingly unified processing elements called artificial neurons, which add up to an elemental computing primitive [82, 102, 150]. Machine learning is a kind of intelligent program which works on example data. It learns from previous experiences and is used to enhance the performances by optimizing a given criterion [5, 156, 178]. Support vector machines (SVM) are known to be the collection of supervised learning techniques. SVMs are very useful in regression and classification analysis [38, 50]. SVMs are fit to handle a number of real-life applications, including text and image classification, or biosequences analysis, to name but a few [38, 50]. Nowadays SVMs are often used as the standard and effective tool for data mining and machine learning activities. Probabilistic reasoning can be defined as the computational method which uses certain logic and probability theory to handle uncertain circumstances [201, 202].

Many researchers utilize the basic features of quantum computing in various evolutionary algorithmic frameworks in the soft computing discipline. The underlying principles of quantum computing are injected into different meta-heuristic structures to develop different quantum inspired techniques. In the context of image analysis, the features are extracted both from pictographic and non-numeric data and are used in these algorithms in different ways [27]. This chapter provides an insight into the various facets of the *quantum computing, image segmentation, image thresholding, and optimization*. This chapter is arranged into a number of relevant sections. Section 1.1 presents an overview of the underlying concepts of image analysis. A brief overview of image segmentation and image thresholding is discussed in this section. Section 1.2 throws light on the basics of quantum computing in detail. Section 1.3 discusses the necessity of optimization in the real world. This section presents different types of optimization procedures with their application in the real world. Apart from the above issues, this chapter also presents a short description of the literature survey on related topics. Different types of approaches in this regard are in detail presented in Section 1.4. The organization of the book is presented in Section 1.5. The chapter concludes in Section 1.6. It also shows the direction of research that can be used as future reference. A brief summary of the chapter is given in Section 1.7. In Section 1.8, a set of questions related to the theme of the chapter is presented.

1.1 Image Analysis

Image analysis has a vital role in extracting relevant and meaningful information from images. There are few automatic or semi-automatic techniques, called *computer/machine vision, pattern recognition, image description, image understanding* to name but a few, used for this purpose. Image segmentation can be thought of as the most fundamental and significant step in several image analysis techniques. A good example of image analysis may involve the organized activities of the human eye with the brain. Computer-based image analysis can be thought of as the best alternative which may reduce human effort in order to make this process faster, more efficient, and automatic. Image analysis has numerous applications in a variety of fields such as medicine, biology, robotics, remote sensing, and manufacturing. It also makes a significant contribution in different industrial activities such as process control, quality control, etc. For example, in the food industry, image analysis plays a significant role to ensure the uniform shape, size and texture of the final food products.

In medical image analysis, clinical images of different views are captured to diagnose and detect diseases in relation to body organs, and study standard physiological procedures for future references. These investigations can be accomplished through images attained from various imaging technologies, such as magnetic resonance imaging (MRI), radiology, ultrasound, etc. For example, image analysis methodology is of the utmost importance in cancer detection and diagnosis [44], thus it helps the physician to ensure accurate treatment for their patient. In the context of cancer treatment, several features like shape, size, and homogeneity of a tumor are taken into consideration when classifying and diagnosing cancer images. Different image analysis algorithms can be introduced that can help radiologists to classify tumor images.

The steps involved in image analysis are presented in Figure 1.1 [112]. Each step is discussed in the following in brief.

1. *Image acquisition*: This is the first step of every vision system. Image acquisition means acquiring a digital image. After obtaining the image successfully, several processing approaches can be used on the image in order to fulfill the different vision

Figure 1.1 Steps in image analysis.

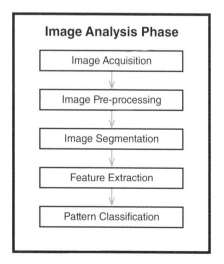

tasks required nowadays. However, if the image cannot be acquired competently, the anticipated tasks may perhaps not be completed successfully by any means.

2. *Image pre-processing*: This step involves improving the image to be suitable for analysis. In this phase, the quality of image is improved by introducing different techniques, such as contrast enhancement, noise contraction, and sharpening of image. The output image can be sent to perform the next step.

3. *Image segmentation*: In this step, the image is partitioned into several regions and the regions of interest are taken out from the input image.

4. *Feature extraction*: This step converts the input image to a number of features on the basis of the traits of the segmented image. Resulting from the discovery of certain facts some data are obtained as the output of this step.

5. *Pattern classification*: This is the final step of image analysis. The extracted features obtained in the last phase are used to classify the given image.

Various techniques can be applied to execute the steps in image analysis depending on the intended application. So, the selected technique for performing each step is of the utmost importance to achieve the desired results from the proposed algorithm.

1.1.1 Image Segmentation

Segmentation separates the patterns into a number of uniform, non-overlapping and homogeneous regions or segments (classes), in which the members of any particular segment are similar to each other and the members in the different segments possess dissimilarity among themselves [9, 22, 87, 133, 166, 194, 203, 252, 258]. The patterns carrying the similar features are known to be clusters/segments. Segmentation is equally effective for localizing and identifying object-specific features both from nonnumeric and pictorial datasets. The challenges lie in the attempt to emulate human perception and intelligence to extract underlying objects accurately. The foremost objective of the segmentation process is to detect the pertinent and meaningful data by taking out the redundant part embedded within. The foundation of a segmentation technique is basically contingent on the assortment of the representation of data elements, the proximity measurement between them and also their grouping. Thus, certain metrics are commonly used for measuring the similarity or difference between the patterns. So far, segmentation has been successfully applied in diverse fields, which may include different engineering disciplines like electrical engineering, mechanical engineering, and others. Apart from that, it has also been widely used in various other fields, such as remote sensing, machine learning, robotic vision, pattern recognition, artificial intelligence, economics, medical sciences, and many others.

Formally, the term "Image segmentation" is defined as follows:

1) $\mathcal{I} = \bigcup_{i=1}^{p} R_i$ (1.1)

2) $R_m \cap R_n = \phi, m \neq n$ (1.2)

where it is assumed that an image, $\mathcal{I}$ is segmented into p number of regions, namely, $R_1, R_2, \ldots, R_p$ [153, 162, 193]. A comprehensive exploration of diverse segmentation methods is available in the literature [27, 97, 231]. Of late, color image segmentation has become a trusted addition in many areas of application [112, 152, 229, 248, 264]. A color pixel is typically manifested as a mixture of different color constituents. The synthesis of different color constituents in color images usually enhances an enormous amount of intrinsic computational complexities in respect of color image processing. As a result, it becomes more challenging to take them up in real-life situations. Some typical examples of color image segmentation may include robotics, object recognition, and data compression to name but a few [27]. A pixel in a color image is recognized in multidimensional color space, which basically enhances the processing complexity in real-life applications. Compared to image segmentation in monochrome image, a higher number of parameters is required to be tuned for optimality in color image segmentation [27].

1.1.2 Image Thresholding

Thresholding is well recognized and probably the most effective tool in the context of image processing (image segmentation) and pattern recognition. From an implementation point of view, it can be considered as the simplest technique among others, which generally provides the most accurate results. Thresholding is basically used in segregating the background and foreground information of the image. This technique is very effective in ascertaining the dissimilar homogeneous components (gray value, color) of the image [97]. Thresholding is equally effective for the images possessing nonhomogeneous components.(textured images).

The threshold can be found by using two approaches, namely, parametric and nonparametric [3, 261, 277]. In the first approach, the distribution of dissimilar gray levels of an object class guides the location of the thresholds. For example, Wang and Haralick [268] used a parametric approach where they divided the pixels of an image into two categories, namely, edge and non-edge pixels. Thereafter, in consonance with the local neighborhoods, the edge pixels are re-classified into two groups, referred to as relatively dark and relatively bright pixels. Afterwards, two histograms are individually and successively drawn from the pixels of each group. The highest peaks are selected from these histograms as the thresholds. Another popular parametric approach is popularly known as moment preserving thresholding, in which the image is segmented on the basis of the condition that the original and thresholded image must have the identical moments [261]. In the latter approach, the concept of optimality is used to find the threshold values, where the threshold divides the gray-level regions of an image on the basis of certain discerning criteria, such as the entropy, cross-entropy, within or between class variance, so on. Typical examples of nonparametric approaches may include Otsu's method [192], Pun's method [206], Kapur's method [140], to name but a few. Otsu's method [192] is clustering-based, where the optimal threshold values are selected by maximizing the between-class variance with a comprehensive search mechanism. Pun's method [206] and Kapur's method [140] are two entropy-based methods, where the gray levels are classified into different classes, and the threshold value is obtained by maximizing the entropy of the histograms of the members of that class. As the consequence of the extensive research over the last few years, a plethora of robust thresholding methods of the parametric or nonparametric type are now available in the literature [220].

Sezgin et al. [231] have presented a comprehensive survey of various thresholding methods.

Any thresholding method is usually of two types: bi-level thresholding and multi-level thresholding. In bi-level thresholding, the image is basically divided into two components, foreground (object) and background, containing different gray-level distribution. Hence, pixels of an image are grouped into two classes in this thresholding method. The pixels with gray levels above a certain threshold are kept in one group, and the other group is formed with the rest of the pixels of that image. The bi-level image thresholding can be computationally extended to its multi-level version, where pixels of an image are segregated into several classes with specific ranges defined by several thresholds.

In both classical and intelligent approaches, the purpose of thresholding also includes reducing the image information complexity by transforming it into monochrome versions, thereby allowing a faithful analysis of the image scene [112, 132, 204, 225, 229]. Thresholding is basically applied to discriminate objects from the background image in an efficient manner. In addition, it can also be used to separate objects from images which comprise of different gray-levels [230, 284]. On the basis of the number of threshold values selected, the thresholding method can be categorized as follows.

1. *Bi-level thresholding*: In this category, the pixel intensity values of the image are grouped into two classes. This kind of thresholding method accepts the gray level/color image as the input and converts it to its corresponding binary image output. The conversion is accomplished on the basis of a predefined pixel intensity value, called the threshold. Based on some criteria, one threshold value as the pixel is chosen from the image, which in turn divides the pixels of the image into two groups. These groups are referred to as object (O) (sometimes called foreground) and background (B). Generally, the pixel intensity values in group (O) are greater than the threshold while the group (B) contains smaller pixels than the threshold or vice versa [220]. To conclude, each element in (O) and (B) is set to be 1 (white) and 0 (black), respectively [278]. Theoretically, for an image (I) and its corresponding threshold value ( ), the subsequent features must be satisfied [25, 41, 251, 261, 278]:

   1) $I \in \{0, 1, 2, \ldots, L-1\}$ $\qquad$ (1.3)

   2) $O = \{I | I > \theta\}$ and $B = \{I | I \leq \theta\}$ $\qquad$ (1.4)

2. *Multi-level thresholding*: When the number of classes of pixels exceeds two, it is called multi-level image thresholding. In this kind of thresholding, multiple number of threshold values as pixels are selected. In this category, the number of groups yielded is one more than the number of threshold selected for image thresholding. As a higher level of thresholding may necessitate more calculations, the time complexity of algorithms increases proportionally with the increase of level of thresholding in multi-level thresholding [10, 39, 119, 120, 279]. This could cause significant difficulties especially when higher level threshold values are required to be evaluated. Hence, multi-level image thresholding possesses more complexity compared to the other one. There exists dissimilar algorithms for bi-level image thresholding in the literature, which can be extended to their respective multi-level versions, if required [120, 185]. Although image thresholding, as stated above, results in a monochrome image, of late, researchers have resorted to multi-level

thresholding [161] to generate multi-point thresholding for the faithful segmentation of gray-level images.

Both of these image thresholding versions can be identified by acclimatizing parametric or nonparametric approaches [161, 186]. So far, different algorithms have been designed to fulfill different purposes. Some distinctive applications of thresholding comprise document image analysis [2], image segmentation [274], or nondestructive quality inspection of materials [230].

Several classical methods [112, 231, 258] have been introduced to attain an apposite thresholding criterion for gray-level images. A score of approaches have been developed to address the problem of image thresholding [231]. A few distinguished methods among them are as follows:

1. *Shape-based methods*: In this category, the peaks, valleys, and curvatures of the smoothed histogram are analyzed [213].
2. *Clustering-based methods*: Here, the gray-level samples are clustered in two sections as background and foreground [192].
3. *Entropy-based methods*: In entropy-based methods, the entropy of the foreground and background regions are used to determine thresholds [158]
4. *Object attribute-based methods*: These methods aim to discover a similarity measure between the gray-level and its binary version. This similarity measure may include edge coincidence, fuzzy shape resemblance, etc.
5. *Spatial methods*: These kinds of methods usually use higher-order probability distribution and/or correlation between pixels [35].
6. *Local methods*: This method acclimatizes the threshold value on every individual pixel to the local image features.

Among the different soft computing approaches in this direction, either a deterministic analysis of the intensity distribution of images or heuristic search and optimization techniques are most extensively used [36, 90, 186, 211]. A survey of classical and non-classical techniques for image thresholding and segmentation is available in [27]. But the inherent problem of these optimization techniques lies in their huge time complexities.

1.2 Prerequisites of Quantum Computing

A quantum computer (QC), as the name implies, fundamentally works on quite a few quantum physical features. In contrast to classical computers, a QC has a faster processing capability (even exponentially), hence, these can be thought of as an immense alternative to today's classical computers. The field of quantum computing [173] has developed to provide a computing speed-up of the classical algorithms by inducing physical phenomena such as, superposition, entanglement, etc. It entails these thoughts to develop a computing paradigm much faster in comparison to the conventional computing. The upsurge in the processing speed is acquired by dint of exploiting the inherent parallelism perceived in the qubits, the building blocks of a quantum computer [57]. Hence, the term *quantum computing* can be primarily considered as a synergistic amalgamation of concepts from quantum physics, computer science, and classical information theory.

Deutsch and Jozsa [58] and, later, Shor [235] exhibited some concrete problems where they proved that such speed-up is possible. The basics of a few QC properties are addressed in the following subsections.

1.2.1 Dirac's Notation

The state of a quantum system is basically described in a complex divergent Hilbert space, symbolized by the notation $\mathcal{H}$. The utility of the "bra-ket" notation is very noteworthy in quantum mechanics. Paul Dirac introduced this standard notation in 1939, and hence, this notation is at times popularly known as Dirac's Notation. The "ket" vector is symbolized as $|\psi\rangle$ and, its Hermitian conjugate (conjugate transpose), referred to as "bra", is denoted by $\langle\phi|$. The "bracket" is formed by uniting these two vectors, and is represented by $\langle\phi|\psi\rangle$ [30, 77, 121].

Formally, a quantum system can be described as follows:

$$|\psi\rangle = \sum_j c_j |\phi_j\rangle \tag{1.5}$$

where, $|\psi\rangle$ is a wave function in $\mathcal{H}$. $|\psi\rangle$ acts as a linear superposition encompassing the basic states ϕ_j. c_j are the complex numbers which satisfies the following equation, as given by

$$\sum_j c_j^2 = 1 \tag{1.6}$$

1.2.2 Qubit

The quantum bit, or in short, qubit can be described as the basic constituent of information in quantum computing. Basically, a qubit state can be represented as a unit vector, defined in 2D complex vector space. Theoretically, a quantum bit can possess an inestimable number of basic states as required which help to provide exponentially augmented information in QC. The basic quantum states are labeled as $\{|0\rangle, |1\rangle\}$, where,

$$|0\rangle = \begin{bmatrix} 1 \\ 0 \end{bmatrix} \text{ and } |1\rangle = \begin{bmatrix} 0 \\ 1 \end{bmatrix} \tag{1.7}$$

Occasionally, $|0\rangle$ and $|1\rangle$ are referred to as "ground state" and "excited state", respectively.

1.2.3 Quantum Superposition

The quantum superposition principle, which expresses the idea that a system can exist simultaneously in two or more mutually exclusive states, is at the heart of the mystery of quantum mechanics. Considering the two state vectors in QC, the superposition between the states is represented by the equation $|\psi\rangle = \alpha|0\rangle + \beta|1\rangle$ where, $(\alpha, \beta) \in \mathbb{C}$ and $|\alpha|^2 + |\beta|^2 = 1$. The superposed quantum states are forced to be collapsed into a single state for quantum measurement. The probability of transforming it into the state $|1\rangle$ is $|\alpha|^2$ and that of $|0\rangle$ is $|\beta|^2$, respectively [30, 77, 173, 266].

1.2.4   Quantum Gates

The quantum gates (sometimes called quantum logic gates) are usually hardware tools which operate on preferred qubits using a fixed unitary operation over a predetermined period of time. Thus, quantum gates are reversible, which means, for n number of inputs, there must be n number of outputs. Some typical examples of quantum gates are NOT gate, C-NOT (Controlled-NOT) gate, controlled phase-shift gate, Hadamard gate, Toffoli gate, and Fredkin gate. Theoretically, for the unitary operator, U, the following equations must hold:

$$U^+ = U^{-1} \text{ and } UU^+ = U^+U = I \tag{1.8}$$

For the Hermitian operator, H

$$U = e^{iHt} \tag{1.9}$$

Quantum gates can be categorized into three categories:

1. One-qubit quantum gates.
2. Two-qubit quantum gates.
3. Three-qubit quantum gates.

A brief summary of popular quantum gates, of each category, is presented below.

1.2.4.1   Quantum NOT Gate (Matrix Representation)

In general, the matrix representation of a quantum gate can be given as

$$\sum_j |input_j\rangle\langle output_j|$$

In this gate, for input $|0\rangle$, output will be $\langle 1|$, and for input $|1\rangle$, output will be $\langle 0|$. The quantum NOT gate can be represented as

$$= |0\rangle\langle 1| + |1\rangle\langle 0| = \begin{bmatrix} 1 \\ 0 \end{bmatrix} \begin{bmatrix} 0 & 1 \end{bmatrix} + \begin{bmatrix} 0 \\ 1 \end{bmatrix} \begin{bmatrix} 1 & 0 \end{bmatrix}$$

$$= \begin{bmatrix} 0 & 1 \\ 0 & 0 \end{bmatrix} + \begin{bmatrix} 0 & 0 \\ 1 & 0 \end{bmatrix} = \begin{bmatrix} 0 & 1 \\ 1 & 0 \end{bmatrix}$$

1.2.4.2   Quantum Z Gate (Matrix Representation)

In this gate, for input $|0\rangle$, output will be $\langle 0|$, and for input $|1\rangle$, output will leads to $\langle -1|$. The quantum Z gate can be represented as

$$= |0\rangle\langle 0| + |1\rangle\langle -1| = \begin{bmatrix} 1 \\ 0 \end{bmatrix} \begin{bmatrix} 1 & 0 \end{bmatrix} + \begin{bmatrix} 0 \\ 1 \end{bmatrix} \begin{bmatrix} 0 & -1 \end{bmatrix}$$

$$= \begin{bmatrix} 1 & 0 \\ 0 & 0 \end{bmatrix} + \begin{bmatrix} 0 & 0 \\ 0 & -1 \end{bmatrix} = \begin{bmatrix} 1 & 0 \\ 0 & -1 \end{bmatrix}$$

1.2.4.3 Hadamard Gate

The most popular quantum gate is known as the Hadamard gate (H). This gate works on a single qubit, and performs the unitary transformation, called the Hadamard transform. It can be defined as follows.

Gate notation Matrix representation

Hadamard gate $|y\rangle$ —$\boxed{H}$— $(-1)^y |y\rangle + |1-y\rangle$ $H = \frac{1}{\sqrt{2}} \begin{bmatrix} 1 & 1 \\ 1 & -1 \end{bmatrix}$

The matrix given here forms the computational basis $\{|0\rangle|1\rangle\}$. The schematic representation of the H gate works on a qubit in state $|y\rangle$, where $y = 0, 1$.

1.2.4.4 Phase Shift Gate

Like the Hadamard gate, the Phase shift gate works on a single qubit, and it is also represented as 2×2 matrices. The gate notation and the matrix representation of this logic gate are given below.

Gate notation Matrix representation

Phase shift gate —$\boxed{\phi}$— $R(\theta) = \begin{bmatrix} 1 & 0 \\ 0 & e^{i\theta} \end{bmatrix}$

1.2.4.5 Controlled NOT Gate (CNOT)

Unlike the Hadamard gate, the CNOT gate possesses two input qubits, called the control and target qubit. The first line is known as the Control qubit, while the second line signifies the target qubit. The CNOT gate works on the basis of the following condition.

1. ***Case 1:*** If control qubit $= 0$, the target qubit is required to be left alone.
2. ***Case 2:*** If control qubit $= 1$, then the target qubit is required to be flipped.

Gate notation Matrix representation

Controlled-NOT gate $|x\rangle$ ———•——— $|x\rangle$ $\begin{bmatrix} 1 & 0 & 0 & 0 \\ 0 & 1 & 0 & 0 \\ 0 & 0 & 0 & 1 \\ 0 & 0 & 1 & 0 \end{bmatrix}$

$|y\rangle$ ——⊕—— $|x\rangle \otimes |y\rangle$

The schematic representation of the *CNOT* gate is given here. The truth table of this gate is presented below.

Input		Output	
$\lvert x\rangle$	$\lvert y\rangle$	$\lvert x\rangle$	$\lvert x\rangle \otimes \lvert y\rangle$
0	0	0	0
0	1	0	1
1	0	1	1
1	1	1	0

1.2.4.6 SWAP Gate

The swap gate is used to swap the states of a pair of qubits. This gate is usually made by using three CNOT gates. This gate works as follows.

First, in this gate input is represented as $\lvert x, y\rangle$. This first CNOT gate has the output of $\lvert x, x \otimes y\rangle$, which acts as the input of the second CNOT gate to produce $\lvert x \otimes (x \otimes y)$, $x \otimes y\rangle = \lvert y, x \otimes y\rangle$. Finally, this is fed in as the input of third CNOT gate, which produces $\lvert y, y \otimes (x \otimes y)\rangle = \lvert y, x\rangle$.

Gate notation   Matrix representation

Swap gate

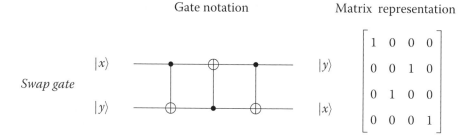

$$\begin{bmatrix} 1 & 0 & 0 & 0 \\ 0 & 0 & 1 & 0 \\ 0 & 1 & 0 & 0 \\ 0 & 0 & 0 & 1 \end{bmatrix}$$

The schematic representation of the Swap gate is presented here. The truth table of the Swap gate is presented below.

Input		Output	
$\lvert x\rangle$	$\lvert y\rangle$	$\lvert y\rangle$	$\lvert x\rangle$
0	0	0	0
0	1	1	0
1	0	0	1
1	1	1	1

1.2.4.7 Toffoli Gate

The Toffoli gate, alias the controlled-controlled-NOT, is a popular universal reversible gate (logic gate). The Toffoli gate can be used to construct any reversible circuit. This gate is composed of 3 input bits and 3 output bits. In this logic gate, any one of the following can occur.

1. ***Case 1:*** If each of first two bits = 1, the third bit is inverted.
2. ***Case 2:*** Otherwise, all bits remains unchanged.

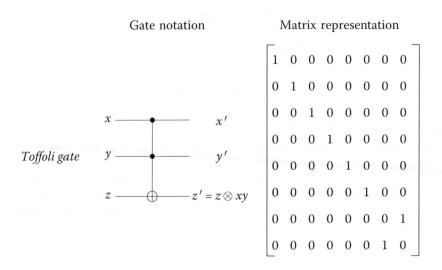

The gate notation and matrix representation of the Toffoli gate are presented here. The truth table of this gate is given below.

Input			Output		
x	y	z	x'	y'	z'
0	0	0	0	0	0
0	0	1	0	0	1
0	1	0	0	1	0
0	1	1	0	1	1
1	0	0	1	0	0
1	0	1	1	0	1
1	1	0	1	1	1
1	1	1	1	1	0

1.2.4.8 Fredkin Gate

Like the Toffoli gate, the Fredkin gate is also a universal gate, appropriate for reversible computing. This logic gate can be effectively used to construct any arithmetic or logical operation. The Fredkin gate, alias the controlled-SWAP gate, is basically a circuit, which has 3 input bits and 3 output bits. It transmits the first bit unaffected and swaps the last pair of bits if and only if the first bit is set to be 1.

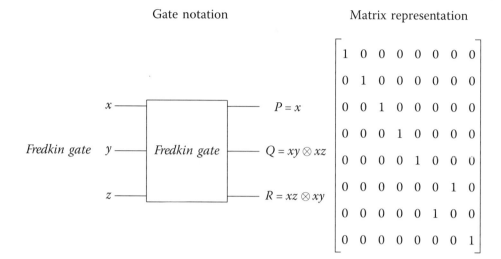

The gate notation and matrix representation of the Fredkin gate are presented here. The truth table of this gate is shown below.

Input			Output		
x	y	z	P	Q	R
0	0	0	0	0	0
0	0	1	0	0	1
0	1	0	0	1	0
0	1	1	0	1	1
1	0	0	1	0	0
1	0	1	1	1	0
1	1	0	1	0	1
1	1	1	1	1	1

1.2.4.9 Quantum Rotation Gate

The quantum rotation gate is a popular quantum gate, generally employed to update any qubit by applying a rotation angle. A distinctive paradigm of this gate used to update any qubit (suppose, k^{th} qubit) (α_k, β_k), can be described as

$$\begin{bmatrix} \alpha'_k \\ \beta'_k \end{bmatrix} = \begin{bmatrix} cos(\theta_k) & -sin(\theta_k) \\ sin(\theta_k) & cos(\theta_k) \end{bmatrix} \begin{bmatrix} \alpha_k \\ \beta_k \end{bmatrix} \tag{1.10}$$

where, θ_k represents the rotation angle of each qubit. The value of θ_k is chosen according to the given problem.

1.2.5 Quantum Register

An anthology of qubits is generally referred to as a quantum register. The size of this register is measured by the size of qubits exercised. The qubit to number conversion can be shown using a typical example given by

$$\underbrace{|1\rangle \otimes |0\rangle \otimes \cdots \otimes |0\rangle \otimes |1\rangle}_{n \text{ qubits}} \equiv \underbrace{|10\cdots01\rangle}_{n \text{ bits}} \equiv |EDN\rangle$$

where EDN stands for equivalent decimal number and $\otimes$ signifies tensor product.

Tensor product: One can construct a new vector space using two given vector spaces. Suppose Y and Z are such two vector spaces. The third vector space is mathematically defined as $Y \otimes Z$, called the tensor product of Y and Z.

Definition: As revealed by quantum mechanics, each and every system S is described by dint of a Hilbert space H. Suppose the system S comprises two subsystems, called S_a and S_b. Symbolically, S can be represented as $S = S_a \cup S_b$. According to the quantum theory, the Hilbert spaces of S, S_a and S_b are correlated by a tensor product, as given by $H = H_1 \otimes H_2$.

Illustration 4: Suppose a quantum register is used to store information as a binary stream. For instance, the decimal number 7 can be represented by a quantum register in state $|1\rangle \otimes |1\rangle \otimes |1\rangle$. Let us describe it in more compact notation. Suppose $b = |b_0\rangle \otimes |b_1\rangle \otimes \ldots |b_{n-2}\rangle \otimes |b_{n-1}\rangle$, where $b_j \in \{0,1\}$. It signifies a quantum register having the value of $b = 2^0 b_0 + 2^1 b_1 + 2^2 b_0 + \ldots + 2^{n-2} b_{n-2} + 2^{n-1} b_{n-1}$. In this representation, there may be 2^n states of this form, which represents n-length binary strings or numbers from 0 to $2^{(n-1)}$. They form a "computational basis," $b \in \{0,1\}^n$, where b composed of n-length binary string indicates that $|b_j|$ is a part of the "computational basis."

A group of n qubits taken together is known as a quantum register of size n. Accordingly, the decimal 1 and 5 can be stored in a quantum register of size 3 as follows:

$$|0\rangle \otimes |0\rangle \otimes |1\rangle \equiv |001\rangle \equiv |1\rangle$$
$$|1\rangle \otimes |0\rangle \otimes |1\rangle \equiv |101\rangle \equiv |5\rangle$$

These two numbers can be stored in chorus as

$$\frac{1}{\sqrt{2}}(|0\rangle + |1\rangle) \otimes |0\rangle \otimes |1\rangle \equiv \frac{1}{\sqrt{2}}(|001\rangle + |101\rangle) = \frac{1}{\sqrt{2}}(|1\rangle + |5\rangle)$$

where, $\frac{1}{\sqrt{2}}(|0\rangle + |1\rangle)$ is called the superposed form of quantum states.

1.2.6 Quantum Entanglement

Quantum entanglement is basically a quantum mechanical phenomenon where the quantum states comprising at least two objects can be described in relation to one another, even though the distinct objects might be spatially isolated. This causes correlations among perceptible physical features in the quantum systems. The quantum entanglement ensures that any alteration made in one object will definitely affect the other one. In real life, there may subsist untold love between a boy and a girl, yet they feel unseen romance, affections and mystical connections for each other. Such connections also happen in the subatomic world, which is known as entanglement between quantum states.

Let there be a bipartite system $S = S_a \cup S_b$, where, S_a and S_b are two subsystems of the system S. A pure state of S is basically a vector $\phi \in H$, where H is known as Hilbert space. The state ϕ is called simple, separable, non-entangled or even factorable if ϕ can be expressed as $\phi = c \otimes d$ for some $c \in H_a$ and $d \in H_b$. Otherwise, the state is called the entangled state. For the non-entangled state, ϕ, S_a and S_b are in states c and d, respectively. It should be noted that if ϕ is entangled, it is not separable. In quantum computing, a pure state comprising two qubits is said to be in the entangled form if the state cannot be represented (not separable) as a tensor product of the participating states of these qubits as $|\vartheta_1\rangle \otimes |\vartheta_2\rangle$.

Quantum entanglement is the utmost notable feature of quantum mechanics. It is the basis of several applications of quantum mechanics, such as quantum teleportation, quantum computation, quantum cryptography, and quantum information so on [190]. The use of innumerable entangled qubits in QC, can accelerate computational capability as compared to its classical version.

Note: We consider two entangled states as follows.

1. *Case 1:* Suppose, in a system, the first qubit is given as $|0\rangle$ and the second qubit is given as $\frac{1}{\sqrt{2}}(|0\rangle + |1\rangle)$. Then the state of this system (two-qubit) can be written as the tensor product of these two qubits as follows:

$$\frac{1}{\sqrt{2}}|0\rangle \otimes (|0\rangle + |1\rangle) = \frac{1}{\sqrt{2}}(|00\rangle + |01\rangle)$$

2. *Case 2:* The entangled states such as $\frac{1}{\sqrt{2}}(|00\rangle + |11\rangle)$ or $\frac{1}{\sqrt{2}}(|01\rangle + |10\rangle)$ cannot be represented as the product of the two-qubits states.

1.2.7 Quantum Solutions of NP-complete Problems

Researchers are sometimes capable of constructing algorithms that can convey a solution for some particular given problem. These kinds of problems are solvable by a number of computational steps delimited by a polynomial m^i where m is the input size and i is a constant. This kind of problem, where a polynomial-time algorithm exists, is categorized as class P problems. It is known that class P problems are solvable with high efficacy. There is one more category where a given arrangement is validated to confirm whether it is a solution to the given problem in polynomial time or not. Such methods exist which can find a solution of this kind. These kinds of problems demand an exhaustive search by configuring all possible arrangements until there is a result of the given problems under a polynomial time. Such problems fall into the class NP. So it is clear that $P \subseteq NP$. NP-complete problems are figured with the subclass of NP. Basically, this subclass surrounds both the NP and NP-hard problems. For a NP-hard problem, there exists an NP-complete problem which can be reduced to an NP-hard problem in polynomial time. Quantum mechanics can store and influence a huge amount of data and processes information using the minimum number of quantum particles due to its rigid architecture. Its hardware is capable of dealing with all possible combinations of solutions simultaneously. It has a fabulous performance in certain kinds of problems like performing the factorial of an integer number as shown in Shor's algorithm [235]. For class P problems, QC has a very high efficacy compared to its classical counterparts, while for other NP-complete problems, it may not live up to its publicity. In [58], the authors have explained that some classically hard problems may be solved efficiently in quantum computing. Later, Grover's algorithm [117] has attracted interest in developing a more efficient search algorithm.

1.3 Role of Optimization

Optimization is referred to as a selection mechanism, used to attain the best combination of feasible solutions from an ample number of acceptable solutions of a specific problem. This selection process is guided by several predefined single/multiple criteria. This is accomplished through a systematic approach by selecting the appropriate values of integer or real variables within an allowable set. The foremost objective of optimization is to get the best possible solutions of any specific problem based on the objective functions for a defined domain. The possible solutions of any certain problem are initially found by exploring its search space, and then they are optimized within a nominal time frame. In principle, the optimization process handles those problems which require one or more of its objectives to be minimized or maximized, where the objectives are basically functions of several integer or real variables. The optimization technique can be categorized as follows.

1. Single-objective optimization
2. Multi-objective optimization

A brief overview of these two kinds of optimization technique is described in the subsequent subsections.

1.3.1 Single-objective Optimization

A single-objective optimization (SOO) problem deals with one objective function. The foremost goal of SOO is to search for the "best" possible solution corresponding to the maximization or minimization of a single objective function that combines all dissimilar objectives into one. This kind of optimization can be effective as a tool which gives decision-makers insight into the type of problem. It is usually unable to deliver the alternative solution sets that trade dissimilar objectives against one another. In this optimization technique, the attention is usually directed towards determining a solution which accomplishes global optima. Formally, the optimization technique is defined as follows.

Let $g : U \rightarrow \mathfrak{R}$ be a given function, which maps a set U to the set of real numbers, $\mathfrak{R}$. The goal of single-objective optimization is to determine an optimum element $y_0 \in U$ such that, $g(y_0) \leq g(y), \forall y \in S$ occurs in performing minimization, while $g(y_0) \geq g(y), \forall y \in U$ occurs in accomplishing maximization

Here, $U \in \mathfrak{R}^n$ comprises an assortment of entities, like equalities, inequalities or constraints and $\mathfrak{R}^n$ denotes the Euclidean space. Here, U signifies a subset of the Euclidean space R^n which is an assortment of entities such as equalities, inequalities or constraints. Each member of U ought to satisfy these said entities. The domain of the above function is known as the search space, and the members of U are referred to as feasible or candidate solutions. The function g is termed the objective function or cost function. Among all feasible solutions, a particular solution which optimizes (minimizes or maximizes) the given objective function is known as an optimal solution.

Different SOO techniques can be classified into the following categories [19, 185, 186].

- *Enumerative techniques*: Enumeration methods are very useful in solving combinatorial optimization problems. These methods encompass assessing every points of the finite, or even discretized infinite search space to attain the optimal solution [19]. Dynamic programming is a popular example of this category of search method. Enumeration methods can be broadly categorized into the following categories.
  (1) Explicit complete enumeration: In this category, all potential alternatives are completely enumerated and a comparison is made among them to get the best possible solution. It is basically impossible as well as very expensive to solve complex problems.
  (2) Implicit complete enumeration: Portions of the solution space that are certainly sub-optimal are left out in this category. This decreases complexity because only the most promising solutions are taken into consideration. In this kind of enumeration, different methods, such as Branch & Bound, dynamic optimization, etc. are generally used.
  (3) Incomplete enumeration: In this category, certain heuristics are applied to select alternatives by only observing portions of the solution space. This does not promise an optimal solution, instead, it delivers estimated solutions.
- *Calculus-based techniques*: The calculus-based methods, also referred to as numerical methods, mean the solution of any optimization problem is found on the basis of a set of necessary and sufficient criteria [19]. This kind of method can be further divided into two different groups, namely, direct and indirect methods. The direct search method is a kind of optimization method used to solve such optimization problems, which does not necessitate any information relating to the gradient of the fitness (objective) function. The traditional optimization methods generally use information about the higher derivatives or gradient for searching an optimal point, whereas a direct search method explores the search space to look for a set of points in all directions of the current point, and selects one of them, which possesses the best objective value compared to the others. The non-continuous or non-differentiable type of objective function can be used to solve problems in the direct search method. Typical applications of this category may include Mesh Adaptive Direct Search (MADS) [11], the Nelder-Mead algorithm (NM) [101, 154], Multidirectional Search (MDS) [259], to name but a few. The indirect search algorithms are based on the derivatives or gradients of the given objective function. The solution of the set of equations is obtained by equating the gradient of the objective function to zero. The calculus-based methods may be very effective in solving the trivial class of unimodal problems [19].
- *Random techniques*: Compared to enumerative methods, random search methods basically use the additional information related to the search space that in turn guides them to possible areas of the search space [24, 112]. This method can be further subdivided into two different categories, namely, single-point search and multiple-point search. In the first category, the aim is to search for a single point, whereas several points are needed to be searched at a time for the multiple-point search method. Some

popular examples of single-point search methods involve Simulated annealing [147], Tabu search [109], etc. Some population-based evolutionary algorithms, such as Particle Swarm Optimization [144], Genetic Algorithms [127], differential Evolution [76], and Ant Colony Optimization [242], are popular examples of the latter category. The guided random search methods are very effective in such problems where the search space is vast, multi-modal, and not continuous. Instead of providing the exact solution, the random search methods deliver a near-optimal solution across an extensive range of problems.

1.3.2 Multi-objective Optimization

Multi-objective optimization (MOO), also known as the multi-attribute or multi-criteria optimization technique [54] that simultaneously optimizes several objectives as one in respect of a set of specific constraints. As opposed to the SOO, the MOO deals with various conflicting objectives and provides no single solution (optimal). The collaboration among diverse objectives presents compromised solutions, which are principally referred to as the trade-off, non-inferior, non-dominated, or even Pareto-optimal solutions.

There are numerous real-life decision-making problems which have several objectives, such as minimizing risks, maximizing reliability or minimizing deviations from anticipated levels, minimizing cost, etc. In MOO, it is not always possible to find a solution in the search space, which produces the best fitness values with regards to each objective. In the given search space, there may be a group of solutions having better fitness with respect to a set of objectives and worse value for the rest as compared to other solutions. From the perspective of MOO, the term "domination" between a pair of solutions can be described as follows [54, 177]:

Suppose a MOO problem finds a set of n solutions, say, $\{y_1, y_2, \ldots, y_n \in Y\}$ for m objectives, say, $\{O_1, O_2, \ldots, O_m \in O\}$, Y and O are known as the solution space and objective space, respectively. A solution (say, y_i) dominates others (say, y_j), where $1 \leq i, j \leq n$ and $i \neq j$, if the following conditions are satisfied:

1. The solution y_i is not inferior to y_j $\forall$ O_k, $1 \leq k \leq m$.
2. The solution y_i must possess better value than y_j in at least one O_k, $1 \leq k \leq m$.

Since, there exists no solution in the Pareto-optimal set, which holds the top spot of all objectives, some specific problem knowledge and decision-making capabilities to select preferred solutions in MOO [54] have to be found. Over the last few years, a number of techniques have been proposed in this literature that have coped with multi-objective optimization problems in different facets [177].

1.3.3 Application of Optimization to Image Analysis

Image analysis appears to be a part of decision support systems in a variety of applications, such as medical, military, industrial and many others. Several techniques have already been introduced in the literature so far to solve different image processing activities. They usually necessitate few method-specific parameters tuning in an optimal way to accomplish the best performance. The obligation to achieve the best results changes the structure of the methods candidate into a corresponding optimization problem. Optimization is an important technique used to solve several problems in image processing. Such a reality is evident as a significant number of researchers use optimization techniques in their research work. Classical optimization techniques habitually face excessive difficulties when dealing with images or systems comprising distortions and noise. In these situations, a variety of evolutionary computation techniques have been shown to be viable alternatives that can efficiently address the challenge of different real-life image processing problems [51]. Image analysis is a dynamic and fast-growing field of research. Apart from that, each novel technique proposed by different researchers in any field is rapidly recognized and -simulated for image processing tasks. Some state-of-the-art techniques that can handle the challenges and image processing problems are available in the literature [191]. The rich amount of information is already available in the literature makes it easy to find the exact optimization technique for a certain image application.

1.4   Related Literature Survey

This section presents a write-up discussion about a brief review of various approaches to handle different optimization problems. This section tries to provide a fundamental overview of present trends to solve different optimization problems. These aforementioned approaches are broadly classified into three categories, namely, quantum-based approaches, meta-heuristic-based approaches, and multi-objective-based approaches. Most of the quantum-based approaches mainly use the basics of quantum computing to serve the optimization purpose. The meta-heuristic-based approaches use the basic anatomy of different meta-heuristics to develop a variety of techniques to handle different optimization problems. Lastly, the multi-objective-based approaches are used to develop different Pareto-based techniques to handle a number of objectives simultaneously for optimization.

This section presents a brief literature survey of the aforesaid trends in single and multi-objective optimization.

1.4.1   Quantum-based Approaches

The field of quantum computing became popular when the notion of a quantum mechanical system was anticipated in the early 1980s [23]. The aforesaid quantum mechanical machine is able to solve some particular computational problems efficiently [117]. In [89], the author has recognized that the classical computer faces a lack of ability while simulating quantum mechanical systems. The author presented a structural framework to build Quantum Computer. Alfares and Esat [4] analyzed how the notion of quantum algorithms can be applied to solve some typical engineering optimization problems. According to them some problems may arise when the features of QC have been applied. These problems can be avoided by using certain kind of algorithms. Hogg [125] presented a framework for a structured quantum search where Grover's algorithm was applied to correlate the cost with the gates, behavior. In [126], the authors extended the work and proposed a new quantum version of combinatorial optimization. Rylander et al. [216] presented a quantum version of genetic algorithm where the quantum principles like superposition and entanglement are employed on a modified genetic algorithm. In [181], Moore and Narayanan proposed a framework for general quantum-inspired algorithms. Later, Han and Kim [121] developed a quantum-inspired evolutionary algorithm which was applied to solve the knapsack problem. Here, qubit is used for the probabilistic representation. A qubit individual is represented by a string of qubits. The authors introduce the quantum gate as a variation operator in order to drive the qubit individuals toward better solutions. A new improved version of this algorithm has been presented in [122]. Here, the authors proposed a termination criterion to accelerate the convergence of qubit individuals. Another version of a quantum genetic algorithm has been proposed by Zhang et al. where the strategies to update the quantum gate by utilizing the best solution and also introducing population catastrophe have been shown. The improved version of the work presented in [121] has been proposed by Zhang et al. where they applied a different approach to get the best solution [287]. Narayan and Moore presented a genetic algorithm where quantum mechanics was used for the modification of a crossover scheme [189]. Moreover, Li and Zhuang developed a modified genetic algorithm using quantum probability representation. They adjusted the crossover and mutation processes to attain the quantum representation [157]. In [188], the authors presented a quantum-inspired neural network algorithm where also the basic quantum principles were employed to symbolize the problem variables. The instinctive compilation of information science with the quantum mechanics constructs the concept of quantum computing. The quantum evolutionary algorithm (QEA) was admired as a probability-based optimization technique. It uses qubits encoded strings for its quantum computation paradigm. The intrinsic principles of QEA help to facilitate maintaining the equilibrium between exploitation and exploration. In recent years, some researchers have presented some quantum evolutionary algorithms to solve particular combinatorial optimization problems. A typical example of this algorithm is filter design by Zhang et al. [285]. A group of heuristic search algorithms have been designed for quantum computers by different researchers. They call these algorithms heuristic search designed quantum algorithms (HDQs) [114, 115, 219, 239]. The capability of HDQs is checked by simulating a quantum computer, such that the efficiency of a quantum algorithm can be judged on classical hardware. Some authors have combined quantum computation with genetic algorithms and developed applications where fitness functions are varied between genetic steps based on the number of outer time-dependent inputs. A few distinctive examples of this category are given in [262, 272]. These papers include some schemes for quantum control processes. Here, genetic algorithms are employed for optimally shaped fields to force a few preferred physical processes [262, 272]. Aytekin et al. [12] developed a quantum-based automatic object extraction technique where quantum mechanical principles were used as the basic constituents of the proposed technique. With reference to Artificial Intelligence (AI), some authors have developed quantum behaved applications on AI. A few of them are presented in [124, 126, 137, 267]. Hogg [125] proposed a new framework for a structured quantum search. In his proposed framework, the author used Grover's algorithm [117] to associate the cost with the activities of the quantum gate. Lukac and Perkowski [164] applied a different approach where they considered each individual in the population as quantum circuits and used the elements of population for the objective quantum circuit. Two popular quantum algorithms developed so far are Quantum Fourier Transform (QFT) [235], and the Grover Search Algorithm [117].

Different problems can be efficiently solved using QFT, for instance, discrete logarithm, factoring and order finding and many others [190]. Research is still going on to create purposeful and scalable quantum computers.

1.4.2 Meta-heuristic-based Approaches

Meta-heuristic optimization techniques are employed heuristically in searching algorithms. They use iterative approaches to yield a better solution by moving from local optima [31, 108]. So they force some basic heuristic to compensate from local optima. The popular meta-heuristic techniques like Genetic Algorithm (GA), Particle Swarm Optimization (PSO), Differential Evolution (DE), Ant Colony Optimization (ACO), Simulated Annealing (SA) and Tabu Search (TS) are applied for optimization in different ways. Holland [127] proposed genetic algorithms (GAs) which impersonate the belief of some natural acts. GAs can be applied efficiently in data mining for classification. In 2006, Jiao et al. presented an organizational coevolutionary algorithm for classification (OCEC) [134]. In OCEC, the bottom-up searching technique has been adopted from the coevolutionary model that can efficiently carry out multi-class learning. Kennedy and Eberhart first proposed PSO in 1995 inspired by the synchronized movement of flocks of birds [144]. In PSO, the population of particles is the particle swarm. In 2004, Sousa et al. projected PSO in data mining [238]. PSO can be skillfully used in rule-based searching processes. In recent years, many researchers have tried to improve the performance of PSO and proposed various alternatives to PSO. In [208], the authors presented an algorithm for parameters settings. Some authors have furnished their efforts for the betterment and variation of PSO by combining diverse techniques [7, 37, 197]. The actions of ant colonies are governed by heuristic algorithms such as ant colony optimization (ACO) [76]. In 2002, Parpinelli et al. proposed the concepts of ant-miner-based ACO to generate the classification rules [196]. Storn and Price [242] presented the differential algorithm in 1997. This is a population-based stochastic meta-heuristic optimization technique. DE has proved its excellence in float-point search space. Numerous modified versions of DE algorithms have been invented so far; most of them are built to solve continuous optimization problems. But they have low efficacy to solve discrete optimization problems and face several problems, such as engineering-related problems especially for routing or scheduling or even combinational type problems. Binary DE was presented in 2006 by Pampara et al. [195] that can solve numerous numerical optimization problems. In [88], a coalesced form of simulated annealing and genetic algorithm (SAGA) has been proposed by Fengjie et al. The authors employed 2D Otsu algorithm in low contrast images. Their algorithm can differentiate the ice-covered cable from its backgrounds. Luo et al. also presented a combined (SAGA) algorithm for colony image segmentation in [165]. According to Luo et al., the collective gives a better result because the individual shortcomings have been eliminated in the proposed algorithm. In [250], SA and an improved Snake model-based algorithm were presented by Tang et al. in image segmentation. In [184], Nakib et al. presented a research work on non-supervised image segmentation based on multi-objective optimization. In addition to this, they have developed another alternative of SA that can resolve the histogram Gaussian curve-fitting problem. Chen et al. presented a segmentation algorithm by combining image entropy and SA. According to their research work, the segmentation process can speed up the contours evaluation.

Apart from that, the combination of this meta-heuristic approach can terminate the contours evolution automatically to apposite boundaries [280]. In another work, Garrido et al. [99] presented an SA-based segmentation approach based on MRI for the left ventricle. The details are presented in [99]. Salavati and Mozafari [221] used a combinational approach based on GA and SA in histogram-based image segmentation. The authors used their proposed approach to find threshold values for different levels of thresholding. In 1997, tabu search was first proposed by Fred Glover to allow hill climbing to arise from local optima [93]. Later, supplementary research has been carried out on this meta-heuristic approach that was reported in [105, 106]. There are many computational experiments that establish the completeness and efficacy as a meta-heuristic approach. Faigle et al. and Fox also investigated the theoretical facets of Tabu Search that have been reported in [86, 93], respectively. Huang and Wang [129] developed an algorithm called the Two-Stage Multithreshold Otsu method (TSMO) to improve the effectiveness of Otsu's method. Zhang et al. presented a Maximum Tsallis entropy Thresholding method (MTT) for image segmentation. They employed Tsallis Entropy using the Artificial Bee Colony Approach for optimization [288]. In 2011, Sathya and Kayalvizhi proposed another multi-level thresholding algorithm for image segmentation based on a histogram, called the Modified Bacterial Foraging (MBF) algorithm [222].

1.4.3 Multi-objective-based Approaches

To date, a large number of researchers have expressed their profound interest in a variety of MOO problems. A detailed review of different evolutionary algorithms was reported in [54]. A number of Pareto-based methods on genetic algorithms are available in the literature. Shaffer [233] has developed a multi-objective evolutionary algorithm (MOEA), called vector-evaluated genetic algorithm (VEGA). VEGA performs efficiently for a number of generations but fails to provide unbiased solution for a few occasions. Fonseca and Fleming [92] proposed a variant MOEA, entitled the Multi-Objective Genetic Algorithm. The concept of MOGA is that all non-dominated chromosomes in the population are allotted the best possible objective value and the remaining chromosomes in the dominated group are reprimanded in accordance with the population density of the respective regions of these chromosomes. Horn et al. [128] proposed another multi-objective optimization method, called the Niched-Pareto Genetic Algorithm (NPGA). Srinivas and Deb [240] developed a variant MOO algorithm, popularly known as the Non-dominated Sorting Genetic Algorithm (NSGA). The NSGA was designed according to the classification theory as proposed by Goldberg [110]. The backbone of NSGA is said to be Pareto ranking of the population at different generations. This makes NSGA a less efficient algorithm compared to other contemporary algorithms. Tanaka and Tanino later integrated the user's association into an evolutionary multi-objective optimization to propose a new system [249]. Instead of searching the entire Pareto front, the user may be interested in only putting effort into a little segment of it. Deb et al. [56] have introduced a second edition of NSGA [240] called Non-dominated Sorting Genetic Algorithm II (NSGA-II). A speedy nondominated sorting strategy was adopted in this algorithm. Unlike the normal selection procedure, a variant operator was used for the selection purpose. The population for the next generation is produced by coalescing chromosomes (equal to the population size) from the parent and offspring pool with respect to best fitness values. Zitzler and Thiele [291] have suggested another multi-objective optimization algorithm, called the Strength Pareto Evolutionary Algorithm (SPEA). This algorithm can be visualized as an approach of amalgamating multiple evolutionary multi-objective optimization algorithms. Zitzler et al. [293] have introduced a second version of SPEA named Strength Pareto Evolutionary Algorithm 2 (SPEA2). This algorithm is more efficient with regards to search capability compared to SPEA. The Pareto Archived Evolution Strategy (PAES) Algorithm is a well-admired MOO algorithm presented by Knowles and Corne [149]. In PAES algorithm, one offspring agent is created from one parent agent. This algorithm uses a crowding method to segregate objective space recursively. Corne et al. [49] introduced a popular MOO algorithm known as the Pareto Envelope-based Selection Algorithm (PESA). They have adopted the good features from both SPEA and PAES to construct PESA. Bandyopadhyay et al. [18] introduced a multi-objective genetic clustering-based algorithm to segregate the pixel values of remote sensing images. The authors applied the NSGA-II [54–56] algorithm to deal with the problem of fuzzy partitioning by simultaneously optimizing several fuzzy clustering indices. Mukhopadhyay and Maulik [182] developed a multi-objective, real coded, genetic fuzzy clustering algorithm to segment the magnetic resonance images. The authors used NSGA-II [54–56] as the MOO algorithm in accordance with the fitness functions (conflicting type). In the literature, a few bio-inspired methods, e.g. ant colony optimization [113] and artificial immune systems [46] have been successfully implemented to solve different MOO problems. Simulated annealing has also been used in different ways to solve various MOO problems. Emmerich et al. [84] used a novel idea to introduce a popular evolutionary multi-objective optimization algorithm, where authors used the hypervolume measure as the selection criterion. Some distinctive examples may be presented in [243, 244] in this regard.

1.5 Organization of the Book

The book presents the exploration and exploitation of the fundamentals of quantum principles in the soft computing paradigm. The main contribution of the book is to introduce a handful of quantum inspired meta-heuristic algorithms in the gray-scale/color domain and address the relevant problem of their classical counterparts. The scope of the book in essence is confined to but not bounded by developing quantum inspired algorithms for binary, multi-level image thresholding for pure and true gray-scale/color images in a single/multi-objective environment. It is also emphasizes the effectiveness of the proposed approaches.

As a whole, the book is arranged into eight chapters. The book encompasses an introductory chapter, and the rest are pertinent chapters. Chapter 2 briefly presents the review of image analysis. Certain things with regards to image analysis are addressed and discussed in this chapter. In Chapter 3, an overview of different meta-heuristics is addressed and discussed separately. Chapter 4 opens up an inimitable horizon in the quantum computing research community. A few novel quantum inspired meta-heuristic algorithms are introduced to deal with binary image thresholding. The benefits achieved are better efficiency and less time complexity. Chapters 5 and 6 stretch out the applicability and effectiveness of these algorithms to the multi-level, gray-scale and color intensity scales. These chapters present the requisite functional amendments of the above-mentioned algorithms, in this regard. Chapter 7 brings in a new idea in the quantum computing research community. We present a quantum behaved meta-heuristic algorithm in the multi-objective framework. Finally, an endnote to the book is drawn in the concluding chapter by emphasizing some of the paths of further research in this field.

1.5.1 Quantum Inspired Meta-heuristics for Bi-level Image Thresholding

In Chapter 4, we exploit the trivial concepts of a quantum computing to introduce a quantum version of two meta-heuristic algorithms by virtue of trimming down the time complexity of the conventional algorithms. The proposed algorithms are inspired by the inherent features of parallelism and time discreteness exhibited by quantum mechanical systems. These algorithms are designed for binary image segmentation. Hence, Chapter 4 centers on introducing efficient quantum inspired algorithms to determine the optimum threshold intensity, which is demonstrated on real-life gray level images. A comparative study is carried out between the proposed algorithms, their conventional counterparts and the quantum evolutionary algorithm (QEA) developed by Han et al. to analyse their performance and the relevant results in this concern are reported thereafter.

1.5.2 Quantum Inspired Meta-heuristics for Gray-scale Multi-level Image Thresholding

In Chapter 5, we present six different quantum inspired meta-heuristic algorithms for performing multi-level thresholding. The proposed algorithms are designed in a different approach exploiting the basic features of quantum mechanism. The efficacy of the proposed algorithms is exhibited in a number of images for multi-level thresholding. After that, the results of the proposed algorithms are used to produce consensus results. The statistical superiority of the six methods is determined using a statistical test, called the Friedman test. As a comparative study, the best performing algorithm among the proposed algorithms is compared with five other algorithms as regards the multi-level image.

1.5.3 Quantum Behaved Meta-heuristics for True Color Multi-level Thresholding

Chapter 6 presents a few popular meta-heuristic algorithms inspired by the intrinsic principles of quantum mechanics. These algorithms are designed to determine the optimal threshold values of color images at different levels of thresholding by exploiting a variety of objective functions. The performance of the proposed algorithms and their conventional counterparts are evaluated and are reported in this chapter.

1.5.4 Quantum Inspired Multi-objective Algorithms for Multi-level Image Thresholding

Chapter 7 deals with the thresholding of multi-level images in multi-objective flavor. Several quantum behaved multi-objective algorithms for bi-level/multi-level image thresholding are introduced. The first algorithm exploits the fundamental properties of quantum computing using a popular multi-objective technique, called the non-dominated sorting genetic algorithm II (NSGA-II) [56]. The other algorithms are designed by using the properties of quantum computing and different meta-heuristic algorithms in the essence of multi-objective optimization. The efficiency of the proposed algorithms as established over other popular methods, is also reported in this chapter.

1.6 Conclusion

According to the central theme of the book, the chapter describes the basics of image analysis followed by image segmentation and image thresholding techniques. The intricate details of quantum computing are touched upon as an indispensable facet of the theme of the book. The different facets of quantum computing are hinted at. In addition, the role of optimization in real-world scenarios and its application parts is also explained. The chapter presents a detailed literature survey based on different approaches, such as quantum-based approach, and so on. The chapter also deals with different evolutionary algorithms and meta-heuristics in detail. These include the differential evolutionary [242], genetic algorithms [127, 210], particle swarm optimization [144], and ant colony optimization [196], to name but a few.

In medical science, medical images, such as Magnetic Resonance Imaging (MRI), PET, CT are very common these days. The application of quantum inspired algorithms may be enhanced in image segmentation in a much more efficient way. Automatic segmentation of images of this category would be a challenging task. The application of these proposed algorithms can be extended tow processing 3D images.

1.7 Summary

- A quantum computer fundamentally works on several quantum physical features.
- The term "quantum computing" is fundamentally a synergistic combination of ideas from quantum physics, classical information theory and computer science.
- Soft computing manages the soft meaning of thoughts.
- Fuzzy logic tenders more elegant alternatives than conventional (Boolean) logic.
- Segmentation separates the patterns into a number of uniform, non-overlapping and homogeneous regions or segments (classes).
- The members in the same segment are similar to each other. The members of different segments possess dissimilarity among themselves.
- Thresholding is well recognized and probably the most effective tool in the context of image processing (image segmentation) and pattern recognition.
- Based on the number of threshold values, thresholding can be categorized as bi-level and multi-level thresholding.
- The quantum bit can be described as the basic constituent of information in quantum computing.
- The quantum gates are usually hardware tools which operate on preferred qubits using a fixed unitary operation over a predetermined period of time.

- Optimization is referred to as a selection mechanism, used to attain the best combination of feasible solutions from an ample number of acceptable solutions to a specific problem.
- The optimization technique can be categorized into two types, namely, single-objective optimization and multi-objective optimization.

Exercise Questions

Multiple Choice Questions

1 A gray level image has
   (a) pixel intensity values between 0 to 255
   (b) pixel intensity values between 1 to 255
   (c) pixel intensity values between 0 to 256
   (d) none of the above

2 Which of the following is not a part of soft computing?
   (a) neural networks
   (b) evolutionary computation
   (c) machine learning
   (d) optimization technique

3 The technique used for image analysis is
   (a) pattern recognition
   (b) computer/machine vision
   (c) image description
   (d) all of the above

4 The segmented image after image segmentation is
   (a) uniform
   (b) non-overlapping
   (c) homogeneous regions
   (d) all of the above

5 Quantum computer works on
   (a) law of Thermodynamics
   (b) quantum mechanical features
   (c) law of genetics
   (d) none of the above

6 The basic constituent of information in QC is known as
   (a) quantum gate
   (b) quantum bit
   (c) quantum superposition
   (d) quantum entanglement

7 Which of the following is a three-qubit quantum gate?
   (a) Hadamard gate
   (b) Controlled NOT gate
   (c) Toffoli gate
   (d) swap gate

8 "Domination" is an important term found in describing
   (a) single-objective optimization
   (b) multi-objective optimization
   (c) both (a) and (b)
   (d) none of the above

Short Answer Questions

1 Describe combinatorial problem with an example. Mention the difficulties in solving these types of problems.

2 What do we mean by image thresholding? How do you categorize the image thresholding method?

3 What is quantum bit? How is it different from classical bit? Illustrate with an example.

4 What is "tensor product?" How is it used to define quantum registers?

5 What do you mean by an NP-complete problem? How do you relate any NP-complete problem to a quantum solution?

Long Answer Questions

1 Describe different categories of quantum gate. Explain with suitable examples.

2 Describe the different steps involved in image analysis. Illustrate with figures and examples.

3 Discuss the role of optimization in the context of real-life scenarios.

4 What are the basic features of quantum computing?

5 What is Dirac's Notation in quantum computing? How is this notation used in quantum computing?

6 What is quantum gate? Discuss 1-input, 2-input and 3-input quantum gate with suitable examples.

2

Review of Image Analysis

2.1 Introduction

This chapter comprises a brief discussion of the review of image analysis. The objective of this technique is to introduce conventional and other new advanced versions of digital image processing methodologies that primarily engage the image analysis class of tasks, namely, image segmentation, feature description (extraction and selection), and object classification. The first stage of image analysis techniques target applying different image segmentation techniques. The most common methodologies in this direction involve thresholding, region-based segmentation, boundary-based segmentation, texture segmentation, and so on. A few popular feature description methods are texture, moments, edge and boundaries, shape features, etc. The typical object classification techniques consist of clustering, similarity measure, neural networks, and many others.

This chapter describes some current trends of image analysis. It also throws light on popular image thresholding techniques in the binary, multi-level and gray-scale domains.

The organization of the chapter is as follows. Section 2.2 presents a brief definition of image analysis. The mathematical formalism of image segmentation technique is presented in Section 2.3. A brief summary of current technologies of image analysis is discussed in Section 2.4. An overview of popular thresholding techniques is given in Section 2.5. In this section, several thresholding techniques are briefly described with mathematical illustration. Nowadays, image analysis has a wide applicability in different areas of application. Section 2.6 discusses some of those areas of image analysis. The chapter ends with a conclusion, in Section 2.7. At the end of the chapter, a short summary is presented. Finally, a set of exercise questions in relation with the theme of the chapter is given.

2.2 Definition

Image analysis can be defined as a popular technique, associated with the extraction of meaningful and useful information from an image. It focuses on making quantifiable measurements of required objects, if possible automatically, from a given image to yield its description [212], and sorts the desired objects thus achieved, into groups, called classes. The image analysis process necessitates the extraction of some specific features that help identify the object. Different segmentation techniques can be employed to isolate the anticipated object from the image in order to measure some useful information on it subsequently. For example, the task in this concern could be identifying a label on a group of item, or assessing the size and location of blood cells in the human body, captured as a medical image. An advanced version of the image analysis technique can be used to measure quantitative information, which is generally applied to make an erudite decision. For instance, a robot can be programmed to regulate its arm to move an object after accurately recognizing it.

2.3 Mathematical Formalism

An overview of the general formulation of the image segmentation technique is demonstrated in this section. Let the entire image be symbolized by $\mathcal{R}_n$. Segmentation can be described as a process that divides $\mathcal{R}_n$ into n number of non-overlapping (disjointed) regions, $R_1, R_2, \dots, R_n$, such that:

1. $\sum_{k=1}^{n} R_k = \mathcal{R}_n$
2. Each R_k is connected with any other region, $k = 1, 2, \dots, n$
3. $R_j \cap R_k = \emptyset \; \forall j, k$ and $j \neq k$
4. $P(R_k) = \text{True}$, for $k = 1, 2, \dots, n$
5. $R_j \cup R_k = \text{FALSE}$ for $j \neq k$

where $P(R_k)$ is called the logical predicate, defined over the members in set R_k. $\emptyset$ is called the empty set.

The brief description of each of the above points is presented below.

Criterion 1 shows that the union of all the segmented regions leads to the entire region, which confirms its completeness. Criterion 2 states that each region is connected. According to criterion 3, all the regions necessarily are separated. Criterion 4 confirms that the existing pixels in any segmented region must hold the presumed predicate. Criterion 5 designates that every two regions must be different with regards to predicate P.

2.4 Current Technologies

The explosive evolution in hardware and software technology has led to several important improvements in imaging technology. Several automatic methods have been developed during the last few years in different fields, which in turn have reduced human efforts by a significant amount with the utmost accuracy. The underlying principles of these methods are to efficiently and automatically detect several image features, which may be consistent with characteristics in relation to sensorial, and other physical properties.

In general, the process starts by segregating the objects of interest from the source image, quantifying numerous features, like shape, size and texture, and finally classifying these objects into different classes in respect of these features. The conventional and advanced image processing procedures that predominantly engage in the image analysis tasks are described in the following subsection.

2.4.1 Digital Image Analysis Methodologies

There are several techniques used for analyzing digital images automatically. Each different technique may be suited to accomplish some specific range of tasks. In this part, the image analysis techniques are discussed in line with its basic class of operations.

2.4.1.1 Image Segmentation

The term "image segmentation" is fundamentally defined as a comprehensive partitioning of an input image into several homogeneous regions, called classes. The homogeneity of different regions is considered in respect of certain image feature of interest, such as pixel intensity, texture, or color. Image segmentation is the most popular, simplest and robust technique among all other image analysis techniques [32]. The most frequently used segmentation techniques are generally categorized into two groups, such as:

1. *Region-based segmentation techniques:* This technique hunts for regions that satisfy a specified homogeneity criterion. In general, this technique shows less sensitivity to noise. Examples include: histogram thresholding; region growing, splitting and merging; watershed and clustering, and so on [232]. Among all the existing approaches, thresholding is one of the most renowned techniques with regard to accuracy, stability, and simplicity. In this category, this is the most used technique, in which a threshold is chosen, and an image is partitioned into two groups. The first group consists of the pixels having values greater than the threshold and the rest of the pixels are selected for the second group or vice versa. Several thresholding techniques exist in the literature [194]. Image thresholding is classified into two categories, namely, parametric and non-parametric approaches. In the first approach, a statistical model is invoked, which is expected to be obeyed by the gray-level distribution of an image. The optimal value is estimated for the model by means of a given histogram. In the second approach, the optimal threshold is found by optimizing a given objective function, for example, entropy [159].

   Region growing, splitting and merging are another type of region segmentation technique. In this class, the adjacent pixels to the similar segment where their image values are very close to each other, are assigned based on some predefined criterion of closeness. These techniques are more suitable on some occasions compared to the thresholding approach, especially when operating in a color space [260].

   Clustering algorithms accomplish region segmentation by dividing the image into different classes, called clusters. The pixels of each individual cluster have some similarity with reference to some features in feature space. This technique is suitable for partitioning of bigger parts, but not that useful for extraction of small regions [217]. The watershed technique utilizes image morphology for segmentation. The morphological watershed technique can be used to divide a binary image into several regions by taking two parameters, namely, inverse distance map and the local minima. The first parameter is treated as the landscape and the second one as markers. The effectiveness of this segmentation technique depends on the selection of the correct markers. This technique has been proved to be effective for dealing with the objects having blurred edges [286].

2. *Edge-based segmentation techniques:* These techniques aim to determine object boundaries and then segment regions surrounded by the boundaries. They represent a large group of techniques based on edge magnitude in the image. Edges

typically consist of a number of points in the image, in which the gray values change pointedly from one pixel to the other. Therefore, edge detection aids in extracting the meaningful information features of the image where abrupt changes occur.

There exist several techniques for detecting edges, which may be categorized into the two following categories.

- Gradient: This technique detects the edges by computing and observing the maximum and minimum value in the first derivative of the image. For instance, some popular operators like Roberts, Prewitt, Sobel, Canny [112] can be applied to detect horizontal and vertical edges.
- Laplacian: In this method, instead of finding the first derivative of the image, this technique looks for zero crossings in the second derivative for detecting edges. Typical examples may include Laplacian or Gaussian and others.

Most edge detectors basically find discontinuous or over-detected edges. A number of edge detection techniques used for image segmentation are available in the literature [166].

Some other popular examples of the edge-based technique are known as graph searching and contour following. In graph searching, each pixel of the image represents a node of the graph, and the paths comprise the potential edge in the image. A cost is associated with each node, which is generally computed by means of the local edge magnitude, its direction, and also a priori knowledge regarding the boundary location. The cost is basically calculated by summing up the associated costs of all nodes that are contained within the path. The optimal border can be defined by exploring the optimal path (low-cost path) in the graph. This is a powerful technique, which is fully dependent explicitly on the application-specific cost function [254]. In the contour-based approaches, first, the edge is detected in local areas. Consequently, efforts are made to find the improved version of results by a global linking procedure that hunts for curvilinear continuity [168].

2.4.1.2   Feature Extraction/Selection

Once the objects of an image have been successfully identified by the image segmentation technique, the second approaches are identification and computation. The features of an image are generally characterized into three basic categories, namely, morphology, color, and texture. Morphological features refer to the appearance of an image object. Color features are very simple in nature, are associated with the object or scene. Textural describes the spatial distribution of the gray-level pixel intensities [168].

There may be quite a lot of image features for any particular instance. It will be difficult to handle a large number of features at once. In addition, modeling of such features will consume huge computational time [42]. The presence of redundant features may hamper the model performance by a significant amount. Furthermore, numerous training samples may be involved for model development. It was seen that the number of training samples may increases, even exponentially, with the increase of features [80]. The decrease of sample dimensionality may be the reason for the improved performance. For any specified set of features, this can be achieved either of the following two ways.

1. *Feature selection:* The feature selection procedure has a vital role in model performance. It is a process of selecting the best possible subset of features from the given feature set.

2. *Feature extraction:* This is also an important approach like the previous one. In this process, new features are being created from the given original features, which are transformed to a lesser dimensional space using a linear/non-linear combination.

Each method is significant in its own way. The selection of the apposite method is dependent on the application domain and also the available training data set [205]. The feature selection procedure may provide the following potential benefits [118]:

- Simplifying data visualization.
- Helping data understanding.
- Minimizing the requirements for storage and measurement.
- Reducing training and usage times.

Feature selection techniques can be categorized into the following two categories:

1. *Supervised feature selection:* This technique measures the correlation among the features and class labels in order to evaluate the feature's relevance. This method entails adequate labeled data samples so as to provide a discerning feature space.
2. *Unsupervised feature selection:* In this technique, the inherent data structure is retained to measure the feature capacity with the purpose of evaluating its relevance.

Since the sampling and labeling procedure proved fussy and expensive, nowadays, in several real-world applications, massive unlabeled data and a very few labeled samples are used. To cope with such situations, researchers devised the idea of developing different semi-supervised feature selection techniques. A new scheme in this approach has recently been developed by Kalakech et al. [138].

Feature extraction is imperative because the curse of high dimensionality is habitually the most important limitations of several hands-on technologies. On the other hand, when the number of training sets becomes very small compared to the numbers of features, it often degrades the effectiveness of the classifiers [276]. Unsupervised feature extraction techniques use raw data to produce representative features, as a consequence, it aids any classifier to attain a more stable solution and achieve a better performance [253].

Feature selection schemes are generally categorized into the following three types [223].

- *Filter techniques:* These techniques measure the significance of features by considering only the inherent features of the data. Mostly, in this approach, a feature relevance score is computed, and the features with low scores are usually eliminated. After that, the remaining features set is taken as input to the subsequent classification algorithm.
- *Wrapper techniques:* In these approaches, a search process in the potential features space is defined, and afterwards, a number of subsets of features are produced and assessed. The assessment of an explicit subset of features is attained by training and testing an explicit classification model.
- *Embedded techniques:* In these techniques, the search procedure for an optimum subset of features is constructed in the classifier structure, and can be understood as a search procedure in the collective features space and hypothesis. Like the wrapper techniques, these techniques are therefore specific to specified learning algorithms.

2.4.1.3 Classification

Once the first two techniques have taken place, the classification techniques are investigated. Classification investigates the numerical features of a variety of features and arranges data into different categories [79]. Once the features have been identified, the class containing identical data is labeled with a common name. Through substantial amount of variations may be observed among various images and objects, very few features are required for classification purposes. The features are separated into classes on the basis of some decision rules, which were set up through the study of prior measurements in a training process. To develop a more complex classifier, a more efficient method that fits this classifier to the data, is also required [81].

Machine learning is the fast-moving developing field of computer science, with wide-ranging applications. Machine learning algorithms can be of two types: supervised and unsupervised. Supervised algorithms deal with the automatic detection of relevant patterns in data. They also deal with the construction of mathematical models, which are then assessed on account of their predictive capacity in the context of measures of change in the data itself. On the other hand, unsupervised algorithms hunt for similarity between chunks of data so as to decide whether they can be considered as creating a group, called clusters [255]. Several taxonomies for classification techniques in pattern recognition are available in the literature [6]. The most common and popular approaches of methods of classification are statistical classification and neural classification.

Statistical methods are suitable for such approaches, in which the discriminants can be calculated in a direct manner. Neural classifiers follow this approach. In this classifier, there is no prerequisite to take up a statistical model since their generated outputs possibly can be transformed to posteriors using a direct approach [142]. Neural classifiers are basically non-parametric, non-linear, and adaptive in nature compared to other conventional classification methods. These classifiers are able to approximate any necessary relationship with arbitrary accurateness. These types of classifier are ideally suitable for problems in which observations are not too hard to attain, but at the same time the data structure or fundamental relationship is unidentified [155].

Neural networks can be acknowledged as the online learning systems, fundamentally the non-parametric type and model-free. The performance of statistical and neural methods usually comes into effect with regards to feature selection, feature extraction and classification. Statistical methods are usually the basis for evaluating the performance of neural classifiers [176].

Different fuzzy techniques have become more popular recently and they have successful applicability in several areas like decision-making, approximate reasoning, control, and pattern recognition to name but a few [143]. Fuzzy theory delivers an effectual framework for dealing with the uncertainty that may be found image processing, and other different forms of textures in several computer vision problems. Theoretically, a fuzzy set comprises objects, which are agreed by an instinctively defined membership function. The grade value of the defined membership of any object varies from 0 to 1. The value of 1 represents full membership, and the value approaching 0 signifies the weaker membership of the object in the fuzzy set [96]. The thought of fuzzy can be applied at the feature/classification level, to demonstrate class membership, and to give an estimation of lost information with regard to membership values. It has also been widely used in solving clustering problems, in which the class labels are given input data in an unsupervised manner on the basis of specified conditions. Applications of fuzzy coupled with image processing and pattern recognition have been presented in several domains in the literature [179]. The theories of fuzzy and neural networks are later integrated to construct a new intelligent processing, called neuro-fuzzy computing, which has enhanced the processing capability compared to traditional fuzzy classifiers or neural networks or can be a more powerful computational paradigm [180].

2.5 Overview of Different Thresholding Techniques

Image thresholding is a well-known, simple and frequently used technique in image segmentation [194]. Using thresholding as a tool is effortless to accomplish bi-level and multi-level image thresholding and it provides robust results in all conditions [34, 161]. There are a variety of image thresholding techniques in the literature, which are typically used mainly for image thresholding optimization. Sezgin and Sankar [231] categorized all the existing techniques into six categories. A host of important applications of image thresholding is available in [231]. Image thresholding appears to be an optimization problem where plenty of objectives like Otsu's function [192], Kapur's function [140], etc., can be used as the optimization functions. Based on the approach used, the output leads to the desired number of threshold values, called threshold. A few imperative, renowned and mostly used thresholding techniques for both categories of thresholding, are addressed below.

To begin with, let us consider the probability distribution of pixels' frequencies of a gray-scale image as given by

$$p_j = \frac{f_j}{\mathcal{E}}, j \in [0, 255] \tag{2.1}$$

and

$$P_T = \sum_{j=0}^{T} p_j \tag{2.2}$$

where f_j and $\mathcal{E}$ denote the frequency of the jth pixel and total pixels in the image, respectively $j \in [0, 255]$. P_T is a measure, computed by summing the probability distributions up to the threshold value, T.

2.5.1 Ramesh's Algorithm

Ramesh's algorithm is a histogram-based thresholding technique [207, 231]. In this technique, some parameters of the histogram, such as valleys, peaks, and curvature are taken into consideration. Ramesh et al. considered a two-step function to find a functional approximation to the probability mass function. This technique determines the optimum threshold value by minimizing the total values obtained by squaring the histogram and bi-level function [207, 231].

The optimum threshold value of the gray-scale image is determined by using an iterative approach, as given by [207].

$$T_{opt} = \min \left(\sum_{t=0}^{T} (b_1(T) - t)^2 + \sum_{t=T+1}^{L} (b_2(T) - t)^2 \right) \tag{2.3}$$

where

$$b_1(T) = \frac{m_f(T)}{P_g} \quad \text{and} \quad b_2(T) = \frac{m_b(T)}{(1 - P_g)} \tag{2.4}$$

and

$$m_f(T) = \sum_{i=0}^{T} ip_i \quad \text{and} \quad m_b(T) = \sum_{i=T+1}^{L} ip_i \tag{2.5}$$

where p_i and P_g are determined by Eqs. (2.1) and (2.2), respectively. $m_f(T)$ and $m_b(T)$ are called the foreground and background mean [207, 231].

2.5.2 Shanbag's Algorithm

This is an entropy-based technology which can be successfully used for bi-level thresholding. A set of axioms has been listed to construct the proposed technology [234]. The pixel intensity values of an image are categorized into two classes, namely, c_1 and c_2. The membership of the threshold, T where $T \in c_1$ is given as [231, 234]

$$Q_{1,T}(T) = 0.5 + d * p(T) \tag{2.6}$$

where, $p(T)$ is the required probability that the gray level i occurs and d is constant, which can be determined by the axioms referred above. In general, it is said that all the pixel intensity values having gray levels $T - i, T - i + 1, T - i + 2, \cdots, T$ may belong to c_1, if the gray level intensity value $T - i$ lies in both of c_1 and $i + 1$ is placed on the left-hand side of the hypothetical gray level T'. So considering the above statement, the following equation can be derived [231, 234]

$$Q_{1,T}(T - i) = 0.5 + d(p(T - i) + p(T - i + 1)$$
$$+ p(T - i + 2) + \cdots + p(T)) \tag{2.7}$$

Now, we have $Q_{1,T}(0) = 0.5 + d(p(1) + p(2) + \cdots + p(T))$ that in turn provides $Q_{1,T}(0) = 0.5 + d(p(c_1))$. Again, using the axioms from [234], we have $1 = 0.5 + d(p(c_1))$ or,

$$d = \frac{0.5}{p(c_1)} \tag{2.8}$$

Hence, Eq. (2.7) can be written again by [234]

$$Q_{1,T}(T - i) = Q_{1,T}(T - i + 1) + d(p(T - i)).$$

If we proceed further in a similar way, the following equation can also be obtained [234]

$$Q_{2,T}(T + i) = 0.5 + \frac{0.5}{p(c_2)}(p(T + i) + p(T + i - 1)$$
$$+ p(T + i - 2) + \cdots + p(T + 1)) \tag{2.9}$$

To normalize all members of c_1, the probability of each pixel is divided by $p(c_1)$. It gives $\frac{p(0)}{p(c_1)}, \frac{p(1)}{p(c_1)}, \cdots, \frac{p(T)}{p(c_1)}$. Summarizing all the information obtained above, for class c_1, the following relation occurs [234]

$$S_1 = -\frac{p(0)}{p(c_1)} \ln Q_{1,T}(0) - \frac{p(1)}{p(c_1)} \ln Q_{1,T}(1)$$

$$- \cdots - \frac{p(T)}{p(c_1)} \ln Q_{1,T}(T) \tag{2.10}$$

Similarly, for the another class, we get [234]

$$S_2 = -\frac{p(N)}{p(c_2)} \ln Q_{2,T}(N) - \frac{p(N-1)}{p(c_2)} \ln Q_{2,T}(N-1)$$

$$- \cdots - \frac{p(T+1)}{p(c_2)} \ln Q_{2,T}(T+1) \tag{2.11}$$

The optimum threshold value is obtained by minimizing $|S_1 - S_2|$ [234].

2.5.3 Correlation Coefficient

The term "correlation" can be termed a measure, which can be used to check the existing relationship between two variables. The correlation coefficient is a real number that gives us a good idea of the closeness between these two variables. Changing one variable requires changing the other in a predictable way. The correlation coefficient can be a either positive or negative number or even zero. The negative number confirms the existence of a negative relationship between these two variables. Hence, a negative relationship signifies that an increase in one variable tends to decrease the value of other in a predictable manner. Similarly, a positive correlation coefficient signifies the opposite phenomena to the previous case. Finally, a zero correlation coefficient indicates that there exists no relationship between these two variables. That means, there is nothing to predict the characteristics of one variable with respect to with the ups and downs of the other [26, 145, 226].

The correlation coefficients varies from -1.00 to $+1.00$. If it is perfectly -1.00, these two variables share a perfectly negative relationship. This means, the increase of the value of one variable directly decreases the value of the other one in a perfectly predictable manner. If it is perfectly $+1.00$, the opposite behavior can be observed as in the previous case.

The correlation coefficient between the foreground and background image can be defined as follows [26, 145, 226]

$$\rho = \frac{cov(X, Y)}{\sigma_X \sigma_Y} \tag{2.12}$$

where X and Y are the set of pixels of the foreground and background image, respectively. $cov(X, Y)$ is the covariance between X and Y and σ_X and σ_Y are the standard deviations of X and Y.

The correlation coefficients can also be calculated using the formula given by [26, 145, 226]

$$\rho = \frac{\sum_{i=1}^{n}(X_i - \overline{X})(Y_i - \overline{Y})}{\sqrt{\sum_{i=1}^{n}(X_i - \overline{X})^2 \sum_{i=1}^{n}(Y_i - \overline{Y})^2}} \tag{2.13}$$

where X_i, Y_i are respectively the pixel intensity values of the ith pixel of the foreground and background image. $\overline{X}$ and $\overline{Y}$ are the mean of the pixel intensity values of the foreground and background image. It is always recommended to have larger correlation coefficients because it signifies a stronger relationship between the two variables. The optimum threshold can be found by [26, 145, 226]

$$t_{opt} = \max(\rho) \tag{2.14}$$

2.5.4 Pun's Algorithm

Pun has proposed an entropy-based technique, called Pun's algorithm [206, 231]. The author has mulled over g-symbol source where $g = 256$ for gray-level histogram, statistically independent in nature. The ratio of entropy [206]

$$H'(T) = -P(T)\log[P(T)] - [1 - P(T)]\log[1 - P(T)] \tag{2.15}$$

to the source entropy [206]

$$H(T) = H_b(T) + H_f(T) \tag{2.16}$$

is

$$\frac{H'(T)}{H(T)} \geq Pu(\beta) = \beta \frac{\log P(T)}{\log(\max(p_0, p_1, p_2, \cdots, p_T))}$$
$$+ (1 - \beta) \frac{\log(1 - P(T))}{\log(\max(p_{T+1}, p_{T+2}, \cdots, p_g))} \tag{2.17}$$

where

$$H_b(T) = \beta H(T) \tag{2.18}$$

where, $H_b(T)$ and $H_f(T)$ are computed by [206].

$$H_b(T) = -\sum_{j=0}^{T} p_j \log(p_j) \tag{2.19}$$

and

$$H_f(T) = -\sum_{j=T+1}^{g} p_j \log(p_j) \tag{2.20}$$

where, p_j is computed using Eq. (2.1). The optimum threshold value is acquired by the particular T which persuades Eq. (2.18) with the argument β that maximizes $Pu(\beta)$ as given in Eq. (2.17) [206].

2.5.5 Wu's Algorithm

Wu, Songde and Hanqing proposed an entropy-based technique, called Wu's algorithm [231, 274]. The authors adopted the same evaluation strategy as suggested in Kapur's algorithm [140] to find the background and foreground entropies using Eqs. (2.19) and (2.20), respectively. The difference of the aforesaid class entropies ($W(T)$) is minimized to obtain the optimum threshold value, as given by [274]

$$W(T) = |H_b(T) - H_f(T)| \tag{2.21}$$

2.5.6 Renyi's Algorithm

This is a entropy-based thresholding technique [220, 231], which employs two probability distributions: one for the object and other for the background for an gray level image. Let us assume that $p_0, p_1, p_2 \cdots, p_{255}$ be the given probability distributions obtained from this image. The object class (D_1) and background class (D_2) are derived from the above probability distributions as given by

$$D_1 : \frac{p_0}{p(D_1)}, \frac{p_1}{p(D_1)}, \cdots, \frac{p_1}{p(D_1)}$$

and

$$D_2 : \frac{p_{T+1}}{p(D_2)}, \frac{p_{T+2}}{p(D_2)}, \cdots, \frac{p_{255}}{p(D_2)}$$

where

$$p(D_1) = \sum_{j=0}^{T} p_j, p(D_2) = \sum_{j=T+1}^{255} p_j \tag{2.22}$$

and

$$p(D_1) + p(D_2) = 1 \tag{2.23}$$

where p_j is obtained by Eq. (2.1). Renyi's entropy can be defined with reference to the foreground and background image distribution with parameter ρ. Formally, it is defined as follows [220]:

$$H_f^\rho = \frac{1}{1-\rho} \ln \left(\sum_{j=0}^{T} \left[\frac{p_j}{p(D_1)} \right]^\rho \right) \tag{2.24}$$

and

$$H_b^\rho = \frac{1}{1-\rho} \ln \left(\sum_{j=T+1}^{255} \left[\frac{p_j}{p(D_2)} \right]^\rho \right) \tag{2.25}$$

The optimum threshold value of a gray-level image is obtained by maximizing $\{H_f^\rho + H_b^\rho\}$.

2.5.7 Yen's Algorithm

Yen, Chang and Chang first introduced an entropy-based technique, called Yen's algorithm [231, 277]. Again the authors used the same approach as Kapur's algorithm [140] to determine the foreground and background image distributions. Renyi's entropy can be defined in relation to the foreground and background image distribution with a parameter (ρ), [277]:

$$H_f^\rho = \frac{1}{1-\rho} \ln \left(\sum_{j=0}^{T} \left[\frac{p_j}{P_T} \right]^\rho \right) \tag{2.26}$$

and

$$H_b^\rho = \frac{1}{1-\rho} \ln \left(\sum_{j=T+1}^{g} \left[\frac{p_j}{1-P_T} \right]^\rho \right)$$

(2.27)

where, p_j and P_T are obtained by Eqs. (2.1) and (2.2), respectively.

The authors defined the *entropic correlation* by [277]

$$EC(T) = Y_b(T) + Y_f(T)$$

$$= -\ln \left(\sum_{j=0}^{T} \left[\frac{p_j}{P_T} \right]^2 \right) - \ln \left(\sum_{j=T+1}^{255} \left[\frac{p_j}{1-P_T} \right]^2 \right)$$

(2.28)

Equations (2.26) and (2.27) are transformed into the functions $Y_b(T)$ and $Y_f(T)$ for $\rho = 2$ [231, 277]. Maximization of $EC(T)$ yields the optimum threshold value of an image.

2.5.8 Johannsen's Algorithm

Johannsen and Bille proposed a popular thresholding technique, called Johannsen's algorithm [135, 231]. To obtain the optimal threshold of a given gray-level image, the method minimizes the expression given by [135]

$$J(T) = J_b(T) + J_f(T)$$

$$= \log(P_T) + \frac{1}{P_T} \{E(p_T) + E(P_{T-1})\} + \log(1 - P_{T-1})$$

$$+ \frac{1}{1-P_{T-1}} \{E(p_T) + E(1 - P_T)\}$$

(2.29)

where, $E(p) = -p\log(p)$, and the values of p_j and P_T are derived from Eqs. (2.1) and (2.2), respectively [135].

2.5.9 Silva's Algorithm

The thresholding technique developed by da Silva, Lins and Rocha [231, 236] performs a statistical measure. For this reason, entropy is calculated between the gray level distributions and its binary version of the test image. Let S be the entropy of the gray-scale image as given by [236]

$$S = \sum_{j=0}^{255} p_j \log_2(p_j)$$

(2.30)

First, the *priori probability distribution*, namely, $\{p_0, p_1, \cdots, p_{255}\}$ is calculated using Eq. (2.1). The value of T is then computed by the *posteriori probability distributions*, namely, $\{P_T, 1 - P_T\}$ for $P_T \leq 0.5$, which is then associated with the entropy-based distribution as given by

$$S'(T) = h(P_T)$$

(2.31)

where $h(p) = -p\log_2(p) - (1-p)\log_2(1-p)$ and the value of P_T is calculated using Eq. (2.2).

A loss factor β is determined by the equation as

$$\beta(S/\log(256)) = -\frac{3}{7}(S/\log(256)) + 0.8,$$

$$if\ S/\log(256) < 0.7$$

$$= (S/\log(256)) - 0.2,\ if\ S/\log(256) \geq 0.7 \qquad (2.32)$$

Finally, the optimum threshold of this image is determined by minimizing $|E(T)|$ as given by

$$|E(T)| = \left| \frac{S'(T)}{S/\log(256)} - \beta(S/\log(256)) \right| \qquad (2.33)$$

2.5.10 Fuzzy Algorithm

The fuzzy measures can be defined as the measurement of degree such that a gray level (g) fits in the object along with its background [231]. The term index of fuzziness has been proposed by Huang and Wang. The authors used this measure by computing the distance between the gray-scale image and the corresponding binary version [130, 257]. Let us consider a gray-scale image of the dimension $C \times D$, which has L gray levels $g \in [0, 1, \cdots , L]$. For a subset $B \subseteq Y$ and the membership function $\mu_Y(g)$, the linear index of fuzziness γ_l can be defined by [130, 257]

$$\gamma_l = \frac{2}{CD} \sum_{g=0}^{L-1} h(g) \times \min\{\mu_B(g), 1 - \mu_B(g)\} \qquad (2.34)$$

where $h(g)$ represents the histogram of the data set. The optimum threshold is obtained by minimizing γ_l [130, 257].

2.5.11 Brink's Algorithm

The cross-entropy-based thresholding technique was first introduced by A. D. Brink and N. E. Pendock [35, 231]. In this technique, the original gray-scale images are used as the preceding distribution with the existing gray-level intensity values to select threshold values (T) of the images. The average values comprising those above and below this threshold value, are considered. These values are represented by the notations $\mu_f(T)$ and $\mu_b(T)$, respectively. These two values are used as the two gray levels to determine the threshold value of binary thresholded image by [35].

$$\mu_b(T) = \sum_{i=0}^{T} ip_i \ \text{ and } \ \mu_f(T) = \sum_{i=T+1}^{255} ip_i \qquad (2.35)$$

where p_i for $i \in [0, 255]$ is determined by Eq. (2.1) [35].

Let us consider two probability distribution functions, called *priori* probability distribution ($p(y)$) and *posteriori* probability distribution ($q(y)$). The cross-entropy distance between these two measures is described as follows [35]

$$Z_{CE}(q, p) = \int q(y) \log \frac{q(y)}{p(y)} dy \qquad (2.36)$$

Using Eq. (2.36), the integration is substituted to the summation of discrete probability distributions. For the discrete cross-entropy, the following equation is satisfied [35]:

$$Z_{CE}(q,p) = \sum_{y=1}^{N} q_y \log \frac{q_y}{p_y} \tag{2.37}$$

$$\text{subject to,} \sum_{y=1}^{N} p_y = \sum_{y=1}^{N} q_y \ (= 1)$$

where N represents the number of discrete points in the discrete probability distribution. By nature, Eq. (2.36) is non-symmetric. For metric distance measure, the non-symmetric equation must be transformed into its equivalent symmetry form using the following equation [34]:

$$Z_{MD}(p,q) = Z_{CE}(q,p) + Z_{CE}(p,q)$$
$$= \sum_{y=1}^{N} q_y \log \frac{q_y}{p_y} + \sum_{y=1}^{N} p_y \log \frac{p_y}{q_y} \tag{2.38}$$

Now, Eqs. (2.35) and (2.38) are turned into the following form [34]

$$\sum_{j=0}^{M} g_j = \left[\sum_{j=0}^{M} \mu_b(T) \right]_{g_j \leq T} + \left[\sum_{j=0}^{M} \mu_f(T) \right]_{g_j > T} \tag{2.39}$$

where M and $g_j \in [0, 255]$ represent the number of pixels and pixel intensity value of an image. In a similar way, Eq. (2.38) can also be written in a different form.

As the size of the image increases, it may take more time to calculate the threshold value using Eq. (2.38) and its equivalent equation (2.39). To improve the computational capability, a gray-level frequency histogram is considered. From the perspective of the histogram, it is easy to calculate the frequency (f) of each gray level (g). This condition leads to transforming Eq. (2.39) into the following form [34]:

$$Z_{CE}(T) = \sum_{j=0}^{T} f\mu_b(T) \log \frac{\mu_b(T)}{j}$$
$$+ \sum_{j=T+1}^{255} f\mu_f(T) \log \frac{\mu_f(T)}{j} \tag{2.40}$$

Similarly, Eq. (2.38) can be transformed into [34].

$$Z_{MD}(T) = \sum_{j=0}^{T} f \left[\mu_b(T) \log \frac{\mu_b(T)}{j} + j \log \frac{j}{\mu_b(T)} \right]$$
$$+ \sum_{j=T+1}^{255} f \left[\mu_f(T) \log \frac{\mu_f(T)}{j} + j \log \frac{j}{\mu_f(T)} \right] \tag{2.41}$$

The optimum threshold value is computed by minimizing $Z_{CE}(T)$ or $Z_{MD}(T)$ using Eqs. (2.40) and (2.41), respectively [34, 231].

2.5.12 Otsu's Algorithm

Otsu's function [192] is a cluster-based approach. It is used as an objective function for underlying optimization problems of bi-level and multi-level image thresholding. For multi-level image thresholding, it finds a set of optimal thresholds $\{\theta_1, \theta_2, \cdots, \theta_{K-1}\}$ that maximize the variance of classes. Formally, it can be defined as follows [192]:

$$F = f\{\theta_1, \theta_2, \cdots, \theta_{K-1}\} = \sum_{j=1}^{K} \omega_j(\mu_j - \mu) \tag{2.42}$$

where K is the number of classes in $C = \{C_1, C_2, \ldots, C_K\}$ and

$$\omega_j = \sum_{i \in C_j} p_i, \ \mu_j = \sum_{i \in C_j} ip_i/\omega_j \tag{2.43}$$

where, p_i is the probability of ith pixel (obtained from Eq. (2.1)). ω_j and μ_j represent the probability and the mean of class C_j, respectively, while μ is the mean of class C. Note that the maximization of F will provide the number of thresholds for an image.

2.5.13 Kittler's Algorithm

Kittler and Illingworth [148] proposed a technique for image thresholding based on error computation. According to their supposition, an image can be described as a mixture distribution between the foreground and the corresponding background pixels. They have considered the histogram and transform it into its parametric form that should be best fitted in Gaussian distribution. Kittler's method can be successfully implemented as a fitness function for multi-level image thresholding. It can find a set of optimal threshold values, $\{\theta_1, \theta_2, \cdots, \theta_{C-1}\}$ for image segmentation where C represents the predefined number of segregated classes. This method computes the error between the original histogram and its corresponding parametric distribution and tries to minimize the error for the optimal solution. Formally, the operative formula of Kittler's method is stated as follows [148]

$$\mathcal{U}_l = \eta^2\{\theta_1, \theta_2, \cdots, \theta_K\}$$

$$= 1 + 2 \times \sum_{j=0}^{K} (\varpi_j(\log \varsigma_j - \log \varpi_j)) \tag{2.44}$$

where $K = C - 1$, $\theta_{op} = \{\theta_1, \theta_2, \cdots, \theta_K\}$ are the optimal solutions where a single threshold value is found from each class in $C = \{C_1, C_2, \ldots, C_K\}$ and

$$\varpi_k = \sum_{i \in C_k} p_i, \ \mu_k = \sum_{i \in C_k} \frac{p_i \times i}{\varpi_k}$$

$$\text{and } \varsigma_k^2 = \sum_{i \in C_k} \frac{p_i \times (i - \mu_k)^2}{\omega_k} \tag{2.45}$$

where, $1 \le k \le C$, p_i represents the probability of ith pixel. p_i is computed using Eq. (2.1). ϖ_k, μ_k and ς_k are called the priori probability, mean, and standard deviation for class C_k. Using Eq. (2.44), the optimum threshold values are determined from the input image by minimizing $\mathcal{U}_l$.

2.5.14 Li's Algorithm

This is a minimum cross-entropy-based thresholding method proposed by Li and Lee [158]. Kullback [151] has proposed the theory of cross-entropy. Let $\mathcal{X}$ be a given data set. Two probability distributions, namely, $\mathcal{A} = a_1, a_2, \ldots, a_r$ and $\mathcal{B} = b_1, b_2, \ldots, b_r$ are generated from $\mathcal{X}$. The cross-entropy between $\mathcal{A}$ and $\mathcal{B}$ can be defined as an information theoretic distance measured between $\mathcal{A}$ and $\mathcal{B}$. Formally, the information theoretic distance is calculated as

$$\mathcal{V}(\mathcal{A}, \mathcal{B}) = \sum_{j=0}^{r} a_j \log \frac{a_j}{b_j} \tag{2.46}$$

Let $\mathcal{I}$ be the given image. For each positive integer $k \in [0, L]$, the histogram, $h(k)$ is computed where, L is the maximum gray-level intensity value in $\mathcal{I}$. This method calculates the cross-entropy between the $\mathcal{I}$ and its thresholded version, say, $\mathcal{I}_T$ to find the optimal threshold values for $\mathcal{I}$. Let, at any level of computation in multi-level thresholding, it find $\{\theta_1, \theta_2, \cdots, \theta_K\}, K = C - 1$ as the threshold values where C denotes the predefined number of classes. The thresholded image is constructed as follows:

$$\begin{aligned}
I_T(u, v) &= \mu(0, \theta_1), \quad I(u, v) \in [0, \theta_1] \\
&= \mu(\theta_1, \theta_2), \quad I(u, v) \in (\theta_1, \theta_2] \\
&= \mu(\theta_2, \theta_3), \quad I(u, v) \in (\theta_2, \theta_3] \\
&\vdots \\
&= \mu(\theta_K, L), \quad I(u, v) \in (\theta_K, L]
\end{aligned} \tag{2.47}$$

where

$$\mu(e, f) = \frac{\sum_{k=e}^{(f-1)} k h(k)}{\sum_{k=e}^{(f-1)} h(k)} \tag{2.48}$$

Formally, the cross-entropy can be defined as

$$\mathcal{V} = \kappa\{\theta_1, \theta_2, \cdots, \theta_K\} = \sum_{j=0}^{K} \sum_{k \in C_j} k h(k) \log \left[\frac{k}{\mu(C_j)} \right] \tag{2.49}$$

Minimization of $\mathcal{V}$ (as given in Eq. (2.49)) leads to the optimum threshold values of the input image.

2.5.15 Kapur's Algorithm

Kapur's function [140] can be used as an objective function in an image thresholding optimization problem. It aims to determine the appropriate number of optimal threshold $\{\theta_1, \theta_2, \cdots, \theta_{K-1}\}$ from gray-scale/color images. Kapur's function is an entropy-based optimization technique which maximizes the sum of entropies for each individual class. Formally, it is described as follows [140]:

$$\mathcal{V}_l = \varsigma^2\{\theta_1, \theta_2, \cdots, \theta_{K-1}\} = \sum_{k=0}^{K-1} Z_k \tag{2.50}$$

where $C = \{C_1, C_2, \ldots, C_K\}$ represents the set of K number of classes and

$$\omega_k = \sum_{i \in C_k} p_i, Z_k = -\sum_{i \in C_k} \frac{p_i}{\omega_k} \ln \frac{p_i}{\omega_k} \tag{2.51}$$

where p_i denotes the probability of ith pixel, computed using Eq. (2.1). The number of optimum thresholds of an image is obtained by maximizing V_l.

2.5.16 Huang's Algorithm

This is a popular image thresholding technique, developed by Liang-Kai Huang and Mao-Jiun J. Wang [130]. Like Kapur's method [140], this method can also be efficiently employed in multi-level image thresholding, which results in a number of optimum threshold values, namely, $\{\theta_1, \theta_2, \cdots, \theta_{K-1}\}$ where K represents a number of selected classes. For an input image, Huang's method minimizes its measures of fuzziness to reach optimality. In this method, the membership function is employed to signify the characteristic relationship between a pixel and the region where it belongs. The details of Huang's method are discussed below.

Let $\mathcal{A}$ represents a set of image of dimension $D_1 \times D_2$ with L levels, where the gray level value of (c, d) pixel is denoted by y_{cd}. Let $\mu_{A(y_{cd})}$ be the membership value in $\mathcal{A}$, which can be described as a characteristic function that expresses the fuzziness of (c, d) pixel in $\mathcal{A}$. Using fuzzy set notation, $\mathcal{A}$ can be defined as follows:

$$\mathcal{A} = \{(y_{cd}, \mu_{A(y_{cd})})\} \tag{2.52}$$

where $\mu_{A(y_{cd})} \in [0, 1], c = 0, 1, \ldots, D_1 - 1$ and $d = 0, 1, \ldots, D_2 - 1$.

Let the pixel intensity values of a gray-scale image be segregated into K number of classes, namely, $C = \{C_1, C_2, \ldots, C_K\}$. The consecutive classes are actually separated by the threshold values, namely, $\{\theta_1, \theta_2, \cdots, \theta_{K-1}\}$, respectively. Let the mean gray levels of the classes in C be denoted by $\mu_0, \mu_1, \ldots, \mu_{K-1}$, respectively, which are calculated using the equations given by

$$\mu_0 = \frac{\sum_{u=0}^{\theta_1} uh(u)}{\sum_{u=0}^{\theta_1} h(u)},$$

$$\mu_1 = \frac{\sum_{u=\theta_1+1}^{\theta_2} uh(u)}{\sum_{u=\theta_1+1}^{\theta_2} h(u)}, \tag{2.53}$$

$$\ldots \ldots \ldots \ldots \ldots \ldots$$

$$\ldots \ldots \ldots \ldots \ldots \ldots$$

$$\mu_{K-1} = \frac{\sum_{u=\theta_{K-1}}^{L} uh(u)}{\sum_{u=\theta_{K-1}}^{L} h(u)}$$

where $h(u)$ denotes the frequency of uth pixel in the image.

The aforementioned membership function for pixel (c, d) is defined as follows:

$$A(y_{cd}) = \frac{1}{1 + |y_{cd} - \mu_0|/ct} , y_{cd} \in [0, \theta_1]$$

$$= \frac{1}{1 + |y_{cd} - \mu_1|/ct} \quad , y_{cd} \in [\theta_1, \theta_2]$$

$$\dots \dots \dots \dots \dots \dots \dots$$

$$\dots \dots \dots \dots \dots \dots \dots \tag{2.54}$$

$$= \frac{1}{1 + |y_{cd} - \mu_{K-1}|/ct} \quad , y_{cd} \in [\theta_{K-1}, L]$$

where ct is constant such that $A(y_{cd}) \in [1/2, 1]$.

Let the complement of fuzzy image set A be denoted by $\overline{A}$. The distance between these two sets is defined by

$$Y_p(A, \overline{A}) = \left[\sum_c \sum_d |A(y_{cd}) - \overline{A}(y_{cd})|^p \right]^{\frac{1}{p}}, \quad p = 1, 2, \dots, L - 1 \tag{2.55}$$

where $\overline{A}(y_{cd}) = 1 - A(y_{cd})$. Accordingly, the measure of fuzziness of A is defined by

$$\mathcal{V} = 1 - \frac{Y_p(A, \overline{A})}{(D_1 \times D_2)^{\frac{1}{p}}} \tag{2.56}$$

For $p = 1$ and 2, Y_p is referred to as the Hamming metric and the Euclidean metric, respectively. The optimum threshold values are determined by minimizing $\mathcal{V}$ as given in Eq. (2.56).

2.6 Applications of Image Analysis

The applications of image analysis are constantly growing in all fields of industry and science. Some typical applications of image analysis are discussed as follows:

1. *Computer vision*: Computer vision is basically defined as an automated system that can be used to recognize and process images as the human vision does, and deliver an apposite output. Computer vision has wide applicability in both science and technology. Typical applications of this process may include the systems to control processes like an autonomous vehicle, an industrial robot, to detect events like visual surveillance, to organize information like indexing image databases, and so on [16].
2. *Robotics*: Among different areas of application of image analysis, Robotics is one of the most important. Robotics can be described as developing automated machines that can be used to replace human beings and imitate human actions, that can reduce significant amounts of human efforts. Many techniques of robotic systems are already available in the literature [214].
3. *Remote sensing*: Remote sensing is an acquisition process used to gather information signals from different sources, such as objects or any other phenomenon. A variety of real-time sensing apparatus, usually of the wireless type, is used to perform the remote sensing operation. In practice, remote sensing is an assortment of diverse data signals acquired from a number of devices, used to gather information on a specified object. Some distinct examples are X-radiation (X-ray), Positron Emission Tomography (PET), MRI, and many others [227, 271].

4. *Biometric verification*: It is a process of automatic identification of humans by their intrinsic behaviors. It is an effective type of recognition and access control process. This process can also be applied to identify any particular individuals from groups of several people. This image analysis technique ensures security of a system by allowing the legitimate users to access its rendered services. It is basically a pattern recognition system which works on the basis of obtaining biometric data from any individual. In this approach, a set of defined features is extracted, based on the given data set, and then this feature set is compared against the stored data in the database [59].

5. *Face detection*: In this process, the essential facial features are figured out and the other parts are ignored. In the true sense, this process is an explicit example of object class detection. The purpose of face recognition is to explore the stated features, such as positions and sizes of different faces. A number of face detection algorithms exist in the literature, which generally focus on detecting frontal human faces [198]. Several attempts have also been made to resolve the more general and challenging problems of multi-view face recognition [198, 256].

6. *Astronomy*: Image analysis has a very effective role in astronomy. The astronomical observations involve raw images that are required to be processed to ascertain the pertinent scientific information. Various metrics like signal to noise ratio and other factors are figured out for different purposes. A variety of methods that can efficiently deal with astronomical images are available in the literature [228].

7. *Medical science*: Image analysis is very important in medical science. Medical imaging, has an effective role in the detection and recovery of different diseases. It is used in various areas of medical science and technology. For example, mammography is the most important procedure for investigating breast cancer, a common malignancy in females. Mostly, cancer can be cured by therapy or surgery, if detected and diagnosed at an early stage. The computerized version of this technique increases its predictive value by pre-reading the mammogram, which may indicate the position of suspicious defects and investigate their features [78].

8. *Optical Character Recognition*: This is a popular and widely used image analysis technique. It is translation process mainly used in artificial intelligence, pattern recognition, and a few others, in which images in the form of handwritten or printed text is converted into machine (computer) editable text. In the literature, some popular and cost-effective types are available [16].

9. *Signature recognition*: Signature recognition is a useful image analysis technique; the signature of a given signer is verified based on the stored image of the signature and a few other samples of it. A variety of features, such as size, orientation, and many others are taken into consideration while verifying the handwritten signature. A robust handwritten signature recognition system has to account for all of these factors [200].

2.7 Conclusion

This chapter focuses on presenting the details of image analysis. The current technologies and a variety of applications of image analysis are also presented in this chapter with the aid of apposite instances. As a sequel, the outline of a number of thresholding techniques of image analysis is also hinted at using apposite illustrations.

2.8   Summary

- The explosive evolution in hardware and software technology has led to several important improvements in imaging technology.
- Several automatic methods have been developed during the last few years in different fields, which in turn have reduced human efforts by a significant amount with the utmost accuracy.
- Image analysis has a vital role in extracting relevant and meaningful data from images.
- There are a few automatic or semi-automatic techniques, called computer/machine vision, pattern recognition, image description, image understanding.
- The stages of image analysis are image acquisition, image pre-processing, image segmentation, feature extraction, and pattern classification.
- Image thresholding is a well-known, simple, and frequently used technique in image segmentation.
- Image thresholding is of two types, namely, bi-level and multi-level thresholding.
- The features of an image are generally characterized into three basic categories, namely, morphology, color, and texture.
- Feature selection techniques are of two types, namely, supervised and unsupervised feature selection.
- Classification investigates the numerical features of a variety of features and arranges data into different categories.

Exercise Questions

Multiple Choice Questions

1   The image analysis class of tasks includes
   - (a) image segmentation
   - (b) feature description
   - (c) object classification
   - (d) all of the above

2   Which of the following is not an example of image segmentation approach?
   - (a) region-based segmentation
   - (b) texture segmentation
   - (c) feature extraction
   - (d) thresholding

3   Some popular object classification techniques include
   - (a) neural networks

(b) fuzzy techniques
(c) machine learning
(d) all of the above

4 Clustering algorithms divide images into several classes, generally known as
(a) field
(b) clusters
(c) areas
(d) none of the above

5 Edge detection technique can be categorized as
(a) gradient
(b) Laplacian
(c) none of the above
(d) both (a) and (b)

6 Feature selection techniques can be described as
(a) supervised feature selection
(b) unsupervised feature selection
(c) none of the above
(d) both (a) and (b)

7 The concepts of fuzzy and neural networks are later combined to form a new intelligent process, called
(a) neuro-fuzzy computing
(b) neuro-based fuzzy computing
(c) fuzzy-based neuro computing
(d) none of the above

Short Answer Questions

1 What do we mean by image analysis? Describe in brief.

2 What is the difference between edge-based and region-based segmentation techniques? Illustrate with suitable examples.

3 What is a cluster? How can you make clusters from an image?

4 Discuss, in brief, several edge detection techniques.

5 What is the difference between feature selection and feature extraction methodologies? Illustrate with examples.

6 Discuss the stages of image analysis with a suitable diagram.

Long Answer Questions

1 Describe the mathematical formulation of the image segmentation technique.

2 Describe, in brief, digital image analysis methodologies with suitable examples.

3 What are the available feature selection schemes? Describe with examples.

4 Give an overview of different thresholding techniques. How can you classify them?

5 Give some application areas of image analysis.

3

Overview of Meta-heuristics

3.1 Introduction

Meta-heuristics are natural stochastics, precisely designed to solve different optimization problems, which may cover a large search space when determining optimum solutions. Meta-heuristics are typically used to expose the appropriate solutions of specific combinatorial optimization problems. Fundamentally, these techniques are recognized as generation-based procedures, which discover optimal solutions by exploring and exploiting their search space within a stipulated computational time. It has been observed that the fields of application of meta-heuristics range between simple local search procedures to complex searching processes. On some occasions, meta-heuristics may necessitate certain added intelligence to obtain a solution to certain kinds of problems. They have been applied as an alternate methodology for underlying optimization problems of multi-level thresholding. Moreover, meta-heuristics have been applied in an extensive range of different applications. Some typical applications of meta-heuristics are presented in [31, 108, 169–171]. Meta-heuristics are notable for the executional time that is needed to find the global optimum.

Evolutionary algorithms (EAs) are stochastic search and optimization methods inspired by the laws of *biological evolution.* These are basically population-based techniques which are inspired by Darwinian principles. These are very useful for solving computational problems in a wide range of applications. In the last few decades, EAs have been successfully employed by a number of researchers in different search and optimization techniques. The rule of nature admits that evolution is generally established by natural selection. To survive in an environment, each individual always remains occupied in the competition to grab environmental resources. In each iteration, the individuals aspire to proliferate their own genetic material. There exists quite a few evolutionary algorithms in this area. These may be categorized as: (i) genetic programming (GP), (ii) evolutionary programming (EP), (iii) evolutionary strategies(ES), and (iv) genetic algorithms (GAs). GP deals with evolving programs. EP is very useful to optimize the continuous functions. It does not have any recombination operation in its schedule. Like EP, ES is employed in optimizing continuous functions but performs the recombination operation. Finally, GA is very popular for optimizing general combinatorial problems. The most widely used evolutionary algorithms in vogue include Evolution Strategies by Rechenberg [209], the Evolutionary Programming by Fogel [91], Genetic Algorithm (GA) by Holland [127], the Simulated Annealing by Kirkpatrick et al. [147], the Tabu Search technique by F. Glover [107, 109, 116],

Quantum Inspired Meta-heuristics for Image Analysis, First Edition.
Sandip Dey, Siddhartha Bhattacharyya, and Ujjwal Maulik.
© 2019 John Wiley & Sons Ltd. Published 2019 by John Wiley & Sons Ltd.

Ant Colony Optimization by Dorigo et al. [76], Differential Evolution by Storn and Price [242], to name a few.

EAs are characterized by the encoding of the solutions, an evaluation function to compute the fitness of the representative solutions, and other parameters such as *population* size, parental *selection* schemes and characteristic *operators*. The *operators* are basically of two types, namely, *reproduction* and *evolution*. The *reproduction* operator include a *selection* mechanism and the *evolution* operator is generally guided by two different operators successively, namely, *crossover* and *mutation* operators. At each iteration, the population is updated using different operators. Generally, an iteration is known as a *generation*. The parent individual produces new solution space in the consecutive generations. The fitness value of each individual is evaluated by a figure of merit, generally known as the *objective function*. Based on fitness values, the participating individuals are stochastically selected for the next generations.

The chapter is arranged as follows. A brief summary of the impact on controlling parameters for EA is discussed in Section 3.1.1. This chapter throws light on six popular meta-heuristics. The fundamentals of these algorithms and their pseudo-codes are presented in brief in Sections 3.2–3.7. The conclusion of this chapter is presented in Section 3.8. Chapter summary and different types of questions are given at the end of this chapter.

3.1.1 Impact on Controlling Parameters

The performance of an EA may suffer due to the inaccurate selection of the control parameters. The best selection of control parameters has a direct influence on its performance. An assortment of studies are already available in this approach. The optimum values of control parameters can be determined as the best use of an EA, particularly for GA. For instance, if *crossover* or/and *mutation* probability is fixed at very high value, GA explores much space to find its solutions. It may possibly resort to losing a few good solutions with a very high probability. Schaffer et al. [224] studied the behavior of control parameters at best. The authors carried out some relevant computational experiments to examine the influence of the control parameters on the performance of GA. As per their assessment, the average performance is susceptible to crossover, mutation rates and population size. The performance also differs on the selection of crossover points (single/multiple) to be used in mating. Two-point crossover is generally superior to single-point crossover. One unpleasant situation may occur in EAs due to the early convergence of requisite solutions. This state of affairs can be avoided in a fairly simple way by adopting two-point crossover instead of single point.

Some popular EA-based algorithms are discussed in the following sections.

3.2 Genetic Algorithms

Genetic algorithms (GAs) [18, 53, 110, 175] are effective, adaptive, and robust optimization and randomized search techniques guided by the Darwinian laws of biological evolution. GAs imitate principles of natural genetics when solving complex optimization problems. These evolutionary techniques fundamentally encompass a set of individual chromosomes, known as population. This technique imitates the natural selection process, in which the fittest individuals (the best chromosomes) are chosen for reproduction to produce the offspring of the subsequent generation. They possess a large amount of implicit parallelism and deliver optimal solutions or near optimal solutions of a given objective function in large, complex and multi-modal scenery.

3.2.1 Fundamental Principles and Features

A genetic algorithm is also characterized by a population of trial solutions and two types of operators, namely, *reproduction* and *evolution* to act on it. The reproduction operator is guided by a selection mechanism. The evolution operator includes the crossover and mutation operators. The search technique is implemented through a series of generations. Two biologically inspired operators called genetic operators are used to create a new set of potentially better population of chromosomes for the next generation by means of a figure of merit referred to as the *objective/fitness* function, which evaluates the fitness of the candidate solutions.

GAs are identified by five distinct components [127, 210] as given by:

- *chromosome encoding*: It encodes the chromosomes to replicate the solution space by binary encoding, value encoding, permutation encoding, or tree encoding methods.
- *fitness evaluation*: It determines the suitability of a chromosome as a solution to a particular problem.
- *selection*: An operator which generates a new population on the basis of a probability proportional to the computed fitness value by adopting one selection strategy from a roulette wheel selection, Boltzmann selection or rank selection.
- *crossover*: An operator which retains the quality of chromosomes of previous generations. Parents having an equal probability of crossover rate are selected for breeding. The basic single-point crossover uses a randomly selected crossover point, based upon which the two chromosomes are divided into heads and tails and swapped and rejoined to produce a new offspring.
- *mutation*: Another genetic operator is used to induce diversity in the population (preventing it from getting stuck to a local optimum) by altering the value at a particular position selected randomly in a chromosome.

Despite these facts, GAs do not guarantee that higher fit solutions will rapidly dominate the population after a certain number of generations.

3.2.2 Pseudo-code of Genetic Algorithms

The working steps of GA are described by the following ***pseudo-code***.

```
begin
   iteration := 1;
   initpop P(iteration);
   evaluate P(iteration);
   repeat
      Z := selectpar P(iteration);
      recombine Z(iteration);
      mutate Z(iteration);
      evaluate Z(iteration);
```

```
    P := selectsurv P, Z(iteration);
    iteration := iteration + 1;
  until iteration>max_gen
end
```

Here, the **initpop** function is used to generate the population of initial (trial) solutions (P) at random. Afterward, the fitness of each member in P is evaluated using the **evaluate** function. The **selectpar** function uses a selection method to produce its descendant (Z). After that, two genetic operators namely, **recombine** and **mutate** are successively used by the selected members of this offspring. Then **evaluate** function evaluates the fitness of the individuals in (Z) to select the best survivors. This practice is continued for a predefined number of generations.

3.2.3   Encoding Strategy and the Creation of Population

In the first step, GAs begins with the chromosomal representation of a set of parameters which is required to be encoded as a fixed size (finite number) string. The string comprises an alphabet of predetermined length. Let us illustrate this encoding strategy with the following examples.

Illustration 1: For binary encoding, a string (binary chromosome) of length eight is taken as follows

1 1 0 1 0 0 0 1

Each chromosome in reality introduces a coded possible solution. A population comprises a set of such chromosomes in a particular generation. Generation-wise, the population size may either be constant or may vary as per the nature of the problem. In general, the initial population is generated either at random or any domain-specific knowledge may be used for that purpose.

3.2.4   Evaluation Techniques

The objective function is selected according to the problem to be solved and optimized. In this way, the strings for possible solutions are representative of good points in the search space that should possess high fitness values. GAs use this information (also called payoff information) to search for possible solutions.

3.2.5   Genetic Operators

Three genetic operators, namely, selection, crossover (reproduction), and mutation operators are successively applied to the population of chromosomes to yield potentially new offspring with diversity. These operators are now discussed below.

3.2.6   Selection Mechanism

The selection mechanism is an obligatory part of GAs. The selected individuals are chosen for mating to produce offspring individuals for the next generation. Several selection scheme ares in vogue. These are illustrated briefly as follows. For selection purposes, the following approaches are mostly used to implement GAs.

- *Roulette wheel selection method*: This is the simplest and most used selection scheme. Each individual in the population is allocated a section of the roulette wheel, proportional to its fitness value [110]. Higher fitness value indicates a larger segment portion of the wheel. The selection process through the roulette wheel selection approach is continued until the required number of individuals are picked to fill the population. The stronger individuals (having high fitness values) have a higher possibility of becoming a member of the population for the next generation. Formally, the selection probability of kth individual can be defined as follows:

$$P_k = \frac{FT_k}{\sum_{j=1}^{ni} FT_j} \tag{3.1}$$

where, FT_k represents the fitness value of kth individual and ni is the size of the population.

- *Ranking selection method*: In this selection approach, the individuals are sorted according to their fitness values. First, the sorted individuals are ranked and thereafter a selection probability is assigned to each individual with regard to its rank [17]. An individual is selected based on its selection probability. The rank selection mechanism does not involve a quick convergence at the earliest. The mode of practice of this method looks more robust compared to the other selection approaches [13, 273]. The implementation of this method is done through the following formula:

$$\mathcal{R}_k = \frac{r_k}{\sum_{j=1}^{ni} r_j} \tag{3.2}$$

where, r_k stands for rank of kth individual and ni is the population size.

- *Tournament selection method*: Tournament selection is a very popular selection procedure and perhaps the simplest method in terms of implementation [111]. This method simply draws a number of individuals at random (without or with replacement) from the existing population and selects the best individual as a parent with respect to fitness. The population for the subsequent generation is filled though this process [241]. With the change of tournament size, the selection pressure can easily be varied. Generally, a larger tournament size indicates higher selection pressure.

A few additional selection methods may include stochastic universal selection and stochastic remainder selection.

3.2.7 Crossover

The aim of applying the crossover operator is to exchange genetic information between two parent chromosomes selected at random. It amalgamates a part of each parent chromosome to create offspring for the subsequent generation. Crossover can be divided into two categories, namely, single-point crossover and multi-point crossover. The former is the most frequently used scheme. In the first step, the members of selected chromosomes in the mating pool are randomly paired. Thereafter, based on a predefined probability μ_{cr}, a pair of parent chromosomes is selected to cross over at a random point, p (also called the crossover point), where $k \in [1, (l-1)]$ and $l > 1$, l is called the string length. A pair of new offspring (chromosomes) is formed by exchanging all information from position $(p + 1)$ to l. Let us illustrate this crossover process with the following examples.

Illustration 2: Two strings, each of length eight, are taken along with the crossover points, as shown below.

$$1 \quad 1 \quad 0 \quad 0 \quad 0 \quad | \quad 1 \quad 0 \quad 0$$

$$0 \quad 1 \quad 1 \quad 0 \quad 1 \quad | \quad 0 \quad 0 \quad 1$$

After crossover, the new offspring will be formed as follow.

$$1 \quad 1 \quad 0 \quad 0 \quad 0 \quad | \quad 0 \quad 0 \quad 1$$

$$0 \quad 1 \quad 1 \quad 0 \quad 1 \quad | \quad 1 \quad 0 \quad 0$$

Some other common crossover techniques are two-point crossover, multiple point crossover, shuffle-exchange crossover and uniform crossover [53].

3.2.8 Mutation

Mutation is a genetic operator which alters the chromosomal genetic structure at random. The foremost objective of mutation operation is to achieve genetic diversity in the population of chromosomes. Sometimes the optimal solution exists in a part of the search space which may not be represented in the existing genetic structure of the population. So, it may not be possible for the process to reach the global optima. Hence, the mutation can be the single approach adopted to direct the population to attain the global optima in the search space in such conditions. Mutation randomly alters the genetic information in a chromosome. Mutation of a binary gene is performed by simply flipping the bit, while there exists multiple ways to mutate real coded genes [85, 175]. Let us describe the mutation process with the following example.

Illustration 3: A binary string can be bit-by-bit mutated with regards to the predefined mutation probability μ_m. Suppose, two positions namely, 5 and 7 satisfy the mutation criteria. The outcome of mutation on these two positions of a chromosome is depicted below.

$$0 \quad 1 \quad 0 \quad 0 \quad 0 \quad 1 \quad 0 \quad 0$$

$$0 \quad 1 \quad 0 \quad 0 \quad 1 \quad 1 \quad 1 \quad 0$$

3.3 Particle Swarm Optimization

Particle Swarm Optimization is a biologically inspired, simple, stochastic, population-based, powerful optimization technique. It was introduced by Kennedy and Eberhart [144] in 1995. The congregation of birds behavior (flocking of bird) or gathering of fish to form a fish school, have been meticulously examined to build this optimization strategy. The panorama of particle swarm is that birds always try to discover some dedicated search space to fly [160]. Their tendency is to pursue some paths which have been visited before with high efficacy [120, 144]. In PSO, the population comprising the possible solutions (feasible solutions) is popularly known as *swarm* and such potential solutions in the swarm are called *particles*. Particle Swarm Optimization (PSO) uses the concept of social communication to solve problems.

The PSO is quite indistinguishable from the GA in many respects. Both these techniques generate population and initialize it with random solutions and look for an optimum solution by repeating through generations. Unlike the GA, PSO does not use any genetic operator like mutation and crossover; instead, it employs particles which modify their components by flying through the search pace (problem space) by following the present optimum particles. Particles employ their fitness values by means of fitness function to judge the quality of their real position in the search space. Particles compare their fitness with their neighbors as well and emulate the best of them.

3.3.1 Pseudo-code of Particle Swarm Optimization

The functioning of PSO is described by the pseudo-code given by

```
begin
   iteration := 0;
   initswarm P(iteration);
   evaluate P(iteration);
   repeat
     begin
     iteration := iteration + 1;
     selectpBest P(iteration);
     selectgBest P(iteration);
     calpVel P(iteration);
     updatePos P(iteration);
     evaluate P(iteration);
     end
   until iteration>max_gen
 end
```

The **initswarm** function generates the swarm of trial solutions (*P*) at random. Then, the **evaluate** function evaluates the fitness of each particle of *P*. The **selectpBest** function finds the best fitness associated with a particle in the solution space and the **selectgBest** function determines the swarm best fitness value achieved hitherto among all particles. **calpVel** and **updatePos** functions are collectively exercised to modify the particle's position in *P*. This process is executed for a predefined number of generations.

3.3.2 PSO: Velocity and Position Update

Each particle updates its position by flying through its search space. A velocity is applied to each particle to move it from one position to other in the problem space. During the operation, each particle continuously keeps track of its position with respect to the best solution (fitness value) it has attained hitherto in the problem space [144]. This value is referred to as pbest (local best value), and the candidate solution that attained this fitness value is called the best position/best candidate solution of the individual. The particle swarm optimizer also keeps track of one more "best value" achieved among all the particles in the entire population (swarm) [144]. This is called gbest (global best) and the candidate solution that attained this fitness is referred to as the global best position/global best candidate solution. The velocity and position of each particle are updated using the following formula:

$$v_{kd}^{new} = \omega.v_{kd}^{old} = c_1.rand(0,1)(p_{kd} - y_{kd}) + c_2.rand(0,1)(p_{gd} - y_{kd})$$

$$y_{kd}^{new} = y_{kd}^{old} + v_{kd}^{new} \tag{3.3}$$

where d is called the dimension, c_1 and c_2 are known to be the cognition parameter and the social parameter, respectively. These two parameters signify how much the particle has faith in its own previous experience and in the swarm, respectively. rand(0,1) is the randomly generated number between (0,1) and ω is called the inertia weight.

In brief, the PSO comprises the following three steps:

1. Evaluate the fitness value of each particle.
2. Update the best solution of each particle.
3. Update the velocity and compute the location of each particle.

These three steps are repeated till any stopping criteria are encountered.

In the last few years, PSO has been efficaciously applied in many fields owing to its ability to achieve better results (outcomes) in a faster. inexpensive way compared to other methods. Furthermore, PSO necessitates very few parameter alterations.

3.4 Ant Colony Optimization

Ant Colony Optimization is a well-known general search, population-based optimization technique, used to find the solution of a variety of combinatorial problems [76]. The communal behavior of real ants is the source of inspiration in designing ACO. The exhilarating source of ACO is obviously the pheromone trail and following the deeds of real ants which spurt pheromone, a chemical substance as a communication channel. This feature of real ant gatherings is exploited in ACO in order to find solutions to different optimization problems.

3.4.1 Stigmergy in Ants: Biological Inspiration

ACO imitates the basic behaviors of real ants to accomplish the solutions of various optimization problems. Ants are socially awkward, but together they can accomplish complex tasks. In real life, ants struggle for food to sustain their existence. They have an ability to discover the shortest route from the colony to the food source and back to their colony by exploiting an indirect communication by means of pheromone. The whole episode is described through the following points.

- Write: The real ants traverse different paths for food and squirt pheromone from their body.
- Read: This chemical helps them to exchange information among themselves and to locate the shortest path to be followed between their nest and a food source. It is observed that a particular path which contains more pheromone is outlined by more number of ants.

- Emergence: This uncomplicated behavior applied by the entire colony can bring about the optimal (shortest) path.

Many scholars have been motivated by the communal behavior of real ants and established numerous algorithms to solve combinatorial optimization problems [196].

3.4.2 Pseudo-code of Ant Colony Optimization

The implementation of ACO is presented by the following pseudo-code.

```
begin
   iteration := 0;
   initPhero Q(iteration);
   initPop P(iteration);
   evaluate P(iteration);
   repeat
      iteration := iteration + 1;
      P: =updatePop P(iteration);
      evaluate P(iteration);
      Q: =updatePhero Q(iteration);
   until iteration>max_gen
 end
```

At the outset, the pheromone matrix (Q) is randomly initialized by the **initPhero** function. Then, the population of trial solutions (P) is randomly generated by the **initPop** function. Afterward, **evaluate** function evaluates fitness of each individual of P. Then, **updatePop** and **updatePhero** functions update the population and pheromone matrix, respectively. Note that, the pheromone matrix (Q) is used to update the population (P) at each successive generation. This procedure is run for a predefined number of generations.

3.4.3 Pheromone Trails

The pheromone trails in ACO can be thought of as disseminated and numerical information, which are used by real ants to probabilistically build solutions to the problem and also adapted by them during the execution of this technique to imitate their search experience. Each individual ant deposits pheromone trails at the time of traveling from the colony, to the colony or perhaps in both ways. This chemical substance slowly evaporates over time. The pheromone trail helps ants to work together and find the shortest path to reach a food source. Every time an ant discovers food, it marks its return journey by means of an amount of pheromone. This trace disappear faster on elongated paths and lasts for a long time on shorter paths. This path could be a viable alternative for most of the agents as the way to the source of food.

3.4.4 Updating Pheromone Trails

Let us illustrate the updating strategy of pheromone trails with the following example. Suppose, ACO technique is applied to solve the Traveling Salesman Problems (TSP).

According to the TSP, a salesman has to visit a number of cities for his work and come back to the city from where he started visiting. The number of cities and the distance between each pair of them is given. The solution of problem is to find shortest possible path with the condition that all the cities must be visited exactly once.

Let us assume that m ants start visiting at random nodes ("city" is considered as node and distance between any two cities is considered as "edge"). Each ant traverses the nodes according to the distance between each pair of nodes. An iteration terminates when all the ants finish their visits to all nodes. After completing each iteration, the pheromone trails need to be updated using the formula given by

$$T'_{cd} = (1 - \rho)T_{cd} + Cng(T_{cd}) \tag{3.4}$$

where, T_{cd} is the pheromone trail of edge (c,d), ρ is a parameter, known as the rate of pheromone decay. $Cng(T_{cd})$ signifies the amount of trails added when an ant traverses an edge (c,d) during the iteration. As we pass through more number of iterations, we reach optimality.

3.5 Differential Evolution

The differential evolution [242] is a population-based, real-valued, heuristic optimization technique used to optimize different non-differentiable and nonlinear continuous functions. DE is simple to use and implement, converges fast and may furnish remarkable results with almost no or a few parameter alterations. It is stochastic in nature and very effective for optimizing unimodal or multi-modal search [281, 289]. In recent years, it became the first choice for the developers to use in various application areas, especially in engineering or statistics. Due to its simplicity and powerfulness, it can easily be implemented to optimize different objective functions, which are very difficult or even analytically impossible to solve. In the last few years, a number of research works have been carried out that testify to the superiority of DE over other meta-heuristics [199].

3.5.1 Pseudo-code of Differential Evolution

The steps of operations of DE are described by the following pseudo-code.

```
begin
   iteration := 0;
   initPop P(iteration);
   evaluate P(iteration);
   repeat
      P := mutate P(iteration);
      P := recombine P(iteration);
      P := selectpar P(iteration);
      iteration := iteration + 1;
      evaluate P(iteration);
   until iteration>max_gen
end
```

Initially, the **initpop** function generates the population of trial solutions (P) randomly. Then, **evaluate** function evaluates the fitness of each individual of P. Each individual in P is operated by **mutate** and **recombine** functions, in succession. These two functions are used to obtain the population diversity for the next generation. Based on a certain condition, **selectpar** function selects the fittest individuals from P to form the population for the next generation. **mutate**, **recombine** and **selectpar** are continued until the predefined stopping criterion is attained.

3.5.2 Basic Principles of DE

There are DE different control parameters, such as the population size (C), the scaling factor (mutation factor) (F) used for mutation operation and the crossover probability (CP) used for crossover operation. A population in DE encompasses C numbers of vectors, each is defined in D-dimension, given by ($\vec{X}_{j,G}, j = 1, 2, \ldots, C$), where optimality is found at G number of generations. Each vector passes through three different steps successively, namely, *mutation*, *crossover* and *selection* to create the population for the subsequent generation. The steps involved in DE are discussed in the next subsections.

3.5.3 Mutation

The mutation operator is used to expand the search space. This operation familiarizes new parameters into the population. The mutation operator produces new vectors, called mutant vectors ($\vec{Y}_{j,G}$) to accomplish it. These mutant vectors are created as follows:

$$\vec{Y}_{j,G+1} = \vec{X}_{a_1,G} + F(\vec{X}_{a_2,G} - \vec{X}_{a_3,G}) \tag{3.5}$$

where $\vec{X}_{a_1,G}$, $\vec{X}_{a_2,G}$ and $\vec{X}_{a_3,G}$ are three vectors selected at random. Using Eq. (3.5), these three vectors are used for perturbation, which creates ($\vec{Y}_{j,G}$). Here, $j, a_1, a_2, a_3 \in \{1, 2, \ldots, C\}$ are selected at random and $j \neq a_1 \neq a_2 \neq a_3, F \in [0, 1]$. The value of C must be greater than 3 to satisfy the aforementioned criteria.

3.5.4 Crossover

The crossover operator starts with creating the trial vectors as given by $\{\vec{Z}_{j,G+1} = (\vec{Z}_{1j,G+1}, \vec{Z}_{2j,G+1}, \ldots, \vec{Z}_{Dj,G+1})\}$. These trial vectors are used to select the vectors for the next generation. A trial vector is created by the amalgamation of a mutant vector with a parent vector on the basis of various distributions like binomial, uniform, exponential distribution, etc. The trial vector is created on the basis of the following criteria:

$$\vec{Z}_{kj,G+1} = \begin{cases} \vec{Y}_{kj,G+1} : if \ (rand(k) \leq CP) \quad or \quad k = q \\ \vec{X}_{kj,G} : if \ (rand(k) > CP) \quad or \quad k = q \end{cases} \tag{3.6}$$

where $k = 1, 2, \ldots, D$; $rand(k) \in [0, 1]$ is a randomly generated number and $CP \in [0, 1]$. $q \in (1, 2, \ldots, D)$ denotes an index selected at random, which confirms that $\vec{Z}_{kj,G+1}$ acquires one or more elements from $\vec{Y}_{kj,G+1}$ after the crossover operation. According to Eq. (3.7), the randomly generated number is compared with CP. Either the mutant vector or parent vector is chosen for the next generation, depending on the condition as given in this equation.

3.5.5 Selection

The selection operator selects the best individuals to create the population for the next generation. The fitness value of the trial vector is compared with the corresponding value of the target vector in this process. In this way, the population is populated with the vectors having better fitness values. The best offspring ($\vec{X}_{j,G+1}$) is chosen based on the following equation:

$$\vec{X}_{kj,G+1} = \begin{cases} \vec{Y}_{j,G+1} : \text{if } gn(\vec{Y}_{j,G+1}) \leq gn(\vec{X}_{j,G+1}) \\ \vec{X}_{j,G} : \text{otherwise} \end{cases} \tag{3.7}$$

where gn is the given objective function.

3.6 Simulated Annealing

Annealing is the process of tempering a certain physical substance, such as glass, alloys of metal, etc. by heating it above their melting point, keeping its temperature on hold, and then cooling it very gradually until it becomes a sturdy crystalline structure. This physical or/and chemical phenomenon is the basis of making high-quality materials. When this process is simulated in a certain approach, it is called simulated annealing [147]. It is a widely used search and optimization technique, which exploits the laws of thermodynamics. SA basically comprises two stochastic processes: the first is to generate solutions and the other is accepting the solutions. SA is a descent technique reformed by random ascent moves with the aim of escaping local minima (which must not be global minima). SA can be described as the mathematical equivalence to a cooling system which has a high acceptability to highly multidimensional and nonlinear functions.

3.6.1 Pseudo-code of Simulated Annealing

The executional steps of SA are demonstrated by the subsequent *pseudo-code*.

```
begin
    initsol xa;
    initialize tmx,tmi,maxiter,r;
    evaluate xa;
    tm=tmx;
    BS=xa;
    repeat
        begin
            repeat
                iteration := 1;
                xf :=neighbor(xa)
                dl=Ft(xf)-Ft(xa);
                if(dl<0)
                xa=xf;
                else if(rand<Power(e,(-dl/tm)))
                xa=xf;
                update BS(iteration);
```

```
            iteration := iteration + 1;
        until iteration>maxiter
        tm=tm*r;
    end
  until tm>tmi
end
```

The **initsol** function randomly generates one trial solution (*xa*) at initial. Then, the values of the required parameters, such as *tmx* (initial temperature), *tmi* (final temperature), *maxiter* (no. of iterations for the inner loop) and *r* (reduction factor) are set. Thereafter, **evaluate** function evaluates the fitness of *xa*. This procedure is continued until the predefined final temperature is reached (through the outer loop). The inner loop is executed for *maxiter* number of times. At each iteration, the neighbor solution of *xa* is found and the best solution is exercised based on a certain criterion. **update** function updates the overall best individual through different iterations. Finally, the temperature is reduced by *r* times to run the outer loop.

3.6.2   Basics of Simulated Annealing

SA generates configurations using the following approach. Let us assume that SA has a current configuration C_j whose energy is referred to as E_j. In the next generation, C_j is passed through a small perturbation to generate a new configuration, say, C_k (with energy E_k). The acceptance of C_k depends on either of the following criteria:

1. Case A: Accept C_k as the present configuration, if $(E_k - E_j) < 0$.
2. Case B: Accept C_k with a probability given by $\exp(-\frac{(E_k - E_j)}{B_\beta T})$, if $(E_k - E_j) \geq 0$.

where T and B_β signify the temperature and Boltzmann's constant, respectively. The Metropolis algorithm (MA) is a popular and simple method which is used to simulate the progress to the thermal equilibrium of a material for a specified temperature [174]. SA is basically a variant of the MA, in which the temperature is lowered from high to low. If the temperature is decreased very slowly, the crystal touches a thermal equilibrium state at each temperature. In the MA, this is accomplished by applying an adequate number of perturbations at every temperature. Geman and Geman [100] proved that SA congregates to the global optima when it is annealed slowly.

The annealing process of an SA technique comprises as follows:

1. Initial temperature (T_{mx})(usually set with a very high value).
2. Schedule for cooling.
3. Number of iterations at each temperature.
4. Stopping condition to end the technique.

So far, SA has been successfully applied in different areas [20, 40]. This technique deals with single objective for optimizing. However, a few attempts have already been made by different researchers to extend SA to multi-objective optimization. The earlier attempts basically involved uniting a number of objectives into a single-objective function by means of a weighted sum approach [52, 123, 187, 246, 263]. The problem is based on choosing the proper weights in advance. A few other approaches have also been applied in this regard. Since SA deals with a single solution at any particular run, the technique uses several SA runs to develop the set of Pareto Optimal (PO) solutions.

3.7 Tabu Search

Tabu search is a local search technique, belonging to the category of present-day heuristics, which are presently used to solve a wide variety of combinatorial optimization problems. TS was first introduced by Glover and Laguna [107, 109, 116]. TS offers a general heuristic stratagem to solve several combinatorial optimization problems of the form given by

$$\min_{y \in Y} f(y)$$

where Y is known to be the solution space comprising a finite number of feasible solutions. f is any arbitrary function s.t. $f : Y \rightarrow \mathcal{R}$. Hitherto this stratagem has been effectively and successfully used in a wide variety of problems and is quickly spreading to several novel fields of applications.

TS undergoes iterative searching to explore its solution space to find the near optimal solution. TS is simple to implement, supple and powerful in nature. A proper mechanism must be employed to free TS from sticking over local optima. The only problem of TS is that one can face parameter adjustment beforehand. Since the parameters used in TS have a direct influence on the solution, a proper knowledge of parameter tuning is an essential and desirable condition to guarantee a good result.

3.7.1 Pseudo-code of Tabu Search

The operational procedure of TS is presented by the following pseudo-code.

```
begin
   iteration := 0;
   initsol xa;
   initTM POP :=null;
   evaluate xa;
   BS=xa;
       repeat
           iteration := iteration+1;
           ns :=neighborSet(xa)
           bs :=findBest ns(iteration);
           updateP POP(iteration);
           BS=bs;
           iteration := iteration + 1;
       until iteration>max_iter
 end
```

The **initsol** function is used to generate one trial solution (xa) at random. Initially, the **initTM** function initializes the tabu memory (POP) as null. Next, the **evaluate** function is utilized to evaluate the fitness of xa. Thereafter, a set of neighbor solutions is determined through **neighborSet** function. The function **findBest** finds the best individual from these sets of solutions. The **updateP** function updates POP. This procedure persists until the predefined stopping criterion is attained.

3.7.2   Memory Management in Tabu Search

Local searches yield a prospective solution to a problem and look at its immediate neighbors in the expectation of discovering an improved solution. These methods have a tendency to get stuck in other suboptimal regions or even on plateaus where numerous solutions are equally acceptable. TS uses memory structures to improve the efficiency of these techniques. The memory represents the visited solutions or user-defined sets of rules. The strategy of using memory is that if a potential solution has already been visited within a specific short period of time or if it has dishonored any defined rule, it is labeled "taboo" so that the technique does not acknowledge that possibility repeatedly.

The memory structures used in TS can be categorized into three classes. These are defined as follows [107, 109, 116]:

- Short-term: This contains the list of solutions considered recently. If any potential solution arrives on this list, it cannot revisit until and unless it reaches an expiry point.
- Intermediate-term: A set of rules deliberately made to bias the exploration towards promising parts of the search space.
- Long-term: A set of rules that endorse assortment in the search procedure. It is basically represented by resets while the search get stuck in a plateau or reaches a suboptimal dead end.

3.7.3   Parameters Used in Tabu Search

Some fundamental ingredients of TS can be described as follows.

1. *Neighborhood*: This consists of collection of adjacent solutions which are found from the existing solution.
2. *Tabu*: To utilize memory in TS the to classify in the neighborhood an entry in the tabu list.
3. *Attributes*: This is referred to as a classification that depends on the recent history.
4. *Tabu list*: Records the tabu moves
5. *Aspiration criterion*: When the tabu classification needs to be overridden using some relevant condition, it is called the aspiration criterion for the tabu search.

3.8   Conclusion

The chapter deals with the overview of six popular meta-heuristics. These comprise the classical genetic algorithm [18, 53, 110, 175], the classical particle swarm optimization [120, 144], the classical ant colony optimization [76], the classical differential evolutionary algorithms [242], classical simulated annealing [147] and finally, the classical tabu search [107, 109, 116] methods. The basics of each with their pseudo-codes are presented in brief in this chapter.

3.9   Summary

- Evolutionary computation is a search and optimization procedure which uses biological evolution inspired by Darwinian principles.

- Genetic algorithms are effective, adaptive and robust optimization and randomized search techniques guided by the Darwinian laws of biological evolution.
- Particle Swarm Optimization is biologically inspired, simple, stochastic, population-based powerful optimization technique.
- Ant Colony Optimization is a well-known general search, population-based optimization technique, used to find the solution for a variety of combinatorial problems.
- The Differential Evolution is population-based, real-valued, heuristic optimization technique used to optimize different non-differentiable and nonlinear continuous functions.
- Simulated Annealing is a widely used search and optimization technique, which exploits the laws of thermodynamics.
- Tabu Search is a local search technique, belongs to the category of present-day heuristics, which are presently extensively used for a wide variety of combinatorial optimization problems.

Exercise Questions

Multiple Choice Questions

1   Meta-heuristics can be suitably used for
   (a) a local search process
   (b) a complex search procedure
   (c) both (a) and (b)
   (d) none of the above

2   EA stands for
   (a) evolutionary arithmetic
   (b) evolutionary algorithm
   (c) evolutionary application
   (d) none of the above

3   Evolution strategies were introduced by
   (a) Fogel
   (b) Darwin
   (c) Rechenberg
   (d) none of the above

4   Pheromone matrix is introduced in developing
   (a) DE
   (b) PSO
   (c) ACO
   (d) SA

5   Elitism is generally found in
   (a) GA

(b) ACO

(c) PSO

(d) DE

6 Which of the following method exploits the laws of thermodynamics?

(a) TS

(b) ACO

(c) SA

(d) DE

7 Memory management can be seen in

(a) GA

(b) ACO

(c) TS

(d) DE

Short Answer Questions

1 Mention two requirements that a problem should satisfy in order to be suitably solved by GA.

2 Discuss in brief, why mutation and crossover operators are important in DE.

3 Describe the idea behind the simulated annealing algorithm making reference to its origins as an optimization methodology.

4 Describe the idea behind the design of SA algorithm.

5 How can you handle division by zero in a program by using a genetic programming methodology?

Long Answer Questions

1 Mention and discuss the basic principles of Genetic Algorithms.

2 What is the difference between single-point crossover and multi-point crossover in GA, illustrate with suitable examples. If you are able to employ a diverse crossover operator, which one you would prefer? Give an example.

3 Write the pseudo-code of a simple DE by describing its main elements, in short. What are the fundamental ingredients of TS?

4 Discuss the impact on the controlling parameters on the performance of EA.

5 How can population diversity can be managed in DE?

4

Quantum Inspired Meta-heuristics for Bi-level Image Thresholding

4.1 Introduction

Several single-point based and population-based evolutionary algorithms were described in detail in Chapter 1. Over the last few decades, these algorithms have been successfully applied as the optimization techniques in different fields of application. These distinguished algorithms are very useful to render the suitable solutions of several combinatorial optimization problems. These approaches explore their respective search space to discover the optimal solutions within the stipulated time frame. Several initiatives with regard to the meta-heuristics in combinatorial optimization are found in [31]. Among the population-based algorithms, Genetic Algorithm [127], Ant Colony Optimization [76], Particle Swarm Optimization [144], and Differential Evolution [242] deserve special mention. The Simulated Annealing [147] and Tabu Search [105–107, 109] are two prevalent paradigms of the single-point search algorithm.

Of late, the field of quantum computing [173] has appeared to offer a speed-up of the classical algorithms. This can be achieved by inducing some physical phenomena of quantum computing into the classical algorithmic framework. Hence, these concepts can be applied in the improvement of a computing paradigm much faster than the conventional computing scenario.

The improvement in the computational speed-up can be achieved by means of intrinsic parallelism perceived in the quantum bits, the building blocks of a quantum computer. This makes quantum computers far more efficient compared to their classical counterparts in different scenarios, such as searching databases [117] and factoring large numbers [235], etc. The first quantum algorithms appeared before the actual fabrication of the first quantum gate [173]. Hence, scientists have resorted to deriving quantum inspired versions of the conventional algorithms in vogue by emulating the features offered by quantum mechanical systems, to achieve time efficiency.

Almost all approaches have focused on the design of quantum-inspired evolutionary algorithms by inducing the physical properties of quantum mechanics into the algorithmic structure [121, 146]. In [146], wave interference was introduced in the traditional crossover operator of a genetic algorithm to solve combinatorial optimization problems. Later, Han et al. developed a quantum-inspired evolutionary algorithm, in which a quantum bit probabilistic representation and superposition of states were introduced [121]. In addition a Q-gate was also used to derive a better solution space. Talbi et al. [247] introduced a new quantum genetic algorithm that can solve the Traveling Salesman Problem using the approaches presented in [121]. A quantum rotation gate was used in

Quantum Inspired Meta-heuristics for Image Analysis, First Edition.
Sandip Dey, Siddhartha Bhattacharyya, and Ujjwal Maulik.
© 2019 John Wiley & Sons Ltd. Published 2019 by John Wiley & Sons Ltd.

each of these approaches, which can aid in embedding interference among the candidate solutions with the help of a look-up table of the probable best solution. However, this cannot be always pertinent because the best solution may not be always known beforehand.

In this chapter, the inadequacy of the aforesaid algorithms is addressed. This chapter introduces two novel quantum inspired meta-heuristic algorithms for bi-level image thresholding [28, 61, 64, 68]. These techniques are associated with the class of quantum inspired evolutionary algorithms (QIEA) [121, 247] which is usually applied to solve a variety of combinatorial optimization problems. The foremost aim of the chapter is to present greatly improved viable alternatives that can persuade uncertainty handling mechanisms in the optimization process. Other distinguished features that prove them to be superior to the prevailing QIEA [121, 247] are that they do not necessitate any a priori knowledge of the best possible solution of the system at hand to perturb the superposition of the system state. Alternatively, it resorts to a chaotic map model if the system is an ergodic dynamical system [33].

Results of the application of the quantum inspired approaches are demonstrated with real-life gray-scale images for the purpose of bi-level image thresholding. The performance of each quantum inspired approach is exhibited by means of several thresholding techniques as objective functions, such as Ramesh's Algorithm [207, 231], Shanbag's Algorithm [231, 234], Pun's Algorithm [206, 231], Wu's Algorithm [231, 274], Renyi's Algorithm [220, 231], Yen's Algorithm [231, 277], Johannsen's Algorithm [135, 231], Silva's Algorithm [231, 236], Brink's Algorithm [35, 231], Otsu's Algorithm [192, 231], Kittler's method [148, 231], Li's Algorithm [158, 231], Kapur's Algorithm [140, 231], Fuzzy Algorithm [130, 257], and the Correlation coefficient [26, 145, 226].

Experiments have been carried out with several real-life images to obtain their optimal threshold value. The specification of the parameters used for experimental purpose has been set after an adequate number of tuning turns. The effectiveness of the quantum inspired techniques, in respect of time efficiency, accuracy, and robustness, has been compared with their respective conventional counterparts for real-life gray-scale images under consideration. The experimental results are presented in the respective sections. The chapter is arranged as follows. Sections 4.2 and 4.3 elucidate two novel quantum inspired meta-heuristic algorithms [64]. The entire methodology of these algorithms is discussed in detail in these sections. The implementation results of quantum inspired procedures are presented in Section 4.4. This section also contains the experimental results of their respective classical counterparts and the results of the algorithm proposed by Han et al. [121]. As a sequel to these experimental results, this chapter also presents a comparative study among the participating competitors, the details of which is demonstrated in Section 4.5. The chapter conclusion is presented in Section 4.6. A chapter summary is presented in Section 4.7. A set of questions in connection with the theme of the chapter is presented. Finally, some sample coding with regard to the basis of experiments of the proposed algorithms ends the chapter.

4.2 Quantum Inspired Genetic Algorithm

The quantum inspired genetic algorithm (QIGA) is a class of quantum inspired evolutionary algorithms (QIEA) [121, 247], which is used to solve combinatorial optimization

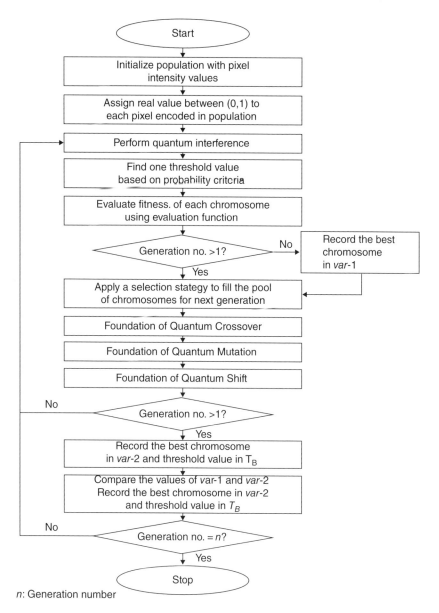

Figure 4.1 QIGA flow diagram.

problems. Instead, it resorts to a chaotic map model considering the system to be an ergodic dynamical system [33]. The proposed QIGA has been achieved in seven phases. The flow diagram of QIGA is presented in Figure 4.1.

4.2.1 Initialize the Population of Qubit Encoded Chromosomes

This phase entails qubit encoded chromosomes for the generation of the initial population of trial solutions. As we know, an L level gray-scale image comprises L number of possible pixel intensity values. For encoding each potential pixel intensity value in the participating image, each chromosome in the initial population is constructed with a $L \times L$ matrix of quantum bits (qubits). Each of the said chromosomes is denoted as follows:

$$|\psi\rangle = \begin{bmatrix} \alpha_{11}|\psi_1\rangle + \beta_{11}|\psi_2\rangle \dots \alpha_{1L}|\psi_1\rangle + \beta_{1L}|\psi_2\rangle \\ \alpha_{21}|\psi_1\rangle + \beta_{21}|\psi_2\rangle \dots \alpha_{2L}|\psi_1\rangle + \beta_{2L}|\psi_2\rangle \\ \dots \dots \dots \dots \dots \dots \dots \dots \dots \dots \\ \dots \dots \dots \dots \dots \dots \dots \dots \dots \dots \\ \alpha_{L1}|\psi_1\rangle + \beta_{L1}|\psi_2\rangle \dots \alpha_{LL}|\psi_1\rangle + \beta_{LL}|\psi_2\rangle \end{bmatrix} \tag{4.1}$$

Hence, in order to encode each pixel, a real number between $(0,1)$ is assigned to it.

4.2.2 Perform Quantum Interference

The quantum interference in the proposed QIGA is performed through the following steps.

4.2.2.1 Generate Random Chaotic Map for Each Qubit State

This is one of the important phases in QIGA as it prepares the foundation for a random interference of the participating qubits based on the representation of the chromosomal quantum states in terms of a real chaotic map. Considering QIGA to be a dynamical time discrete and random quantum mechanical system exhibiting transformations C and D, which possess the properties of automorphism, endomorphism, flow and semi-flow [33], the *qubit* chromosomal representation can be approximated by a real chaotic map. Given a measure space $(U; A; \mu)$, if $v \in L^1(U; A; \mu)$, then for μ-a.a. and $u \in U$, the following limits exist and are equal to each other.

1. Automorphism: If C is an automorphism, then [33]

$$\lim_{n \to \infty} \frac{1}{n} \sum_{k=0}^{n-1} v(C^k u) = \lim_{n \to \infty} \frac{1}{n} \sum_{k=0}^{n-1} v(C^{-k} u) \tag{4.2}$$

i.e.,

$$\frac{1}{2n+1} \sum_{k=-n}^{n} v(C^k u) \overset{def}{=} \bar{v}(u) \tag{4.3}$$

2. Endomorphism: If C is an endomorphism, then [33]

$$\lim_{n \to \infty} \frac{1}{n} \sum_{k=0}^{n-1} v(C^k u) \overset{def}{=} \bar{v}(u) \tag{4.4}$$

3. Flow: If $\{C^t\}_{t \in R}$ (R being the measurement space) is a flow, then [33]

$$\lim_{t \to \infty} \frac{1}{t} \int_0^t v(C^\tau u) d\tau = \lim_{t \to \infty} \frac{1}{t} \int_0^t v(C^{-\tau} u) d\tau \tag{4.5}$$

i.e.,

$$\lim_{t \to \infty} \frac{1}{2t} \int_{-t}^t v(C^\tau u) d\tau \overset{def}{=} \bar{v}(u) \tag{4.6}$$

4. Semiflow: If $\{D^t\}_{t \in R^+}$ (R being the measurement space) is a semiflow, then [33]

$$\lim_{t \to \infty} \frac{1}{2t} \int v(D^t u) d\tau \overset{def}{=} \bar{v}(u) \tag{4.7}$$

Moreover,

$$\bar{v}(Cu) = \bar{v}(u) \;\; or \;\; \bar{v}(C^t u) = \bar{v}(u) \;\; or \;\; \bar{v}(D^t u) = \bar{v}(u) \tag{4.8}$$

Any transformation which possesses properties stated in Eqs. (4.2), (4.4), (4.5) and (4.7) is said to be invariant w.r.t. these properties. A dynamical system is said to be ergodic if the measure $\mu(A)$ of any invariant set A equals 0 or 1 [33].

On similar lines, the participating *qubit* chromosomes ($\langle \psi| = \alpha_j |\psi_j\rangle + \beta_j |\psi_j\rangle, j = 1, 2$) may be considered to be in a dynamical ergodic system and according to Birkhoff's Ergodic theorem [33], they can be transformed by some invariant point transformation C to a collection of chaotic maps $C = \{t_1, t_2, I, p_1, p_2, 1 - p_1, 1 - p_2\}$ [33]. $\tau_k (k = 1, 2) = h \circ C \circ h^{-1}$ are nonlinear point transformations differentially conjugated to C, where $\circ$ may be any composition operator. I represents the identity map, t_1, t_2 are non-linear point transformations and p_1, p_2 [33] are respective weighting probabilities of τ_1 and τ_2 respectively. $p_k, k = 1, 2$ can be determined by [33]

$$p_k = (b_k h_k) / \sum_{k=1}^{2} b_k h_k \tag{4.9}$$

where, b_k are positive constants and $h_k = |\psi_k\rangle\langle\psi_k| = \psi_k^* \psi_k$ are the probability density function (pdf) corresponding to τ_k.

4.2.2.2 Initiate Probabilistic Switching Between Chaotic Maps

In this phase, a simple probabilistic switching between the maps results in the change of states of the maps of the corresponding *qubits* and yields the expected interference between them, which is merely the result of the superposition of two wave functions in some orderly fashion. Thus, the novelty of the present algorithm lies in the fact that the interference process does not follow any predefined order/fashion and is simplistically random by nature arising out of the probabilistic transformation C achieved in the previous phase.

In practice, considering a two-state QC, the population of chromosomes is formed as the superposition between two states as specified in Eq. (1.5). Each chromosome in the population undergoes a quantum interference satisfying the following relation.

$$|\alpha_{ij}|^2 + |\beta_{ij}|^2 = 1 \tag{4.10}$$

where, $i, j \in \{1, 2, \cdots, L\}$.

A typical example of quantum interference among the participating qubits is given below. Figure 4.2 depicts $(0.2395)^2 + (0.9708)^2 = 1$, $(-0.1523)^2 + (0.9883)^2 = 1$, and so on.

Figure 4.2 Quantum interference.

0.2395	−0.1523	−0.0987	0.2232	0.9937	−0.9881
0.9708	0.9883	0.9950	0.9747	0.1119	−0.1532

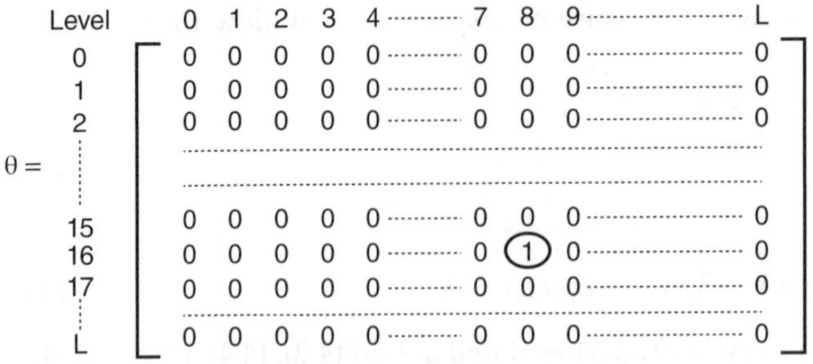

Level	0	1	2	3	4	⋯	7	8	9	⋯	L
0	0	0	0	0	0	⋯	0	0	0	⋯	0
1	0	0	0	0	0	⋯	0	0	0	⋯	0
2	0	0	0	0	0	⋯	0	0	0	⋯	0
⋮											
15	0	0	0	0	0	⋯	0	0	0	⋯	0
16	0	0	0	0	0	⋯	0	(1)	0	⋯	0
17	0	0	0	0	0	⋯	0	0	0	⋯	0
⋮											
L	0	0	0	0	0	⋯	0	0	0	⋯	0

$\theta =$ (matrix above)

Figure 4.3 The solution matrix for bi-level image thresholding.

4.2.3 Find the Threshold Value in Population and Evaluate Fitness

Quantum measurement entails the measurement of the quantum states leading to the destruction of the superposition between the states. In this phase, the best solutions are obtained by means of a comparison between a randomly generated number (N_m) and $|p_2|^2$. All those τ_2 positions within the ($L \times L$) chromosomes having $|p_2|^2 > N_m$ indicate possible solutions and the states are preserved for further searching. For a gray-level image with L possible gray levels, a typical possible solution may be represented where, the proposed algorithm finds a solution at position $(16, 8)$ as shown in Figure 4.3. The solution represents a threshold value of $(16 \times 8) = 128$ (shown by circle).

4.2.4 Apply Selection Mechanism to Generate a New Population

Since a quantum measurement results in chaos of quantum states, in QIGA, the previous states are insured from loss by means of a backup of the states for further evolution of the algorithm. The measurement procedure computes the fitness value of each chaotic map. This phase of the proposed algorithm applies an elitist selection strategy to select only those previously preserved chromosomes corresponding to the best solutions with the best fitness values. These selected chromosomes form the population for the next iteration. The entire process is repeated until the best solution obtained at any stage ceases to exceed the overall best solution and the pool of chromosome is filled with population.

4.2.5 Foundation of Quantum Crossover

This phase introduces diversity in the population of chaotic maps. In this phase, two parent chromosomes are selected randomly for the crossover operation. A crossover probability (p_{cr}) determines a position within these participating chromosomes and the parts on either side of the participating chromosomes are swapped to generate two individuals referred to as offspring. Hence, this operation augments the number of chromosomes in the population. Figure 4.4 depicts a typical illustration of a quantum crossover.

4.2.6 Foundation of Quantum Mutation

This phase emulates the mutation operator in QIGA thereby inducing diversity in the population. Initially, a chromosome is selected randomly from the augmented

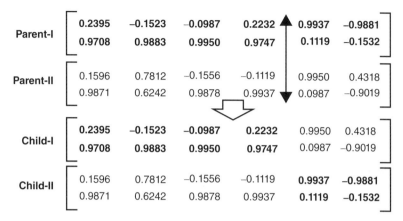

Figure 4.4 The quantum crossover.

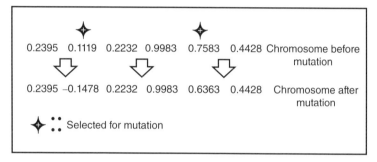

Figure 4.5 The quantum mutation.

population. Subsequently, the chaotic map at a random position (r_p) in the chromosome is muted, based on a predefined mutation probability, P_m. A typical example of quantum mutation is given in Figure 4.5.

4.2.7 Foundation of Quantum Shift

This phase is analogous to the above two phases which cause the population diversity of the participating chromosomes. Quantum shifting is applied at each chromosome based on a predefined shift probability, P_s. The details of quantum shifting are presented in Algorithm 1.

4.2.8 Complexity Analysis

The worst case time complexity analysis of the proposed QIGA is given below.

1. In QIGA, the time complexity to produce the initial chromosome in the population is $O(\mathcal{P} \times \mathcal{X})$, where $\mathcal{P}$ and $\mathcal{X}$ signify the population size and length of each chromosome in QIGA. Note that $\mathcal{X} = L \times L$ where L represents the maximum pixel intensity value of the image.
2. The complexity to assign a positive real number < 1 to each pixel encoded in the population becomes $O(\mathcal{P} \times \mathcal{X})$.

Algorithm 1: Quantum Shift

Require: Qubit encoded population of chromosomes, **cr**
   and L level gray-scale image.
Ensure: Quantum shift for some elected chromosomes
1: **for** each chromosome $y \in cr$ **do**
2:    generate first real random number, $r_s \in [0, 1]$
3:    **if** $(r_s \leq P_s)$ **then**
4:       generate second real random number, $r_1 \in [0, L/3]$
5:       generate a random position, $r_2 \in [1, L - r_1]$
6:       generate forth random number, $r_3 \in [1, L - r_1 - r_2]$
7:       Shift r_1 lines start with the r_{2th} position according to r_3
8:    **end if**
9: **end for**

---

3. Each chromosome in the population undergoes a quantum interference to create another population matrix, the complexity for this process is $O(P \times \mathcal{X})$.
4. To find the threshold value based on some probability criteria, the time complexity turns into $O(P \times \mathcal{X})$.
5. The fitness of each chromosome is computed using different fitness measures, the time complexity for this computation is $O(P)$.
6. For selection, the time complexity becomes $O(P)$.
7. For each mutation and crossover, the time complexity is $O(P \times \mathcal{X})$.
8. Hence, the time complexity to run the proposed QIGA for a predefined number of generations turns into $O(P \times \mathcal{X} \times \mathcal{G})$. Here, $\mathcal{G}$ represents the number of generations.

Therefore, recapitulating the above discussion, it can be said that the overall worst case time complexity for QIGA for bi-level image thresholding becomes $O(P \times \mathcal{X} \times \mathcal{G})$.

4.3 Quantum Inspired Particle Swarm Optimization

In this section, like QIGA, a variant version of the quantum inspired particle swarm optimization (QIPSO), is presented. The first five steps of QIPSO are identical with QIGA, which have already been discussed in Section 4.2. The remaining parts of the proposed algorithm are illustrated in this section. Based on a probability criteria, QIPSO searches and discovers a threshold value as the pixel intensity of the image at each generation. The particle swarm is intended to move to find its best location keeping record of the particle's individual as well as its global best position in a D dimensional space. Let $Y = (y_{k1}, y_{k2}, \cdots, y_{kD})$ be the kth particle swarm where its best position is recorded as $Pbest_k = (p_{k1}, p_{k2}, \cdots, p_{kD})^T$. $Gbest$ represents the swarm best particle. A velocity is applied on each particle to move it into the new position. The position of each particle is updated using the formulae given by

$$v_k^{t+1}(D) = v_k^t(D) + a_1 * rand(0, 1) * (Pbest_k^t(D) - y_k^t(D))$$
$$+ a_2 * rand(0, 1) * (Gbest^t(D) - y_k^t(D))$$
$$y_k^{t+1}(D) = y_k^t(D) + v_k^{t+1}(D) \tag{4.11}$$

where, $v_k^t(D)$ and $y_k^t(D)$ are the velocity and current position of kth particle at t^{th} generation. $rand \in (0, 1)$ represents a real random number. The flow diagram of QIPSO is presented in Figure 4.6.

4.3.1 Complexity Analysis

The worst case time complexity of QIPSO is discussed below. Since the first five steps of QIPSO and QIGA are identical, the proposed algorithms possess the same time complexity for the first five steps, which have already been discussed in the respective subsection of QIGA. The time complexity for the remaining parts of QIPSO is illustrated below.

1. Each particle in the swarm is given an injecting velocity to move into the search space. The time complexity to perform this process in QIPSO turns into $O(\mathcal{P} \times \mathcal{X})$.
2. QIPSO is executed for $\mathcal{G}$ number of generations. Hence, the time complexity to execute QIPSO for $\mathcal{G}$ number of times becomes $O(\mathcal{P} \times \mathcal{X} \times \mathcal{G})$.

Therefore, summarizing the above discussion, it may be concluded that the overall time complexity (worse case) for QIPSO happens to be $O(\mathcal{P} \times \mathcal{X} \times \mathcal{G})$.

4.4 Implementation Results

Application of the proposed algorithms for the determination of the optimum threshold (θ) of gray-level images has been demonstrated with a number of real-life gray-level images of dimensions 256×256. The original test images, namely, Lena (intensity dynamic range of $[23, 255]$), Peppers (intensity dynamic range of $[0, 239]$), Woman (intensity dynamic range of $[0, 216]$), Baboon (intensity dynamic range of $[0, 236]$), Corridor (intensity dynamic range of $[0, 255]$) and Barbara (intensity dynamic range of $[2, 248]$) are presented in Figure 4.7. A group of thresholding methods has been used as the objective functions in the present treatment. The results of image thresholding by the proposed algorithms are demonstrated with the aforementioned gray-scale images shown in Figure 4.7. The experimental results are presented in three different phases in this chapter.

First, results of the proposed algorithms with the five different fitness functions given by Otsu's Algorithm [192], Li's Algorithm [158], Ramesh's Algorithm [207], Shanbag's Algorithm [234] and the Correlation coefficient [26, 145, 226], are reported for Lena, Peppers, and Woman in the first phase. Subsequently, the set of results of the proposed QIGA are presented using six different fitness functions, called Wu's Algorithm [274], Renyi's Algorithm [220], Yen's Algorithm [277], Johannsen's Algorithm [135], Silva's Algorithm [236] and Linear Index of Fuzziness [257] for Baboon, Peppers, and Corridor in the next phase. Results of QIGA are finally listed with three different objective functions, named Brink's Algorithm [35], Kapur's Algorithm [140], and Pun's Algorithm [206] for Lena and Barbara.

The efficiency of the proposed approaches over their respective classical counterparts and the method proposed by Han et al. [121] as regards to the image thresholding of bi-level images, is also depicted in this chapter.

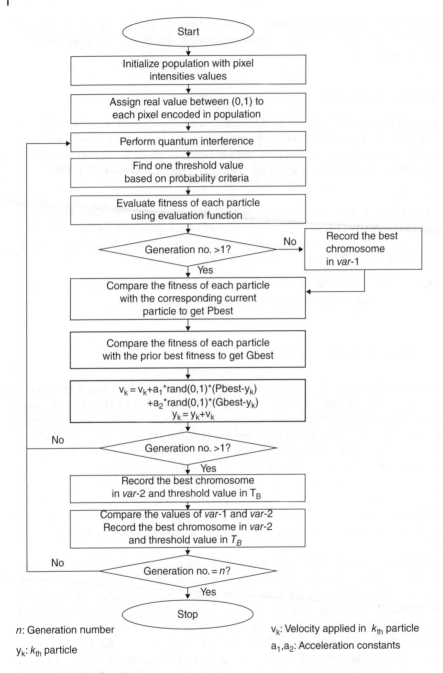

Figure 4.6 QIPSO flow diagram.

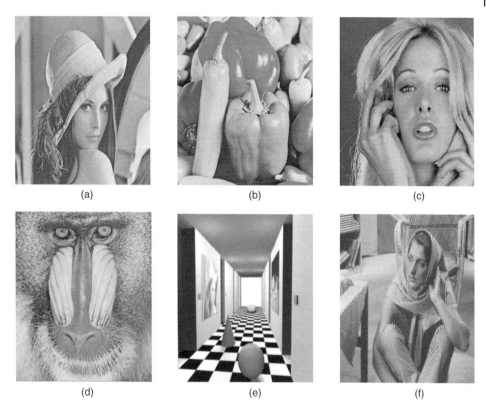

Figure 4.7 Original test images (a) Lena, (b) Peppers, (c) Woman, (d) Baboon, (e) Corridor, and (f) Barbara.

As the part of the comparative study, at the outset, each of the proposed algorithms is compared with its respective classical counterpart. Afterwards, the performance of the proposed approaches is also judged against the quantum evolutionary algorithm developed by Han et al. [121] for all test images.

4.4.1 Experimental Results (Phase I)

The proposed quantum-inspired algorithms are demonstrated using three gray-scale real-life images (Figures 4.7 (a)–(c)) for binary image thresholding. The implementation results of the proposed QIGA and its classical counterpart are listed in Tables 4.1–4.5. In addition, the implementation results of the proposed QIPSO and its classical version are presented in Tables 4.6–4.10. Each table lists the optimum threshold value (θ_{op}), the fitness value (ν_{op}) and the execution time (t) (in seconds) obtained for the three test images taking the best results from 10 different runs for each approach. Moreover, the standard deviation of fitness $\sigma(\nu)$ and the standard deviation of executional time $\sigma(t)$ over all runs for QIGA, GA, QIPSO, and PSO are also reported in Tables 4.1–4.10 for all test images.

Let us assume that the dimension and pixel size of each test image are $m \times n$ and s, respectively. Each chromosome or particle in the population is represented by two state

Table 4.1 Optimum results of QIGA, GA [127, 210] for three gray-scale images Lena, Peppers, and Woman using Otsu's method [192].

SN	QIGA			GA		
		Lena				
	θ_{op}	v_{op}	t	θ_{op}	v_{op}	t
1	118	1424.307	343	118	1424.307	665
$\sigma(v)$	0			0.182		
$\sigma(t)$	4.11			16.03		
SN		Peppers				
	θ_{op}	v_{op}	t	θ_{op}	v_{op}	t
1	130	1813.048	344	130	1813.048	686
$\sigma(v)$	0			0.282		
$\sigma(t)$	3.16			19.01		
SN		Woman				
	θ_{op}	v_{op}	t	θ_{op}	v_{op}	t
1	124	1298.236	343	124	1298.236	660
$\sigma(v)$	0			0.229		
$\sigma(t)$	3.43			17.08		

Table 4.2 Optimum results of QIGA, GA [127, 210] for three gray-scale images Lena, Peppers, and Woman using Ramesh's method [207].

SN	QIGA			GA		
		Lena				
	θ_{op}	v_{op}	t	θ_{op}	v_{op}	t
1	124	533792.559	343	124	533792.559	676
$\sigma(v)$	0			0		
$\sigma(t)$	2.11			14.19		
SN		Peppers				
	θ_{op}	v_{op}	t	θ_{op}	v_{op}	t
1	129	467278.548	344	129	467278.548	680
$\sigma(v)$	0			0		
$\sigma(t)$	2.56			16.28		
SN		Woman				
	θ_{op}	v_{op}	t	θ_{op}	v_{op}	t
1	117	521443.038	342	117	521443.038	676
$\sigma(v)$	0			0.229		
$\sigma(t)$	3.11			17.29		

qubits in the proposed algorithms. Hence, the total size required to produce an individual in the population is calculated as $P = 2 \times m \times n \times s \times 8 = 16mns$ bits. Unlike the proposed algorithms, the binary encoding scheme is used in the conventional algorithms. So, the total size required to produce an individual in the population of each conven-

Table 4.3 Optimum results of QIGA, GA [127, 210] for three gray-scale images Lena, Peppers, and Woman using Li's method [158].

SN	QIGA			GA		
	θ_{op}	v_{op}	t	θ_{op}	v_{op}	t
			Lena			
θ_{op}	v_{op}	t	θ_{op}	v_{op}	t
1	111	−612.412	343	111	−612.412	670
$\sigma(v)$	0			0.070		
$\sigma(t)$	2.20			16.11		
SN			**Peppers**			
	θ_{op}	v_{op}	t	θ_{op}	v_{op}	t
1	123	−633.227	343	123	−633.227	675
$\sigma(v)$	0			0.009		
$\sigma(t)$	3.36			18.30		
SN			**Woman**			
	θ_{op}	v_{op}	t	θ_{op}	v_{op}	t
1	118	−676.120	342	118	−676.120	674
$\sigma(v)$	0			0.015		
$\sigma(t)$	4.19			18.11		

Table 4.4 Optimum results of QIGA, GA [127, 210] for three gray-scale images Lena, Peppers, and Woman using the correlation coefficient [26, 145, 226].

SN	QIGA			GA		
			Lena			
	θ_{op}	v_{op}	t	θ_{op}	v_{op}	t
1	119	0.830751	343	119	0.830751	4652
$\sigma(v)$	0			0.025		
$\sigma(t)$	4.32			29.30		
SN			**Peppers**			
	θ_{op}	v_{op}	t	θ_{op}	v_{op}	t
1	131	0.850624	343	131	0.850624	4448
$\sigma(v)$	0			0.012		
$\sigma(t)$	3.84			25.24		
SN			**Woman**			
	θ_{op}	v_{op}	t	θ_{op}	v_{op}	t
1	123	0.856899	342	123	0.856899	4685
$\sigma(v)$	0			0.016		
$\sigma(t)$	3.91			26.34		

tional approach is $Q = v \times 8 = 8v$ bits where v represents the size of the image pixel. Therefore, size of Q/size of $P=8v/16mns = v/2mns$. In practice, the size of a test image has been considered as 256×256. The qubit encoded individuals in a population are implemented using double data type in the proposed approaches. Therefore, the size of

Table 4.5 Optimum results of QIGA, GA [127, 210] for three gray-scale images Lena, Peppers, and Woman using Shanbag's method [234].

SN	QIGA			GA		
			Lena			
	θ_{op}	v_{op}	t	θ_{op}	v_{op}	t
1	124	0.000002	343	124	0.000002	1329
$\sigma(v)$	0			0		
$\sigma(t)$	3.81			18.24		
SN			**Peppers**			
	θ_{op}	v_{op}	t	θ_{op}	v_{op}	t
1	131	0.000009	342	131	0.000009	1077
$\sigma(v)$	0			0		
$\sigma(t)$	4.30			17.34		
SN			**Woman**			
	θ_{op}	v_{op}	t	θ_{op}	v_{op}	t
1	129	0.000031	342	129	0.000031	1136
$\sigma(v)$	0			0		
$\sigma(t)$	3.84			17.20		

Table 4.6 Optimum results of QIPSO and PSO [144] for three gray-scale images Lena, Peppers, and Woman using Otsu's method [192].

SN	QIPSO			PSO		
			Lena			
	θ_{op}	v_{op}	t	θ_{op}	v_{op}	t
1	118	1424.307	214	118	1424.307	411
$\sigma(v)$	0			0.171		
$\sigma(t)$	3.73			13.18		
SN			**Peppers**			
	θ_{op}	v_{op}	t	θ_{op}	v_{op}	t
1	130	1813.048	213	130	1813.048	453
$\sigma(v)$	0			0.119		
$\sigma(t)$	4.13			14.31		
SN			**Woman**			
	θ_{op}	v_{op}	t	θ_{op}	v_{op}	t
1	124	1298.236	210	124	1298.236	438
$\sigma(v)$	0			0.180		
$\sigma(t)$	3.84			16.22		

each individual in the proposed algorithms turns into $2 \times 256 \times 256 \times 8 \times 8 = 8388608$ bits. As the intensity dynamic range varies from 0 and 255 in the gray level image, the length of each individual is taken as 8 (as $2^8 = 256$). So, each individual in the population can be generated as an integer data type. Hence, the required size to represent

Table 4.7 Optimum results of QIPSO and PSO [144] for three gray-scale images Lena, Peppers, and Woman using Ramesh's method [207].

SN	QIPSO			PSO		
	θ_{op}	v_{op}	t	θ_{op}	v_{op}	t
	Lena					
1	124	533792.559	215	124	533792.559	423
$\sigma(v)$	0			0		
$\sigma(t)$	3.53			18.14		
SN	**Peppers**					
	θ_{op}	v_{op}	t	θ_{op}	v_{op}	t
1	129	467278.548	217	129	467278.548	435
$\sigma(v)$	0			0		
$\sigma(t)$	2.43			16.44		
SN	**Woman**					
	θ_{op}	v_{op}	t	θ_{op}	v_{op}	t
1	117	521443.038	212	117	521443.038	420
$\sigma(v)$	0			0		
$\sigma(t)$	3.05			17.28		

Table 4.8 Optimum results of QIPSO and PSO [144] for three gray-scale images Lena, Peppers, and Woman using Li's method [158].

SN	QIPSO			PSO		
	θ_{op}	v_{op}	t	θ_{op}	v_{op}	t
	Lena					
1	111	−612.412	209	111	−612.412	436
$\sigma(v)$	0			0.007		
$\sigma(t)$	3.13			17.48		
SN	**Peppers**					
	θ_{op}	v_{op}	t	θ_{op}	v_{op}	t
1	123	−633.227	212	123	−633.227	450
$\sigma(v)$	0			0.003		
$\sigma(t)$	4.01			19.22		
SN	**Woman**					
	θ_{op}	v_{op}	t	θ_{op}	v_{op}	t
1	118	−676.120	210	118	−676.120	441
$\sigma(v)$	0			0.005		
$\sigma(t)$	2.91			19.33		

an individual in conventional algorithms is calculated as $8 \times 4 \times 8 = 256$ bits. In practice, 16 individuals for the proposed algorithms and 8192 individuals for their respective classical counterparts have been considered for experimental purposes. Thus, the size of each of the proposed algorithm becomes $16 \times 2 \times 256 \times 256 \times 8 \times 8 = 134217728$ bits and the size of their counterparts is $8192 \times 8 \times 4 \times 8 = 2097152$ bits. Therefore, the size

Table 4.9 Optimum results of QIPSO and PSO [144] for three gray-scale images Lena, Peppers, and Woman using the correlation coefficient.

SN	QIPSO			PSO		
	Lena					
	θ_{op}	v_{op}	t	θ_{op}	v_{op}	t
1	119	0.830751	325	119	0.830751	3342
$\sigma(v)$	0			0.006		
$\sigma(t)$	3.67			27.20		
SN	Peppers					
	θ_{op}	v_{op}	t	θ_{op}	v_{op}	t
1	131	0.850624	330	131	0.850624	3124
$\sigma(v)$	0			0.001		
$\sigma(t)$	3.73			28.38		
SN	Woman					
	θ_{op}	v_{op}	t	θ_{op}	v_{op}	t
1	123	0.856899	328	123	0.856899	3063
$\sigma(v)$	0			0.002		
$\sigma(t)$	3.97			26.38		

Table 4.10 Optimum results of QIPSO and PSO [144] for three gray-scale images Lena., Peppers, and Woman using Shanbag's method [234].

SN	QIPSO			PSO		
	Lena					
	θ_{op}	v_{op}	t	θ_{op}	v_{op}	t
1	124	0.000002	198	124	0.000002	1052
$\sigma(v)$	0			0		
$\sigma(t)$	3.70			16.25		
SN	Peppers					
	θ_{op}	v_{op}	t	θ_{op}	v_{op}	t
1	131	0.000009	201	131	0.000009	955
$\sigma(v)$	0			0		
$\sigma(t)$	3.96			18.34		
SN	Woman					
	θ_{op}	v_{op}	t	θ_{op}	v_{op}	t
1	129	0.000031	204	129	0.000031	1004
$\sigma(v)$	0			0.002		
$\sigma(t)$	4.10			15.39		

of each of the proposed approaches is $134217728/2097152 = 64$ times bigger than that of its classical counterpart. The superiority of the proposed approaches over the conventional algorithms is also established in terms of fitness value, computational time, accuracy and robustness, as reported in Tables 4.1–4.10.

(a) Thresholded Lena image with θ_{op} of 118

(b) Thresholded Peppers image with θ_{op} of 130

(c) Thresholded Woman image with θ_{op} of 124

(d) Thresholded Lena image with θ_{op} of 124

(e) Thresholded Peppers image with θ_{op} of 129

(f) Thresholded Woman image with θ_{op} of 117

(g) Thresholded Lena image with θ_{op} of 111

(h) Thresholded Peppers image with θ_{op} of 123

(i) Thresholded Woman image with θ_{op} of 118

(j) Thresholded Lena image with θ_{op} of 119

(k) Thresholded Peppers image with θ_{op} of 131

(l) Thresholded Woman image with θ_{op} of 123

(m) Thresholded Lena image with θ_{op} of 124

(n) Thresholded Peppers image with θ_{op} of 131

(o) Thresholded Woman image with θ_{op} of 129

Figure 4.8 Thresholded images of Lena, Peppers and Woman using Otsu's method [192] in Figures (a)–(c), Ramesh's method [207] in Figures (d)–(f), Li's method [158] in Figures (g)–(i), correlation coefficient [26, 145, 226] in Figures (j)–(l), and Shanbag's method [234] in Figures (m)–(o) with θ_{op} for QIGA.

The convergence curves for the proposed QIGA and QIPSO for different objective functions are presented in Figures 4.9–4.13 and Figures 4.14–4.18, respectively. Furthermore, Figures 4.19–4.23 show convergence curves for the conventional GA and PSO. The superiority of the proposed approaches is also visually established with

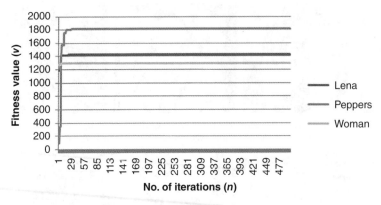

Figure 4.9 Convergence curve for Otsu's method [192] for QIGA.

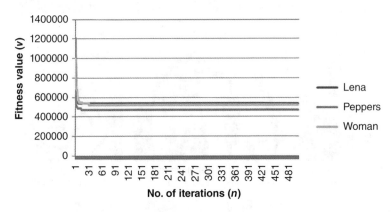

Figure 4.10 Convergence curve for Ramesh's method [207] for QIGA.

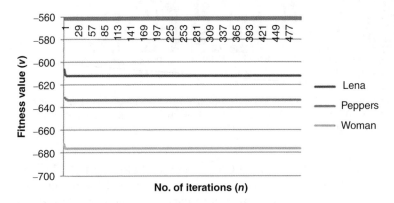

Figure 4.11 Convergence curve for Li's method [158] for QIGA.

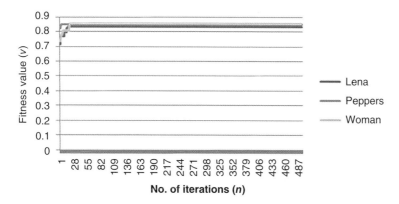

Figure 4.12 Convergence curve for correlation coefficient [26, 145, 226] for QIGA.

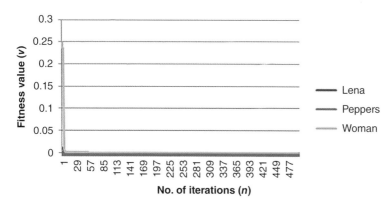

Figure 4.13 Convergence curve for Shanbag's method [234] for QIGA.

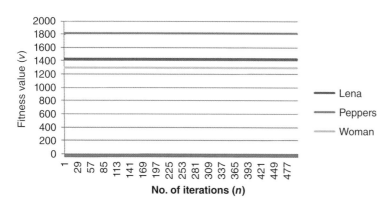

Figure 4.14 Convergence curve for Otsu's method [192] for QIPSO.

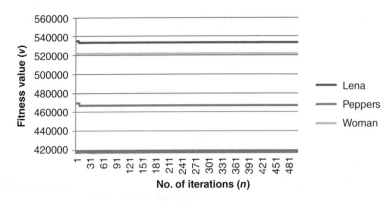

Figure 4.15 Convergence curve for Ramesh's method [207] for QIPSO.

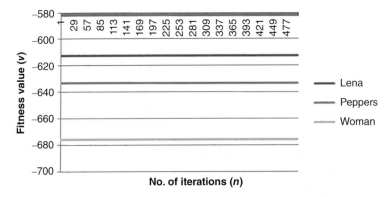

Figure 4.16 Convergence curve for Li's method [158] for QIPSO.

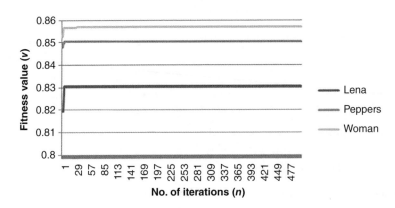

Figure 4.17 Convergence curve for correlation coefficient [26, 145, 226] for QIPSO.

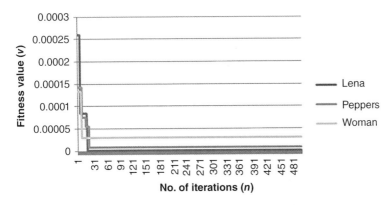

Figure 4.18 Convergence curve for Shanbag's method [234] for QIPSO.

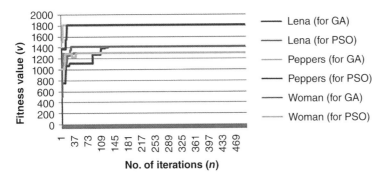

Figure 4.19 Convergence curve for Otsu's method [192] for GA and PSO.

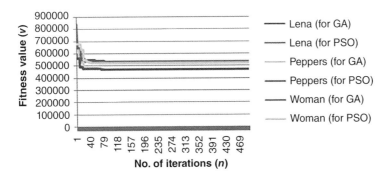

Figure 4.20 Convergence curve for Ramesh's method [207] for GA and PSO.

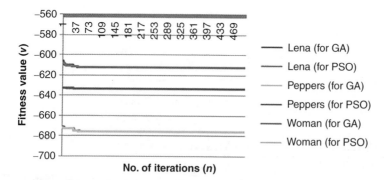

Figure 4.21 Convergence curve for Li's method [158] for GA and PSO.

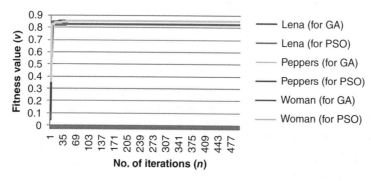

Figure 4.22 Convergence curve for correlation coefficient [26, 145, 226] for GA and PSO.

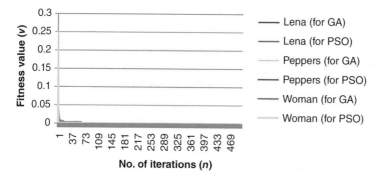

Figure 4.23 Convergence curve for Shanhag's method [234] for GA and PSO.

regards to the nature of these curves. As the thresholds of QIGA and QIPSO are the same or close to each other, one set of thresholded images for QIGA is shown in Figures 4.8(a)–(c) using Otsu's method [192]; in Figures 4.8(d)–(f) using Ramesh's method [207]; in Figures 4.8(a)–(c) using Li's method [158]; in Figures 4.8(d)–(f) using correlation coefficient [26, 145, 226] and in Figures 4.8(a)–(c) using Shanbag's method [234].

4.4.1.1 Implementation Results for QEA

For comparative purposes, 10 different runs of the algorithm introduced by Han et al. [121] with the objective functions offered by Otsu's Algorithm [192], Li's Algorithm [158], Ramesh's Algorithm [207], Shanbag's Algorithm [234], and the correlation coefficient [26, 145, 226], have been executed for the test images (Lena, Peppers, and Woman). A couple of look-up tables have been designed separately for each test image for experimental purpose (for random interference). Two distinct threshold values are randomly picked from the intensity dynamic range of each image (preferably in close proximity to the average values of the given range) to build the look-up table. The selected threshold vales for Lena are 108 (12×9) and 116 (4×29), for Peppers are 120 (3×40) and 125 (5×25), and for Woman are 108 (6×18) and 112 (8×14). The look-up tables for the test images are shown in Figures 4.24–4.26, respectively. The average and best result among all runs are listed in Tables 4.11–4.15. The convergence curves of the test images using the above objective functions are presented for this method in Figures 4.27–4.31. The respective thresholded images using this method are also shown in Figures 4.32(a)–(c) for Otsu's method [192]; in Figures 4.32(d)–(f) for Ramesh's method [207]; in Figures 4.32(a)–(c) for Li's method [158]; in Figures 4.32(d)–(f) for the correlation coefficient [26, 145, 226], and in Figures 4.32(a)–(c) for Shanbag's method [234].

Figure 4.24 Look-up table of Han et al. for Lena for $\theta = 108$ (12×9) and $\theta = 116$ (4×29).

$$\begin{bmatrix} 0 \cdots 0 \cdots 0 \cdots 0 \\ \cdots \cdots \cdots \cdots \cdots \\ 0 \cdots 1 \cdots 0 \cdots 0 \\ \cdots \cdots \cdots \cdots \cdots \\ 0 \cdots 0 \cdots 0 \cdots 0 \end{bmatrix} \quad \begin{bmatrix} 0 \cdots 0 \cdots 0 \cdots 0 \\ \cdots \cdots \cdots \cdots \cdots \\ 0 \cdots 0 \cdots 1 \cdots 0 \\ \cdots \cdots \cdots \cdots \cdots \\ 0 \cdots 0 \cdots 0 \cdots 0 \end{bmatrix}$$

Figure 4.25 Look-up table of Han et al. for Peppers for $\theta = 120$ (3×40) and $\theta = 125$ (5×25).

$$\begin{bmatrix} 0 \cdots 0 \cdots 0 \cdots 0 \\ \cdots \cdots \cdots \cdots \cdots \\ 0 \cdots 1 \cdots 0 \cdots 0 \\ \cdots \cdots \cdots \cdots \cdots \\ 0 \cdots 0 \cdots 0 \cdots 0 \end{bmatrix} \quad \begin{bmatrix} 0 \cdots 0 \cdots 0 \cdots 0 \\ \cdots \cdots \cdots \cdots \cdots \\ 0 \cdots 1 \cdots 0 \cdots 0 \\ \cdots \cdots \cdots \cdots \cdots \\ 0 \cdots 0 \cdots 0 \cdots 0 \end{bmatrix}$$

Figure 4.26 Look-up table of Han et al. for Woman for $\theta = 108$ (6×18) and $\theta = 112$ (8×14).

$$\begin{bmatrix} 0 \cdots 0 \cdots 0 \cdots 0 \\ \cdots \cdots \cdots \cdots \cdots \\ 0 \cdots 1 \cdots 0 \cdots 0 \\ \cdots \cdots \cdots \cdots \cdots \\ 0 \cdots 0 \cdots 0 \cdots 0 \end{bmatrix} \quad \begin{bmatrix} 0 \cdots 0 \cdots 0 \cdots 0 \\ \cdots \cdots \cdots \cdots \cdots \\ 0 \cdots 1 \cdots 0 \cdots 0 \\ \cdots \cdots \cdots \cdots \cdots \\ 0 \cdots 0 \cdots 0 \cdots 0 \end{bmatrix}$$

Table 4.11 Experimental results of Han et al. [121] for three gray-scale images Lena, Peppers, and Woman using Otsu's method [192].

SN	108			116		
	Lena					
	θ	v	t	θ	v	t
Average	118	1424.307	347	118	1424.307	347
Best	118	1424.307	346	118	1424.307	346
θ_{op}	118			118		
v_{op}	1424.307			1424.307		
SN	120			125		
	Peppers					
	θ	v	t	θ	v	t
Average	130	1813.048	346	130	1813.048	347
Best	130	1813.048	345	130	1813.048	346
θ_{op}	130			130		
v_{op}	1813.048			1813.048		
SN	108			112		
	Woman					
	θ	v	t	θ	v	t
Average	124	1298.236	347	124	1298.236	346
Best	124	1298.236	347	124	1298.236	346
θ_{op}	124			124		
v_{op}	1298.236			1298.236		

Table 4.12 Experimental results of Han et al. [121] for three gray-scale images Lena, Peppers, and Woman using Ramesh's method [207].

SN	108			116		
	Lena					
	θ	v	t	θ	v	t
Average	124	533792.559	345	124	533792.559	346
Best	124	533792.559	344	124	533792.559	345
θ_{op}	124			124		
v_{op}	533792.559			533792.559		
SN	120			125		
	Peppers					
	θ	v	t	θ	v	t
Average	129	467278.548	345	129	467278.548	345
Best	129	467278.548	345	129	467278.548	345
θ_{op}	129			129		
v_{op}	467278.548			467278.548		
SN	108			112		
	Woman					
	θ	v	t	θ	v	t
Average	117	521443.038	346	117	521443.038	345
Best	117	521443.038	345	117	521443.038	344
θ_{op}	117			117		
v_{op}	521443.038			521443.038		

Table 4.13 Experimental results of Han et al. [121] for three gray-scale images Lena, Peppers, and Woman using Li's method [158].

	108			116		
SN	Lena					
	θ	v	t	θ	v	t
Average	111	−612.412	345	111	−612.412	347
Best	118	1424.307	346	118	1424.307	346
θ_{op}	111			111		
v_{op}	−612.412			−612.412		
	120			125		
SN	Peppers					
	θ	v	t	θ	v	t
Average	123	−633.227	346	123	−633.227	346
Best	123	−633.227	345	123	−633.227	345
θ_{op}	123			123		
v_{op}	−633.227			−633.227		
	108			112		
SN	Woman					
	θ	v	t	θ	v	t
Average	118	−676.120	345	118	−676.120	345
Best	118	−676.120	345	118	−676.120	345
θ_{op}	118			118		
v_{op}	−676.120			−676.120		

Table 4.14 Experimental results of Han et al. [121] for three gray-scale images Lena, Peppers, and Woman using correlation coefficient [26, 145, 226].

	108			116		
SN	Lena					
	θ	v	t	θ	v	t
Average	119	0.830751	357	119	0.830751	356
Best	119	0.830751	357	119	0.830751	356
θ_{op}	119			119		
v_{op}	0.830751			0.830751		
	120			125		
SN	Peppers					
	θ	v	t	θ	v	t
Average	131	0.850624	356	131	0.850624	355
Best	131	0.850624	356	131	0.850624	355
θ_{op}	131			131		
v_{op}	0.850624			0.850624		
	108			112		
SN	Woman					
	θ	v	t	θ	v	t
Average	123	0.856899	355	123	0.856899	356
Best	123	0.856899	355	123	0.856899	356
θ_{op}	123			123		
v_{op}	0.856899			0.856899		

Table 4.15 Experimental results of Han et al. [121] for three gray-scale images Lena, Peppers, and Woman using Shanbag's method [234].

SN	108			116		
	Lena					
	θ	v	t	θ	v	t
Average	124	0.000002	347	124	0.000002	347
Best	124	0.000002	346	124	0.000002	346
θ_{op}	124			124		
v_{op}	0.000002			0.000002		
	120			125		
SN	**Peppers**					
	θ	v	t	θ	v	t
Average	131	0.000009	346	131	0.000009	347
Best	131	0.000009	345	131	0.000009	346
θ_{op}	131			131		
v_{op}	0.000009			0.000009		
	108			112		
SN	**Woman**					
	θ	v	t	θ	v	t
Average	129	0.000031	345	129	0.000031	346
Best	129	0.000031	344	129	0.000031	345
θ_{op}	129			129		
v_{op}	0.000031			0.000031		

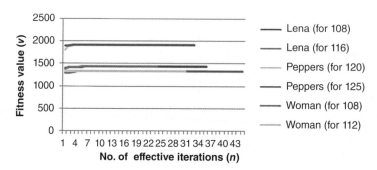

Figure 4.27 Convergence curve of Otsu's method [192] for Han et al. [121].

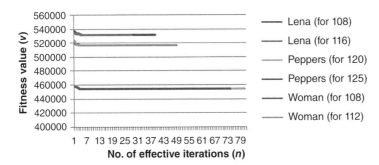

Figure 4.28 Convergence curve of Ramesh's method [207] for Han et al. [121].

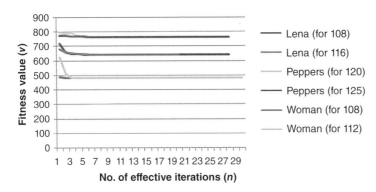

Figure 4.29 Convergence curve of Li's method [158] for Han et al. [121].

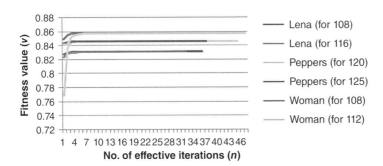

Figure 4.30 Convergence curve of correlation coefficient [26, 145, 226] for Han et al. [121].

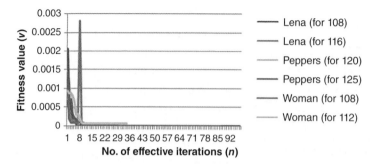

Figure 4.31 Convergence curve of Shanbag's method [234] for Han et al. [121].

4.4.2 Experimental Results (Phase II)

4.4.2.1 Experimental Results of Proposed QIGA and Conventional GA

In this phase, the proposed quantum-inspired genetic algorithm is exhibited using three gray-scale real-life images (Figures 4.7 (b), (d) and (e)) for bi-level thresholding. The proposed QIGA and its classical version have been executed 10 times each using six different thresholding methods as objective functions. The respective test results are listed in Tables 4.16–4.21. Each table comprises a group of 10 results (threshold value (θ_r), the fitness value (v_r), and the execution time (t) (in seconds)) at 10 different runs. Each table also contains the optimum threshold (θ_{op}) and its fitness value v_r among all runs.

Thresholded images for QIGA are shown in Figures 4.33(a)–(c) using Wu's method [274]; in Figures 4.33(d)–(f) using Renyi's method [220]; in Figures 4.33(g)–(i) using Yen's method [277]; in Figures 4.33(j)–(k) using Johannsen's method [135]; in Figures 4.34(a)–(c) using Silva's method [236], and in Figures 4.33(d)–(f) using the Linear Index of Fuzziness [130, 257]. Moreover, the convergence curves of QIGA for different objective functions are depicted in Figures 4.35–4.40.

4.4.2.2 Results Obtained with QEA

For comparative analysis, the QEA, proposed by Han et al. [121], has been applied in the same arrangement as QIGA. Six different heuristically made predefined look-up tables (two for each image) have been used for its quantum random interference. For this purpose, two different thresholds have randomly been selected from the dynamic intensity ranges of each of the test images (possibly, in close proximity to the middle value of the given range). The required look-up tables have been produced accordingly. The average and best results of Baboon, Peppers, and Corridor are listed in Tables 4.22–4.27. The thresholded images of these images are presented in Figures 4.41–4.42. Finally, the convergence curves for QEA for these six different objective functions are separately depicted in Figures 4.43–4.48.

In another approach, a convex combination of two different indices of fuzziness [104, 231] is used for fitness computation. These include the linear and quadratic indices of fuzziness of a fuzzy set (considering the intensity images to be fuzzy sets of brightness/darkness). The general form of the index of fuzziness [104, 231] is given by

$$v(A) = \frac{2}{n^k} d(A, \underline{A}) \tag{4.12}$$

(a) Thresholded Lena image with θ_{op} of 118

(b) Thresholded Peppers image with θ_{op} of 130

(c) Thresholded Woman image with θ_{op} of 124

(d) Thresholded Lena image with θ_{op} of 124

(e) Thresholded Peppers image with θ_{op} of 129

(f) Thresholded Woman image with θ_{op} of 117

(g) Thresholded Lena image with θ_{op} of 111

(h) Thresholded Peppers image with θ_{op} of 123

(i) Thresholded Woman image with θ_{op} of 118

(j) Thresholded Lena image with θ_{op} of 119

(k) Thresholded Peppers image with θ_{op} of 131

(l) Thresholded Peppers image with θ_{op} of 123

(m) Thresholded Lena image with θ_{op} of 124

(n) Thresholded Peppers image with θ_{op} of 131

(o) Thresholded Woman image with θ_{op} of 129

Figure 4.32 Thresholded images of Lena, Peppers and Woman using Otsu's method [192] in Figures (a)–(c), Ramesh's method [207] in Figures (d)–(f), Li's method [158] in Figures (g)–(i), correlation coefficient [26, 145, 226] in Figures (j)–(l), and Shanbag's method [234] in Figures (m)–(o) with θ_{op} for QEA [121].

Table 4.16 Optimum results of QIGA, GA [127, 210] for three gray-scale images Baboon, Peppers, and Corridor using Wu's method [274].

SN	QIGA			GA		
	θ_r	v_r	t	θ_r	v_r	t
	Baboon					
1	134	0.0480	350	134	0.0480	1329
2	134	0.0480	349	134	0.0480	1209
3	134	0.0480	348	134	0.0480	997
4	134	0.0480	346	134	0.0480	967
5	134	0.0480	346	134	0.0480	1001
6	134	0.0480	347	134	0.0480	1116
7	134	0.0480	349	134	0.0480	897
8	134	0.0480	349	134	0.0480	1018
9	134	0.0480	348	134	0.0480	884
10	134	0.0480	350	134	0.0480	1109
θ_{op}	134			134		
v_{op}	0.0480			0.0480		
SN	**Peppers**					
	θ_r	v_r	t	θ_r	v_r	t
1	125	0.0179	348	125	0.0179	886
2	125	0.0179	353	125	0.0179	1077
3	125	0.0179	349	125	0.0179	908
4	125	0.0179	348	125	0.0179	895
5	125	0.0179	347	125	0.0179	991
6	125	0.0179	347	125	0.0179	1055
7	125	0.0179	345	125	0.0179	1011
8	125	0.0179	346	125	0.0179	943
9	125	0.0179	348	125	0.0179	1072
10	125	0.0179	347	125	0.0179	994
θ_{op}	125			125		
v_{op}	0.0179			0.0179		
SN	**Corridor**					
	θ_r	v_r	t	θ_r	v_r	t
1	136	0.0084	346	136	0.0084	556
2	136	0.0084	346	136	0.0084	766
3	136	0.0084	345	136	0.0084	689
4	136	0.0084	347	136	0.0084	747
5	136	0.0084	347	136	0.0084	802
6	136	0.0084	345	136	0.0084	739
7	136	0.0084	346	136	0.0084	741
8	136	0.0084	346	136	0.0084	752
9	136	0.0084	347	136	0.0084	809
10	136	0.0084	345	136	0.0084	840
θ_{op}	136			136		
v_{op}	0.0084			0.0084		

Table 4.17 Optimum results of QIGA, GA [127, 210] for three gray-scale images Baboon, Peppers, and Corridor using Renyi's method [220].

SN	QIGA			GA		
	Baboon					
	θ_r	v_r	t	θ_r	v_r	t
1	108	8.9564	346	108	8.9564	3329
2	108	8.9564	343	108	8.9564	3102
3	108	8.9564	345	108	8.9564	3401
4	108	8.9564	345	108	8.9564	2806
5	108	8.9564	344	108	8.9564	2924
6	108	8.9564	343	108	8.9564	3278
7	108	8.9564	344	108	8.9564	3124
8	108	8.9564	344	108	8.9564	3178
9	108	8.9564	346	108	8.9564	3233
10	108	8.9564	344	108	8.9564	3212
θ_{op}	108			108		
v_{op}	8.9564			8.9564		
SN	Peppers					
	θ_r	v_r	t	θ_r	v_r	t
1	98	9.3140	344	98	9.3140	3978
2	98	9.3140	345	98	9.3140	3554
3	98	9.3140	345	98	9.3140	2997
4	98	9.3140	344	98	9.3140	3447
5	98	9.3140	343	98	9.3140	3100
6	98	9.3140	346	98	9.3140	3248
7	98	9.3140	344	98	9.3140	3278
8	98	9.3140	345	98	9.3140	3079
9	98	9.3140	346	98	9.3140	3002
10	98	9.3140	343	98	9.3140	3212
θ_{op}	98			98		
v_{op}	9.3140			9.3140		
SN	Corridor					
	θ_r	v_r	t	θ_r	v_r	t
1	138	9.3174	343	138	9.3174	3334
2	138	9.3174	344	138	9.3174	3014
3	138	9.3174	345	138	9.3174	2864
4	138	9.3174	344	138	9.3174	2678
5	138	9.3174	344	138	9.3174	3421
6	138	9.3174	345	138	9.3174	3178
7	138	9.3174	344	138	9.3174	3328
8	138	9.3174	344	138	9.3174	3452
9	138	9.3174	346	138	9.3174	3125
10	138	9.3174	344	138	9.3174	2907
θ_{op}	138			138		
v_{op}	9.3174			9.3174		

Table 4.18 Optimum results of QIGA, GA [127, 210] for three gray-scale images Baboon, Peppers, and Corridor using Yen's method [277].

SN	QIGA			GA		
	θ_r	v_r	t	θ_r	v_r	t
	Baboon					
1	108	8.5512	345	108	8.5512	410
2	108	8.5512	344	108	8.5512	425
3	108	8.5512	344	108	8.5512	509
4	108	8.5512	345	108	8.5512	512
5	108	8.5512	343	108	8.5512	424
6	108	8.5512	344	108	8.5512	434
7	108	8.5512	345	108	8.5512	463
8	108	8.5512	344	108	8.5512	455
9	108	8.5512	346	108	8.5512	439
10	108	8.5512	345	108	8.5512	430
θ_{op}	108			108		
v_{op}	8.5512			8.5512		
SN	**Peppers**					
	θ_r	v_r	t	θ_r	v_r	t
1	95	8.9983	346	95	8.9983	444
2	95	8.9983	348	95	8.9983	431
3	95	8.9983	346	95	8.9983	425
4	95	8.9983	347	95	8.9983	426
5	95	8.9983	345	95	8.9983	501
6	95	8.9983	346	95	8.9983	486
7	95	8.9983	345	95	8.9983	437
8	95	8.9983	347	95	8.9983	419
9	95	8.9983	348	95	8.9983	440
10	95	8.9983	345	95	8.9983	416
θ_{op}	95			95		
v_{op}	8.9983			8.9983		
SN	**Corridor**					
	θ_r	v_r	t	θ_r	v_r	t
1	151	7.3945	345	151	7.3945	409
2	138	9.3174	345	138	9.3174	432
3	138	9.3174	344	138	9.3174	429
4	138	9.3174	344	138	9.3174	430
5	138	9.3174	345	138	9.3174	437
6	138	9.3174	343	138	9.3174	416
7	138	9.3174	344	138	9.3174	424
8	138	9.3174	345	138	9.3174	437
9	138	9.3174	345	138	9.3174	513
10	138	9.3174	343	138	9.3174	471
θ_{op}	138			138		
v_{op}	9.3174			9.3174		

Table 4.19 Optimum results of QIGA, GA [127, 210] for three gray-scale images Baboon, Peppers, and Corridor using Johannsen's method [135].

SN	QIGA			GA		
	θ_r	v_r	t	θ_r	v_r	t
	Baboon					
1	118	0.1484	345	118	0.1484	407
2	118	0.1484	344	118	0.1484	412
3	118	0.1484	343	118	0.1484	441
4	118	0.1484	343	118	0.1484	437
5	118	0.1484	344	118	0.1484	490
6	118	0.1484	344	118	0.1484	431
7	118	0.1484	344	118	0.1484	430
8	118	0.1484	345	118	0.1484	442
9	118	0.1484	344	118	0.1484	416
10	118	0.1484	343	118	0.1484	422
θ_{op}	118			118		
v_{op}	0.1484			0.1484		
SN	**Peppers**					
	θ_r	v_r	t	θ_r	v_r	t
1	137	0.0714	342	137	0.0714	444
2	137	0.0714	343	137	0.0714	421
3	137	0.0714	344	137	0.0714	434
4	137	0.0714	342	137	0.0714	408
5	137	0.0714	344	137	0.0714	434
6	137	0.0714	344	137	0.0714	405
7	137	0.0714	344	137	0.0714	419
8	137	0.0714	343	137	0.0714	451
9	137	0.0714	343	137	0.0714	431
10	137	0.0714	342	137	0.0714	410
θ_{op}	137			137		
v_{op}	0.0714			0.0714		
SN	**Corridor**					
	θ_r	v_r	t	θ_r	v_r	t
1	47	0.0276	345	47	0.0276	409
2	47	0.0276	346	47	0.0276	405
3	47	0.0276	346	47	0.0276	419
4	47	0.0276	345	47	0.0276	425
5	47	0.0276	345	47	0.0276	436
6	47	0.0276	347	47	0.0276	471
7	47	0.0276	345	47	0.0276	469
8	47	0.0276	346	47	0.0276	409
9	47	0.0276	347	47	0.0276	461
10	47	0.0276	346	47	0.0276	467
θ_{op}	47			47		
v_{op}	0.0276			0.0276		

Table 4.20 Optimum results of QIGA, GA [127, 210] for three gray-scale images Baboon, Peppers, and Corridor using Silva's method [236].

SN	QIGA			GA		
	Baboon					
	θ_r	v_r	t	θ_r	v_r	t
1	135	0.3391	346	135	0.3391	485
2	135	0.3391	347	135	0.3391	497
3	135	0.3391	347	135	0.3391	413
4	135	0.3391	348	135	0.3391	455
5	135	0.3391	345	135	0.3391	462
6	135	0.3391	346	135	0.3391	461
7	135	0.3391	348	135	0.3391	424
8	135	0.3391	347	135	0.3391	429
9	135	0.3391	347	135	0.3391	431
10	135	0.3391	345	135	0.3391	430
θ_{op}	135			135		
v_{op}	0.3391			0.3391		
SN	Peppers					
	θ_r	v_r	t	θ_r	v_r	t
1	126	0.4242	346	126	0.4242	511
2	126	0.4242	348	126	0.4242	502
3	126	0.4242	349	126	0.4242	490
4	126	0.4242	347	126	0.4242	477
5	126	0.4242	348	126	0.4242	469
6	126	0.4242	348	126	0.4242	470
7	126	0.4242	346	126	0.4242	523
8	126	0.4242	347	126	0.4242	512
9	126	0.4242	345	126	0.4242	448
10	126	0.4242	346	126	0.4242	453
θ_{op}	126			126		
v_{op}	0.4242			0.4242		
SN	Corridor					
	θ_r	v_r	t	θ_r	v_r	t
1	128	0.3611	342	128	0.3611	449
2	128	0.3611	343	128	0.3611	483
3	128	0.3611	344	128	0.3611	486
4	128	0.3611	344	128	0.3611	476
5	128	0.3611	342	128	0.3611	503
6	128	0.3611	343	128	0.3611	482
7	128	0.3611	344	128	0.3611	506
8	128	0.3611	344	128	0.3611	439
9	128	0.3611	342	128	0.3611	470
10	128	0.3611	343	128	0.3611	481
θ_{op}	128			128		
v_{op}	0.3611			0.3611		

Table 4.21 Optimum results of QIGA, GA [127, 210] for three gray-scale images Baboon, Peppers, and Corridor using Linear Index of Fuzziness [130, 257].

	QIGA			GA		
SN	**Baboon**					
	θ_r	v_r	t	θ_r	v_r	t
1	128	0.2963	350	128	0.2963	2511
2	128	0.2963	353	128	0.2963	2636
3	128	0.2963	352	128	0.2963	2914
4	128	0.2963	354	128	0.2963	2007
5	128	0.2963	354	128	0.2963	2714
6	128	0.2963	353	128	0.2963	2466
7	128	0.2963	352	128	0.2963	2435
8	128	0.2963	351	128	0.2963	2676
9	128	0.2963	355	128	0.2963	2778
10	128	0.2963	354	128	0.2963	2586
θ_{op}	128			128		
v_{op}	0.2963			0.2963		
SN	**Peppers**					
	θ_r	v_r	t	θ_r	v_r	t
1	128	0.2397	356	128	0.2397	2763
2	128	0.2397	354	128	0.2397	2802
3	128	0.2397	353	128	0.2397	2337
4	128	0.2397	352	128	0.2397	2642
5	128	0.2397	354	128	0.2397	2843
6	128	0.2397	354	128	0.2397	2774
7	128	0.2397	352	128	0.2397	2559
8	128	0.2397	353	128	0.2397	2502
9	128	0.2397	354	128	0.2397	2364
10	128	0.2397	353	128	0.2397	2548
θ_{op}	128			128		
v_{op}	0.2397			0.2397		
SN	**Corridor**					
	θ_r	v_r	t	θ_r	v_r	t
1	128	0.1666	348	128	0.1666	2614
2	128	0.1666	347	128	0.1666	2473
3	128	0.1666	346	128	0.1666	2531
4	128	0.1666	346	128	0.1666	2782
5	128	0.1666	347	128	0.1666	2701
6	128	0.1666	345	128	0.1666	2146
7	128	0.1666	345	128	0.1666	2348
8	128	0.1666	346	128	0.1666	2635
9	128	0.1666	347	128	0.1666	2443
10	128	0.1666	346	128	0.1666	2475
θ_{op}	128			128		
v_{op}	0.1666			0.1666		

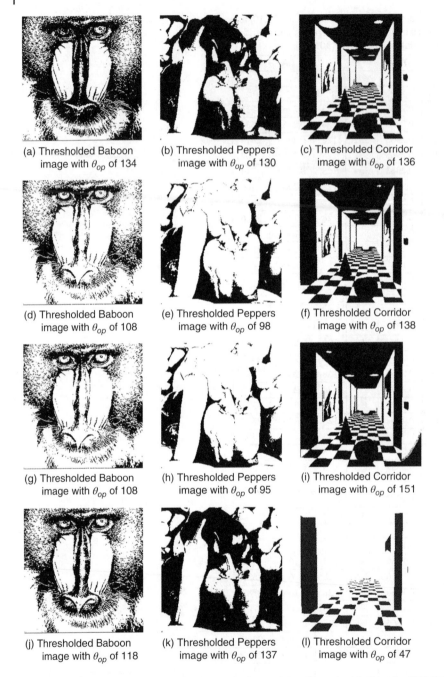

(a) Thresholded Baboon image with θ_{op} of 134

(b) Thresholded Peppers image with θ_{op} of 130

(c) Thresholded Corridor image with θ_{op} of 136

(d) Thresholded Baboon image with θ_{op} of 108

(e) Thresholded Peppers image with θ_{op} of 98

(f) Thresholded Corridor image with θ_{op} of 138

(g) Thresholded Baboon image with θ_{op} of 108

(h) Thresholded Peppers image with θ_{op} of 95

(i) Thresholded Corridor image with θ_{op} of 151

(j) Thresholded Baboon image with θ_{op} of 118

(k) Thresholded Peppers image with θ_{op} of 137

(l) Thresholded Corridor image with θ_{op} of 47

Figure 4.33 Thresholded images of Baboon, Peppers and Corridor using Wu's method [274] in Figures (a)–(c), Renyi's method [220] in Figures (d)–(f), Yen's method [277] in Figures (g)–(i), and Johannsen's method [135] in Figures (j)–(l) with θ_{op} for QIGA.

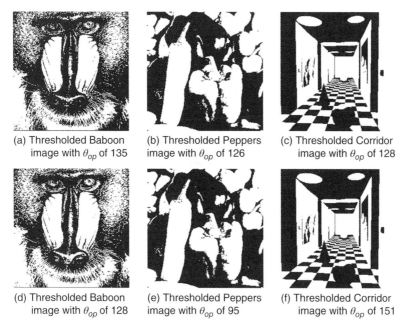

(a) Thresholded Baboon image with θ_{op} of 135

(b) Thresholded Peppers image with θ_{op} of 126

(c) Thresholded Corridor image with θ_{op} of 128

(d) Thresholded Baboon image with θ_{op} of 128

(e) Thresholded Peppers image with θ_{op} of 95

(f) Thresholded Corridor image with θ_{op} of 151

Figure 4.34 Thresholded images of Baboon, Peppers, and Corridor using Silva's method [236] in Figures (a)–(c) and Linear Index of Fuzziness [130, 257] in Figures (d)–(f) with θ_{op} for QIGA.

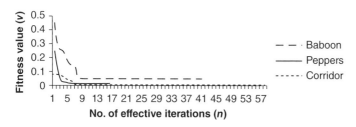

Figure 4.35 Convergence curve of Wu's method [274] for QIGA.

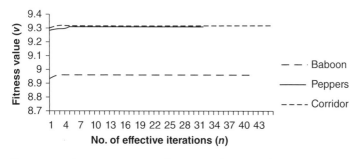

Figure 4.36 Convergence curve of Renyi's method [220] for QIGA.

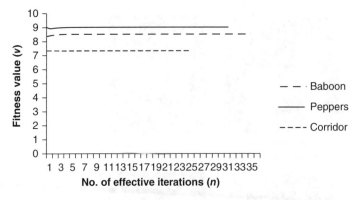

Figure 4.37 Convergence curve of Yen's method [277] for QIGA.

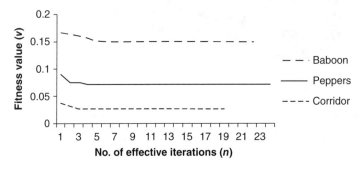

Figure 4.38 Convergence curve of Johannsen's method [135] for QIGA.

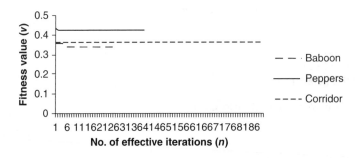

Figure 4.39 Convergence curve of Silva's method [236] for QIGA.

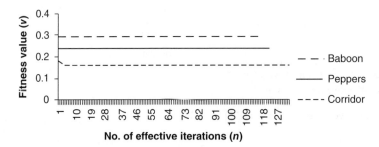

Figure 4.40 Convergence curve of Linear Index of Fuzziness [130, 257] for QIGA.

Table 4.22 Experimental results of Han et al. [121] for three gray-scale images Baboon, Peppers, and Corridor using Wu's method [274].

	118			120		
SN	**Baboon**					
	θ	v	t	θ	v	t
Average	134	0.0480	346	134	0.0480	346
Best	134	0.0480	344	134	0.0480	344
θ_{op}	134			134		
v_{op}	0.0480			0.0480		
	120			124		
SN	**Peppers**					
	θ	v	t	θ	v	t
Average	125	0.0179	345	125	0.0179	344
Best	125	0.0179	343	125	0.0179	342
θ_{op}	125			125		
v_{op}	0.0179			0.0179		
	125			128		
SN	**Corridor**					
	θ	v	t	θ	v	t
Average	136	0.0084	344	136	0.0084	345
Best	136	0.0084	343	136	0.0084	344
θ_{op}	136			136		
v_{op}	0.0084			0.0084		

Table 4.23 Experimental results of Han et al. [121] for three gray-scale images Baboon, Peppers, and Corridor using Renyi's method [220].

	118			120		
SN	**Baboon**					
	θ	v	t	θ	v	t
Average	108	8.9564	346	108	8.9564	345
Best	108	8.9564	343	108	8.9564	344
θ_{op}	108			108		
v_{op}	8.9564			8.9564		
	120			124		
SN	**Peppers**					
	θ	v	t	θ	v	t
Average	98	9.3140	344	98	9.3140	344
Best	98	9.3140	342	98	9.3140	343
θ_{op}	98			98		
v_{op}	9.3140			9.3140		
	125			128		
SN	**Corridor**					
	θ	v	t	θ	v	t
Average	138	9.3174	344	138	9.3174	345
Best	138	9.3174	343	136	9.3174	344
θ_{op}	138			138		
v_{op}	9.3174			9.3174		

Table 4.24 Experimental results of Han et al. [121] for three gray-scale images Baboon, Peppers, and Corridor using Yen's method [277].

SN	118			120		
	Baboon					
	θ	v	t	θ	v	t
Average	108	8.5512	345	108	8.5512	344
Best	108	8.5512	344	108	8.5512	343
θ_{op}	108			108		
v_{op}	8.5512			8.5512		
	120			124		
SN	**Peppers**					
	θ	v	t	θ	v	t
Average	95	8.9983	346	95	8.9983	344
Best	95	8.9983	344	95	8.9983	342
θ_{op}	95			95		
v_{op}	8.9983			8.9983		
	125			128		
SN	**Corridor**					
	θ	v	t	θ	v	t
Average	151	7.3945	345	151	7.3945	344
Best	151	7.3945	343	151	7.3945	342
θ_{op}	151			151		
v_{op}	7.3945			7.3945		

Table 4.25 Experimental results of Han et al. [121] for three gray-scale images Baboon, Peppers, and Corridor using Johannsen's method [135].

SN	118			120		
	Baboon					
	θ	v	t	θ	v	t
Average	118	0.1484	344	118	0.1484	345
Best	118	0.1484	343	118	0.1484	343
θ_{op}	118			118		
v_{op}	0.1484			0.1484		
	120			124		
SN	**Peppers**					
	θ	v	t	θ	v	t
Average	137	0.0714	344	137	0.0714	344
Best	137	0.0714	343	137	0.0714	342
θ_{op}	137			137		
v_{op}	0.0714			0.0714		
	125			128		
SN	**Corridor**					
	θ	v	t	θ	v	t
Average	47	0.0276	345	47	0.0276	344
Best	47	0.0276	343	47	0.0276	342
θ_{op}	47			47		
v_{op}	0.0276			0.0276		

Table 4.26 Experimental results of Han et al. [121] for three gray-scale images Baboon, Peppers, and Corridor using Silva's method [236].

SN	118			120		
	\multicolumn{6}{**Baboon**}					
	θ	v	t	θ	v	t
Average	135	0.3391	346	135	0.3391	344
Best	135	0.3391	345	134	0.3391	342
θ_{op}	135			135		
v_{op}	0.3391			0.3391		
	120			124		
SN	**Peppers**					
	θ	v	t	θ	v	t
Average	126	0.4242	345	126	0.4242	345
Best	126	0.4242	344	126	0.4242	342
θ_{op}	126			126		
v_{op}	0.4242			0.4242		
	125			128		
SN	**Corridor**					
	θ	v	t	θ	v	t
Average	128	0.3611	344	128	0.3611	344
Best	128	0.3611	343	128	0.3611	343
θ_{op}	128			128		
v_{op}	0.3611			0.3611		

Table 4.27 Experimental results of Han et al. [121] for three gray-scale images Baboon, Peppers, and Corridor using the Linear Index of Fuzziness [130, 257].

SN	118			120		
	Baboon					
	θ	v	t	θ	v	t
Average	128	0.2963	348	128	0.2963	346
Best	128	0.2963	346	128	0.2963	344
θ_{op}	128			128		
v_{op}	0.2963			0.2963		
	120			124		
SN	**Peppers**					
	θ	v	t	θ	v	t
Average	128	0.2397	345	128	0.2397	346
Best	128	0.2397	344	128	0.2397	344
θ_{op}	128			128		
v_{op}	0.2397			0.2397		
	125			128		
SN	**Corridor**					
	θ	v	t	θ	v	t
Average	128	0.1666	346	128	0.1666	345
Best	128	0.1666	345	128	0.1666	342
θ_{op}	128			128		
v_{op}	0.1666			0.1666		

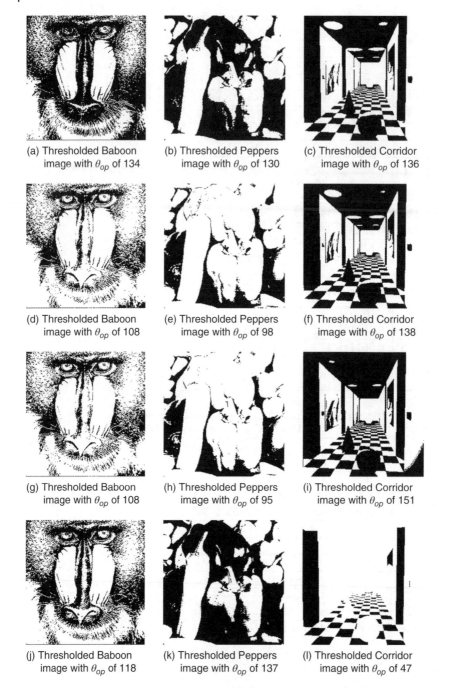

(a) Thresholded Baboon image with θ_{op} of 134

(b) Thresholded Peppers image with θ_{op} of 130

(c) Thresholded Corridor image with θ_{op} of 136

(d) Thresholded Baboon image with θ_{op} of 108

(e) Thresholded Peppers image with θ_{op} of 98

(f) Thresholded Corridor image with θ_{op} of 138

(g) Thresholded Baboon image with θ_{op} of 108

(h) Thresholded Peppers image with θ_{op} of 95

(i) Thresholded Corridor image with θ_{op} of 151

(j) Thresholded Baboon image with θ_{op} of 118

(k) Thresholded Peppers image with θ_{op} of 137

(l) Thresholded Corridor image with θ_{op} of 47

Figure 4.41 Thresholded images of Baboon, Peppers, and Corridor using Wu's method [274] in Figures (a)–(c), Renyi's method [220] in Figures (d)–(f), Yen's method [277] in Figures (g)–(i) and Johannsen's method [135] In Figures (j)–(l) with θ_{op} for QEA.

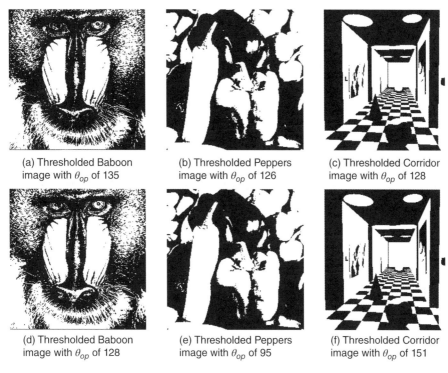

(a) Thresholded Baboon image with θ_{op} of 135

(b) Thresholded Peppers image with θ_{op} of 126

(c) Thresholded Corridor image with θ_{op} of 128

(d) Thresholded Baboon image with θ_{op} of 128

(e) Thresholded Peppers image with θ_{op} of 95

(f) Thresholded Corridor image with θ_{op} of 151

Figure 4.42 Thresholded images of Baboon, Peppers, and Corridor using Silva's method [236] in Figures (a)–(c) and the Linear Index of Fuzziness [130, 257] in Figures (d)–(f) with θ_{op} for QEA.

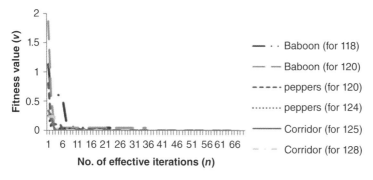

Figure 4.43 Convergence curve of Wu's method [274] for Han et al. [121].

where $d(A, \underline{A})$ denotes the distance between a fuzzy set A and its nearest ordinary set $\underline{A}$ [104]. If $k = 1$ and $d(A, \underline{A})$ represents the linear index of fuzziness $v_q(A)$. $k = 2$ stands for the quadratic index of fuzziness $v_q(A)$ [104].

The convex combination of these linear and quadratic indices of fuzziness used in this treatment in a multiple criteria scenario is given by

$$v_c(A)) = \lambda_1 v_l(A) + \lambda_2 v_q(A) \tag{4.13}$$

where λ_1 and λ_2 are Lagrange's multipliers.

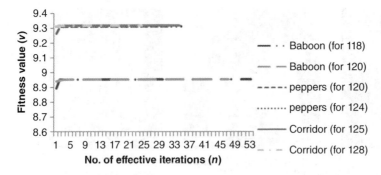

Figure 4.44 Convergence curve of Renyi's method [220] for Han et al. [121].

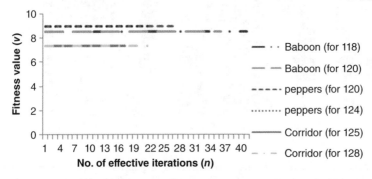

Figure 4.45 Convergence curve of Yen's method [277] for Han et al. [121].

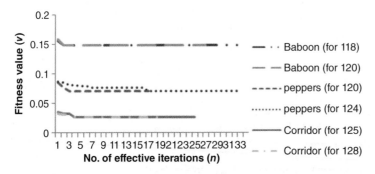

Figure 4.46 Convergence curve of Johannsen's method [135] for Han et al. [121].

The proposed QIGA algorithm has been tested for the optimization of the threshold intensity of two gray-level intensity images, namely, Lena and Peppers (Figures 4.7 (a)–(b)). Ten different runs of the algorithm with the fitness function given by Eq. (4.13) with $\lambda_1 = 0.5$ and $\lambda_2 = 0.5$ have been executed. The average threshold values for both images have been obtained as 128.

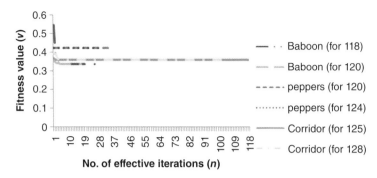

Figure 4.47 Convergence curve of Silva's method [236] for Han et al. [121].

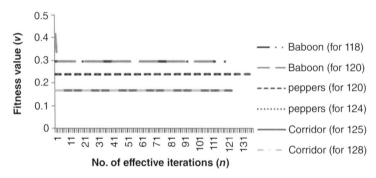

Figure 4.48 Convergence curve of Linear Index of Fuzziness [130, 257] for Han et al. [121].

Table 4.28 Comparative performance analysis of QIGA and GA.

	QIGA			GA		
Image	Lena					
	v_c	n_c	t	v_c	n_c	t
Lena	0.722982	40	2554.4	0.722982	3500	6559.0
Peppers	0.675204	43	2666.4	0.675204	3500	6384.0

Table 4.28 lists the average fitness values (v_c), the average number of iterations (n_c), and the average time (t) (in seconds) over 10 different runs of the QIGA algorithm for the test images. The conventional genetic algorithm (GA) has also been run for the same number of times on the test images with the same fitness function yielding the same threshold value of 128 for both the test images. Table 4.28 also lists the corresponding v_c, n_c, and t (in seconds) of the runs conducted with GA for the test images. From Table 4.28 it is evident that QIGA outperforms its conventional counterpart as far as the average number of iterations and the average time complexity are concerned. This is solely attributed to the inherent parallelism and random interference induced by QIGA in the search heuristics.

4.4.3 Experimental Results (Phase III)

4.4.3.1 Results Obtained with Proposed QIGA and Conventional GA

The proposed QIGA has also been applied to binary thresholding of gray-scale real life images (Figures 4.7 (a) and (f)). Comparative results of image thresholding are presented using these two images in this section. The implementation results of QIGA and GA are reported in Tables 4.29–4.31. Each table has the similar attributes as the tables reported in Section 4.4.2.1.

Thresholded images for the proposed QIGA are shown in Figures 4.49(a)–(b) using Brink's method [34]; in Figures 4.49(c)–(d) using Pun's method [206], and in Figures 4.49(e)–(f) using Kapur's method [140]. Thereafter, the convergence curves of QIGA for these objective functions are portrayed in Figures 4.50–4.52.

Table 4.29 Optimum results of QIGA, GA [127, 210] for three gray-scale images Lena and Barbara using Brink's method [34].

	QIGA			GA			QIGA			GA		
SN		Lena						Barbara				
	θ_r	v_r	t	θ_r	v_r	t	θ_r	v_r	t	θ_r	v_r	t
1	131	−41.910	344	131	−41.910	422	115	−37.606	343	115	−37.606	440
2	131	−41.910	344	131	−41.910	401	115	−37.606	344	115	−37.606	428
3	131	−41.910	344	131	−41.910	489	115	−37.606	343	115	−37.606	445
4	131	−41.910	343	131	−41.910	481	115	−37.606	344	115	−37.606	443
5	131	−41.910	344	131	−41.910	470	115	−37.606	344	115	−37.606	433
6	131	−41.910	344	131	−41.910	468	115	−37.606	343	115	−37.606	451
7	131	−41.910	344	131	−41.910	405	115	−37.606	344	115	−37.606	439
8	131	−41.910	344	131	−41.910	401	115	−37.606	343	115	−37.606	437
9	131	−41.910	344	131	−41.910	422	115	−37.606	344	115	−37.606	440
10	131	−41.910	344	131	−41.910	434	115	−37.606	344	115	−37.606	442
θ_{op}	131			131			115			115		
v_{op}	−41.910			−41.910			−37.606			−37.606		

Table 4.30 Optimum results of QIGA, GA [127, 210] for three gray-scale images Lena and Barbara using Pun's method [206].

	QIGA			GA			QIGA			GA		
SN		Lena						Barbara				
	θ_r	v_r	t	θ_r	v_r	t	θ_r	v_r	t	θ_r	v_r	t
1	130	0.154	343	130	0.154	389	113	0.151	343	113	0.151	463
2	130	0.154	343	130	0.154	406	113	0.151	342	113	0.151	478
3	130	0.154	344	130	0.154	416	113	0.151	343	113	0.151	494
4	130	0.154	344	130	0.154	421	113	0.151	344	113	0.151	474
5	130	0.154	343	130	0.154	410	113	0.151	343	113	0.151	468
6	130	0.154	343	130	0.154	385	113	0.151	342	113	0.151	466
7	130	0.154	344	130	0.154	438	113	0.151	344	113	0.151	472
8	130	0.154	343	130	0.154	413	113	0.151	342	113	0.151	482
9	130	0.154	344	130	0.154	396	113	0.151	343	113	0.151	488
10	130	0.154	344	130	0.154	422	113	0.151	342	113	0.151	490
θ_{op}	130			130			113			113		
v_{op}	0.154			0.154			0.151			0.151		

Table 4.31 Optimum results of QIGA, GA [127, 210] for three gray-scale images Lena and Barbara using Kapur's method [140].

	QIGA			GA			QIGA			GA		
	Lena						Barbara					
SN	θ_r	v_r	t	θ_r	v_r	t	θ_r	v_r	t	θ_r	v_r	t
1	125	8.846	343	125	8.846	444	117	8.992	343	117	8.992	420
2	125	8.846	343	125	8.846	456	117	8.992	344	117	8.992	426
3	125	8.846	344	125	8.846	436	117	8.992	344	117	8.992	436
4	125	8.846	342	125	8.846	436	117	8.992	342	117	8.992	428
5	125	8.846	342	125	8.846	451	117	8.992	343	117	8.992	430
6	125	8.846	344	125	8.846	429	117	8.992	344	117	8.992	431
7	125	8.846	344	125	8.846	439	117	8.992	342	117	8.992	419
8	125	8.846	343	125	8.846	431	117	8.992	344	117	8.992	425
9	125	8.846	343	125	8.846	450	117	8.992	343	117	8.992	447
10	125	8.846	344	125	8.846	449	117	8.992	343	117	8.992	442
θ_{op}	125			125			117			117		
v_{op}	8.846			8.846			8.992			8.992		

(a) Thresholded Lena image with θ_{op} of 131

(b) Thresholded Barbara image with θ_{op} of 115

(c) Thresholded Lena image with θ_{op} of 130

(d) Thresholded Barbara image with θ_{op} of 113

(e) Thresholded Lena image with θ_{op} of 125

(f) Thresholded Barbara image with θ_{op} of 117

Figure 4.49 Thresholded images of Lena and Barbara using Brink's method [34] in Figures (a)–(b), Thresholded images of Lena and Barbara using Pun's method [206] in Figures (c)–(d), and Thresholded images of Lena and Barbara using Kapur's method [140] in Figures (e)–(f) with θ_{op} for QIGA.

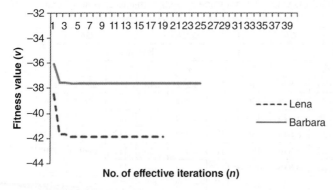

Figure 4.50 Convergence curve of Brink's method [34] for QIGA.

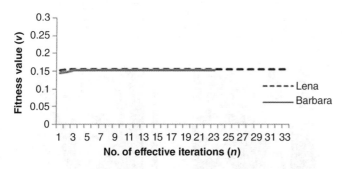

Figure 4.51 Convergence curve of Pun's method [206] for QIGA.

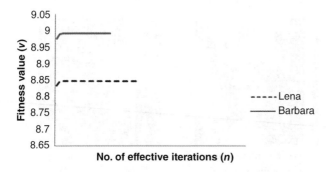

Figure 4.52 Convergence curve of Kapur's method [140] for QIGA.

4.4.3.2 Results obtained from QEA

To compare the performance of QIGA and QEA, proposed by Han et al., QEA has been executed for 10 different runs using Brink's method [34], Pun's method [206], and Kapur's method [140] as the objective functions on Lena and Barbara. The requisite set-up has been taken care of adequately to run QEA (as discussed in Section 4.4.2). The average and best results among 10 individual runs are listed in Tables 4.32–4.34. Furthermore, the thresholded images for QEA are shown in Figures 4.53(a)–(b) using Brink's method [34]; in Figures 4.53(c)–(d) using Pun's method [206], and in Figures 4.53(e)–(f) using Kapur's method [140]. Thereafter, the convergence curves for QEA for the above methods are presented in Figures 4.54–4.56.

Table 4.32 Experimental results of Han et al. [121] for three gray-scale images Lena and Barbara using Brink's method [34].

	108			116		
SN	**Lena**					
	θ	v	t	θ	v	t
Average	131	−41.910	344	131	−41.910	344
Best	131	−41.910	343	131	−41.910	343
θ_{op}	131			131		
v_{op}	−41.910			−41.910		
	120			124		
SN	**Barbara**					
	θ	v	t	θ	v	t
Average	115	−37.606	343	115	−37.606	343
Best	115	−37.606	343	115	−37.606	342
θ_{op}	115			115		
v_{op}	−37.606			−37.606		

Table 4.33 Experimental results of Han et al. [121] for three gray-scale images Lena and Barbara using Pun's method [206].

	108			116		
SN	**Lena**					
	θ	v	t	θ	v	t
Average	130	0.154	344	130	0.154	345
Best	130	0.154	342	130	0.154	342
θ_{op}	130			130		
v_{op}	0.154			0.154		
	120			124		
SN	**Barbara**					
	θ	v	t	θ	v	t
Average	113	0.151	344	113	0.151	344
Best	113	0.151	343	113	0.151	343
θ_{op}	113			113		
v_{op}	0.151			0.151		

Table 4.34 Experimental results of Han et al. [121] for three gray-scale images Lena and Barbara using Kapur's method [140].

	108			116		
SN	**Lena**					
	θ	v	t	θ	v	t
Average	125	8.846	343	125	8.846	343
Best	125	8.846	343	125	8.846	342
θ_{op}	125			125		
v_{op}	8.846			8.846		
	120			124		
SN	**Barbara**					
	θ	v	t	θ	v	t
Average	117	8.992	343	117	8.992	344
Best	117	8.992	342	117	8.992	343
θ_{op}	117			117		
v_{op}	8.992			8.992		

(a) Thresholded Lena image with θ_{op} of 131

(b) Thresholded Barbara image with θ_{op} of 115

(c) Thresholded Lena image with θ_{op} of 130

(d) Thresholded Barbara image with θ_{op} of 113

(e) Thresholded Lena image with θ_{op} of 125

(f) Thresholded Barbara image with θ_{op} of 117

Figure 4.53 Thresholded images of Lena and Barbara using Brink's method [34] in Figures (a)–(b), Thresholded images of Lena and Barbara using Pun's method [206] in Figures (c)–(d), and Thresholded images of Lena and Barbara using Kapur's method [140] in Figures (e)–(f) with θ_{op} for QEA [121].

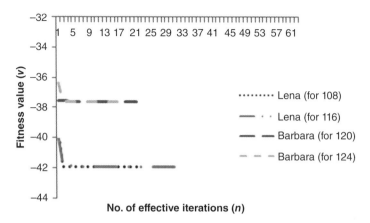

Figure 4.54 Convergence curve of Brink's method [34] for Han et al. [121].

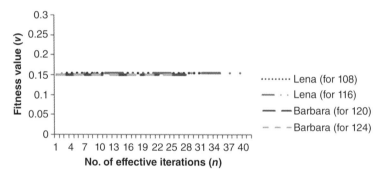

Figure 4.55 Convergence curve of Pun's method [206] for Han et al. [121].

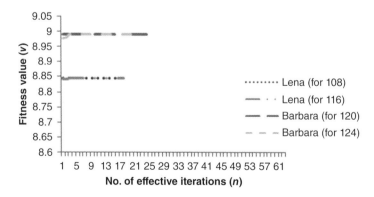

Figure 4.56 Convergence curve of Kapur's method [140] for Han et al. [121].

4.5 Comparative Analysis among the Participating Algorithms

The proposed algorithms, their classical versions and the method introduced by Han et al. [121] with the above-mentioned fitness functions, were executed for 10 different runs. The performance of the proposed algorithms is influenced by their inherent attribute of parallelism and their random interference for search heuristics [8]. The time complexity for each of the proposed algorithms has been discussed in their respective sections. The main disadvantage of Han et al.'s algorithm is that for the random interference, it depends on a predefined look-up table. But no such methods exist so far that can provide a look-up table beforehand. In addition, a rotation gate needs to update each qubit individual for the population diversity. This rotation gate works in consultation with the aforesaid predefined look-up table that makes a fuss for the entire system. In the proposed algorithms, we have used a probabilistic approach to determine the random interference which grants the success of our algorithms. Moreover, QEA [121] proposed by Han et al. starts with a random search until the probability of the best fit Q-bit individual converges to 1. This implies that QEA [121] starts with a global search followed by a local search, thereby benefiting from the features of exploration and exploitation [121]. In the proposed approaches, howsoever, the tenure of the global search mechanism is very small compared to that of Han et al.'s QEA [121]. This is evident from the convergence curves shown in Figures 4.9–4.18, 4.35–4.40, 4.50–4.52 and Figures 4.27–4.31, 4.43–4.48, 4.54–4.56. Whereas, Figures 4.27–4.31, 4.43–4.48, 4.54–4.56 depict the behavioral pattern of Han et al.'s QEA [121] algorithm applied for the determination of the optimal threshold of the test images using the five evaluation metrics under consideration, Figures 4.9–4.13, 4.14–4.18, 4.35–4.40 and 4.50–4.52 illustrate the mechanism of the proposed algorithms in the determination of the optimal threshold of the images using those metrics. Therefore, the local search mechanism prevails in the proposed approaches as compared to QEA [121] using heuristically chosen look-up tables. This feature also helps QIGA and QIPSO to outperform QEA [121] when it comes to thresholding optimization. The outputs obtained with Han et al.'s algorithm were discussed in the respective results sections in detail.

4.6 Conclusion

In this chapter, two distinct quantum inspired meta-heuristics, called quantum inspired genetic algorithm and quantum inspired particle swarm optimization, are discussed and introduced. These two novel algorithms are a class of quantum inspired evolutionary algorithm (QIEA), designed to determine the binary threshold value from gray level real-life images. The proposed algorithms resort to a time discrete and random interference between participating qubit encoded superposed solutions based on Birkhoff's Ergodic theorem. The chaotic model is determined by applying nonlinear point transformations on the qubits in the quantum chromosomes of the population. The use of a random interference method removes the need for any predefined best solution for the algorithm. The performance of the proposed algorithms as regards to bi-level image thresholding of real gray-scale images is compared with their respective classical counterparts and later with QEA proposed by Han et al. The novelty of the proposed algorithms lies in the fact that it does not require any look-up table [121] to decide upon the interference of the constituent qubits, which is characteristic of any QIEA. In contrast, the interference is decided in a random fashion by means of the generation of a chaotic map-based point transformation of the constituent wave functions. Results of the application of the proposed algorithms on the determination of the optimal thresholds of real-life gray-level images indicate encouraging avenues. The results of the application of these proposed algorithms indicate the superiority over their respective classical counterparts and QEA as proposed by Han et al. in terms of its reduced time complexity. Moreover, as far as the accuracy and robustness are concerned, it is found that the proposed QIPSO outperforms the participating algorithms.

4.7 Summary

- The field of quantum computing offers a speed-up of the classical algorithms.
- The speed-up occurs by inducing some physical phenomena of quantum computing into the classical algorithmic framework.
- The improvement in the computational speed-up can be achieved by means of intrinsic parallelism perceived in the quantum bits, the building blocks of a quantum computer.
- The first quantum algorithms appeared before the actual fabrication of the first quantum gate.
- Two novel quantum inspired meta-heuristic algorithms for bi-level image thresholding are introduced in this chapter.
- These proposed techniques resort to a chaotic map model because the system is an ergodic dynamical system.

Exercise Questions

Multiple Choice Questions

1 The speed-up in different classical algorithms can be achieved by inducing some physical phenomena of quantum computing, such as
   (a) interference
   (b) superposition
   (c) entanglement
   (d) all of the above

2 The advance in the computational speed-up can be obtained by means of intrinsic parallelism perceived in
   (a) quantum bit
   (b) quantum gate
   (c) quantum registrar
   (d) none of these

3 A pure quantum algorithm is found by
(a) searching databases
(b) factoring large numbers
(c) both (a) and (b)
(d) none of these

4 In the quantum inspired evolutionary algorithm developed by Han et al., which of the following has been used?
(a) quantum bit probabilistic representation
(b) superposition of states
(c) quantum gates
(d) all of the above

5 In a popular quantum genetic algorithm, proposed by Talbi et al., which of the following gates have been used?
(a) quantum CNOT gate
(b) quantum rotation gate
(c) quantum Hadamard gate
(d) none of the above

6 In the proposed QIGA, the size of initial population was taken as
(a) $L \times L$ matrix of quantum bits
(b) $L \times 1$ matrix of quantum bits
(c) $1 \times L$ matrix of quantum bits
(d) there is no specific size
Here, L denotes the number of possible pixel intensity values of a gray-scale image.

7 The proposed QIGA and QIPSO have been designed to find
(a) two or more optimal threshold values
(b) a single optimal threshold value
(c) any number of optimal threshold values
(d) none of the above

Short Answer Questions

1 What do you mean by bi-level image thresholding? How does it differ from multi-level image thresholding?

2 What do you mean by a quantum inspired evolutionary algorithm? Illustrate with suitable examples.

3 Differentiate between single-point-based and population-based evolutionary algorithms with suitable examples.

4 How can the computational speed-up be improved in classical algorithms?

5 Describe, in brief, the automorphism, endomorphism, flow and semiflow properties in relation to a real chaotic map.

6 Discuss briefly, with suitable examples, how the quantum crossover, quantum mutation and quantum shift occur in QIGA.

7 How does the population diversity occur in QIPSO?

Long Answer Questions

1 Discuss the algorithmic steps of QIGA with proper illustrations. Derive its time complexity.

2 Discuss the algorithmic steps of QIPSO with proper illustrations. Derive its time complexity.

3 How is the random chaotic map for each qubit state generated in QIGA?

4 Discuss in detail, how the quantum measurement occurs in QIGA.

Coding Examples

In this chapter, two quantum inspired evolutionary techniques have been introduced and discussed. The code for the proposed QIGA is presented in Figures 4.57–4.59.

```
Func
function [tr,fv]=QIGA(n,mu,cr)
%n-Number of chromosomes, Mu-Mutation probability, Cr-Crossover
probability
tic ; %Computational time calculation
t = 0;
ntrials = 0 ;
L=256;
His=insert(L);%Histogram of the gray scale image is calculated
M=Generate(L,n);%Function used for generating initial population
s=M;
maxf=0;
po=1;
T=zeros(1,1);
for i=1:50
  [tr,fv,fb,pos]=Fitness(L,s,His,n); %Function used for computing
  fitness of individuals
  if(i==1)
      po=pos;
  end
  if(maxf<fb)
      maxf=fb;
      po=pos;
      T=tr;
      t = toc;
  end
  p=s(po,:);
  S=selection(s,n,fv); %Function used for selecting best
  individuals
  C=cross(S,cr,L,n); %Function used for crossover operation
  MU=mute(C,mu,L,n,p); %Function used for mutation operation
  s=MU;
  ntrials = ntrials + 1 ;
  t = toc ;
  t = t / ntrials
end
```

```
Func
function His=insert(L)
img=imread('D:\test\image1.bmp'); %Read pixels of the image
val=zeros(1,L);
c=0;
d=1;
while(d<=L*L)
    if(img(d)==0)
        c=c+1;
    end
    for j=1:(L-1)
        if(img(d)==j)
            val(j)=val(j)+1;
        end
    end
    d=d+1;
end
val(L)=c;
for i=1:L
    im(i)=val(i)/(L*L);
end
His=im;
end

Func

function M=Generate(L,n)
for k=1:n
    for i=1:L
        a(k,i)=rand;
    end
end
b=sqrt(1-a.^2);
M=b;
end
```

Figure 4.57 Sample code of Func[1], Func[2], and Func[3] for QIGA.

```
Func⁴

function [tr,fv,fb,pos]=Fitness(L,s,His,n)
[tr,fv,fb,pos]=thcal(L,s,His,n);
end


Func⁵

function [tr,fv,fb,pos]=thcal(L,s,His,n,K)
mx=0;
j=1;
p=1;
P=partition(s,L); %Function used for partitioning the pixel
values
T=zeros(1,K);
for i=1:n
  [tr1,fv1]=thli(L,His,P); %Function used for threshold value
  f(j)=fv1(1);              calculation using Li's method
  if(mx<f(j))
    mx=fv1(1);
    T=tr1;
    p=j;
  end
  j=j+1;
end
tr=T;
fv=f;
fb=mx;
pos=p;
end


Func⁶

function P=partition(s,L)
f=0;
st=zeros(1,K-1);
j=1;
```

```
while(1)
  p=randint(1,1,[1,L-1]);
  pm=rand; %probability factor
  if(pm>s(p)*s(p) && p~=f)
    st(j)=p;
    f=p;
    j=j+1;
  end
  if(j==2)
    break;
  end
end
P=st;
end


Func⁷

function [tr1,fv1]=thli(L,His,P)
im=His;
w=zeros(1,2);
muk=zeros(1,2);
mu=0;
P=sort(P);
T=P(1);
m0a= m0b= m1a= m1b= mua= mub= rs= 0;

for i=1:(T-1)
  m0a=m0a+im(i);
  m1a=m1a+i*im(i);
end
for i=T:L
  m0b=m0b+im(i);
  m1b=m1b+i*im(i);
end

mua=m1a/m0a;
mub=m1b/m0b;
```

```
if(m0a~=0 && m0b~=0)
  rs=-m1a*log(mua)-m1b*log(mub);
else
  rs=1000;
end
  fv1=rs;
  tr1=P;
end


Func⁸

function S=selection(s,n,fv)
for i=1:n
while 1
  m1=randint(1,1,[1,n]);
  m2=randint(1,1,[1,n]);
if(m1~=m2)
  break;
end
end
if(fv(m1)>fv(m2))
  s1(i,:)=s(m1,:);
else
  s1(i,:)=s(m2,:);
end
end
S=s1;
end
```

Figure 4.58 Sample code of Func⁴, Func⁵, Func⁶, Func⁷, and Func⁸ for QIGA.

```
Func⁹

function C=cross(S,cr,L,n)
s1=zeros(n,L);
C=S;
for i=1:floor(n/2)
while 1
  m1=randint(1,1,[1,n]);
  m2=randint(1,1,[1,n]);
if(m1~=m2)
 break;
end
end
if(rand<cr)
  pos=randint(1,1,[1,L]);
  s1(m1,1:pos)=S(m1,1:pos);
  s1(m1,pos+1:L)=S(m2,pos+1:L);
  s1(m2,pos+1:L)=S(m1,pos+1:L);
  s1(m2,1:pos)=S(m2,1:pos);
end
  S(m1,:)=s1(m1,:);
  S(m2,:)=s1(m2,:);
end
  C=S;
end
```

```
Func¹⁰

function MU=mute(C,mu,L,n,p)
s1=zeros(1,L);
for i=1:n
for j=1:L
if(rand<mu)
C(i,j)=rand;
end
end
end
  D=sqrt(1-C.^2);
  D(n,:)=p;
  MU=D;
end
```

Figure 4.59 Sample code of Func⁹ and Func¹⁰ for QIGA.

5

Quantum Inspired Meta-Heuristics for Gray-Scale Multi-Level Image Thresholding

5.1 Introduction

Different types of image thresholding techniques [231], both for bi-level and multi-level approaches, were used over the past few decades to drive image segmentation. Two different bi-level image thresholding techniques were introduced and discussed [28, 61, 64, 68] in Chapter 4. In this chapter, the bi-level image thresholding technique is extended to its multi-level version [63, 67]. However, the complexity increases with the increase of level of thresholding. The multi-level thresholding computationally takes more time when the level of thresholding increases.

The proposed techniques, as discussed in Chapter 4, are constrained in their function-alities to act in response to bi-level image information. The foremost focus of Chapter 4 was to develop more efficient quantum inspired techniques starting from the conventional algorithms. These techniques take into cognizance both the quantum mechanical features and the popular meta-heuristics in their operation.

In this chapter, the functionality of the quantum inspired meta-heuristic algorithms has been extended to the gray-scale domain so as to enable them to determine optimum threshold values from synthetic/real-life gray-scale images. As regards gray-scale images, pixel intensity information is exhibited in all possible levels, starting from 0 (perfectly black) to 255 (perfectly white). This chapter is focused on addressing the use of quantum mechanisms to introduce six different quantum inspired meta-heuristic algorithms to perform multi-level thresholding faster. The proposed algorithms are Quantum Inspired Genetic Algorithm, Quantum Inspired Particle Swarm Optimization, Quantum Inspired Differential Evolution, Quantum Inspired Ant Colony Optimization, Quantum Inspired Simulated Annealing, and Quantum Inspired Tabu Search [63, 67].

As a sequel to the proposed algorithms, we have also conducted experiments with the Two-Stage Multithreshold Otsu method (TSMO) [129], the Maximum Tsallis entropy Thresholding (MTT) [288], the Modified Bacterial Foraging (MBF) algorithm [222], the classical particle swarm optimization [144, 167], and the classical genetic algorithm [127, 210]. The effectiveness of the proposed algorithms is demonstrated on fifteen images (ten real-life gray-scale images and five synthetic images) at different levels of thresholds quantitatively and visually. Thereafter, the results of six quantum meta-heuristic algorithms are considered to create consensus results. Finally, a statistical test, called the Friedman test, was conducted to judge the superiority of the algorithms among themselves. Quantum Inspired Particle Swarm Optimization

Quantum Inspired Meta-heuristics for Image Analysis, First Edition.
Sandip Dey, Siddhartha Bhattacharyya, and Ujjwal Maulik.
© 2019 John Wiley & Sons Ltd. Published 2019 by John Wiley & Sons Ltd.

is found to be superior among the proposed six quantum meta-heuristic algorithms and the five other algorithms used for comparison. A Friedman test has also been conducted between the Quantum Inspired Particle Swarm Optimization and all the other algorithms to justify the statistical superiority of the proposed Quantum Inspired Particle Swarm Optimization. Finally, the computational complexities of the proposed algorithms have been elucidated to find the time efficiency of the proposed algorithms.

The chapter is organized as follows. Six different quantum inspired meta-heuristics for multi-level image thresholding are introduced and discussed in Sections 5.2–5.7. These techniques use Otsu's method as the objective function to find the optimal threshold values of the input images. Experimental results of the proposed techniques for multi-level thresholding of several real-life images and synthetic images, are reported in Section 5.8. As a sequel to the performance of the proposed algorithms for multi-level thresholding, the focus of this chapter is also centered on the performance of the comparable algorithms. The best performing algorithm among the proposed algorithms is compared with the other five comparable algorithms to find the overall best algorithm. A comparative result is presented in Section 5.9. Section 5.10 ends this chapter drawing relevant conclusions. A brief summary of the chapter is presented in Section 5.11. A set of exercise questions of different types with regard to the theme of the chapter is presented, along with sample coding of different techniques.

5.2 Quantum Inspired Genetic Algorithm

In this section, we have linked the genetic algorithm, a meta-heuristic optimization technique with quantum mechanical principles to develop a new Quantum Inspired Genetic Algorithm for multi-level thresholding (QIGA). The proposed QIGA is depicted step-wise in Algorithm 2. The different phases of the QIGA are discussed in detail in this section.

5.2.1 Population Generation

This is the initial part of the proposed QIGA. Here, a number of qubit encoded chromosomes are employed to generate an initial population. Initially, an $n \times L$ matrix is formed using two superposed quantum states, as given below

$$|\psi\rangle = \begin{bmatrix} \alpha_{11}|\psi_1\rangle + \beta_{11}|\psi_2\rangle & \alpha_{12}|\psi_1\rangle + \beta_{12}|\psi_2\rangle \cdots \alpha_{1L}|\psi_1\rangle + \beta_{1L}|\psi_2\rangle \\ \alpha_{21}|\psi_1\rangle + \beta_{21}|\psi_2\rangle & \alpha_{22}|\psi_1\rangle + \beta_{22}|\psi_2\rangle \cdots \alpha_{2L}|\psi_1\rangle + \beta_{2L}|\psi_2\rangle \\ \cdots\cdots\cdots\cdots\cdots\cdots\cdots\cdots\cdots\cdots\cdots\cdots\cdots\cdots\cdots\cdots\cdots\cdots \\ \cdots\cdots\cdots\cdots\cdots\cdots\cdots\cdots\cdots\cdots\cdots\cdots\cdots\cdots\cdots\cdots\cdots\cdots \\ \alpha_{n1}|\psi_1\rangle + \beta_{n1}|\psi_2\rangle & \alpha_{n2}|\psi_1\rangle + \beta_{n2}|\psi_2\rangle \cdots \alpha_{nL}|\psi_1\rangle + \beta_{nL}|\psi_2\rangle \end{bmatrix} \quad (5.1)$$

Here, $|\alpha_{ij}|^2$ and $|\beta_{ij}|^2$ are the probabilities for the state $|\psi_1\rangle$ and $|\psi_2\rangle$, respectively where, $i = 1, 2, \cdots, n, j = 1, 2, \cdots, L$ where L represents the maximum pixel intensity value of the input gray-scale image. Each row in Eq. (5.1) signifies the qubit representation of a single chromosome. This is the possible scheme for encoding participating chromosomes for a required number of solutions using the superposition principle. $|\psi_1\rangle$ and $|\psi_2\rangle$ represent the "0" state and "1" state, respectively where $|\psi_1\rangle + |\psi_2\rangle$ is the superposition of these two states for two-state quantum computing.

Algorithm 2: Steps of QIGA for multi-level thresholding

    Input: Number of generation: *MxGen*
          Size of the Population: $\mathcal{V}$
          Crossover probability: P_{cr}
          Mutation probability: P_m
          No. of thresholds: K
    Output: Optimal threshold values: θ

1: Selection of pixel values randomly to generate $\mathcal{V}$ number of initial chromosomes (*POP*) where length of eachchromosome is $\mathcal{L} = \sqrt{L}$, where L represents maximum gray value of a selected image.
2: Using the concept of qubits to assign real value between (0,1) to each pixel encoded in *POP*. Let it produce *POP′*.
3: Using quantum rotation gate as given in Eq. (1.10), update *POP′*.
4: *POP′* undergoes *quantum orthogonality* to generate *POP″*.
5: Find K number of thresholds as pixel values from each chromosome in *POP*. It should satisfy the condition greater than ($>$) $rand(0,1)$ with its corresponding value in *POP″*. Let it give *POP**.
6: Compute fitness of each chromosome in *POP** using Eq. (2.42).
7: Record the best chromosome $b \in POP″$ and its threshold values in $T_B \in POP^*$.
8: Apply tournament selection strategy to replenish the chromosome pool.
9: **repeat**
10:     Select two chromosomes, k and m at random from $[1, \mathcal{V}]$ where $k \neq m$.
11:     **if** ($rand(0,1) < P_{cr}$) **then**
12:         Select a random position, $pos \in [1, \mathcal{L}]$.
13:         Perform crossover operation between two chromosomes, k and m at the position *pos*.
14:     **end if**
15: **until** the pool of chromosomes are filled up
16: **for** all $k \in POP″$ **do**
17:     **for** all position in k **do**
18:         **if** ($rand(0,1) < P_m$) **then**
19:             Flip the corresponding position with random real number.
20:         **end if**
21:     **end for**
22: **end for**
23: Repeat steps 3, 4 and 5.
24: Save the best chromosome in $c \in POP″$ and its corresponding threshold values in $T_B \in POP″$.
25: Compare the fitness value of the chromosomes of b and c.
26: Store the best chromosome in $b \in POP″$ and its corresponding threshold values in $T_B \in POP″$ (elitism).
27: Repeat steps 8 to 26 for a fixed number of generations, *MxGen*.
28: Report the optimal threshold values, $\theta = T_B$.

5.2.2 Quantum Orthogonality

The second step of the proposed QIGA entails the quantum orthogonality of the population of chromosomes as shown in Eq. (5.1). This step preserves the basic constraint of the qubit individual in Eq. (5.1) as given by

$$|\alpha_{ij}|^2 + |\beta_{ij}|^2 = 1 \tag{5.2}$$

where $i = 1, 2, \cdots, n$ and $j = 1, 2, \cdots, L$.

    Each qubit experiences a quantum orthogonality to validate Eq. (5.2). A typical example of quantum orthogonality is shown in Figure 5.1. Note that, $(0.2395)^2 + (0.9708)^2 = 1, (-0.1523)^2 + (0.9883)^2 = 1$, etc.

| 0.1119 | 0.2395 | 0.2232 | 0.9883 | 0.9950 | −0.1532 |
| 0.9937 | 0.9708 | 0.9747 | −0.1523 | 0.0987 | 0.9881 |

Figure 5.1 Quantum orthogonality.

5.2.3 Determination of Threshold Values in Population and Measurement of Fitness

In quantum measurement, a special treatment in measurement policy is espoused that leads to the optimal solution. Since the proposed algorithm is executed on conventional computers, the superposed quantum states must be destroyed in order to find a coherent solution and the states are conserved for the iteration of next generation. Depending on the level of computation in multi-level thresholding, the number of states having exactly a single qubit are determined. This phase resorts to quantum measurement to determine optimal thresholds. First, $\mathcal{V}$ number of pixel intensity values are randomly selected from the test image to form a population, POP. The length of each participating chromosome in POP is taken as $\mathcal{L} = \sqrt{L}$, where L signifies the maximum pixel intensity value of the selected image. Afterwards, the concept of qubits is applied to assign a real number at random between (0, 1) to each pixel encoded in POP, which produces POP'. Next, POP' passes through a quantum orthogonality to generate POP''. A predefined number of threshold values as pixel intensities are obtained using a probability criteria. Next, another population matrix, POP^* is created by applying the condition $rand(0, 1) > |\beta_i|^2$, $i = 1, 2, \cdots, L$ in POP. Finally, each chromosome in POP^* is evaluated to derive the fitness value using Eq. (2.42). The entire process is repeated for a predefined number of generations and the global best solution is updated and reported.

A typical example is presented below to demonstrate the measurement strategy. Let, at a particular generation, the condition $rand(0, 1) > |\beta_i|^2$ be satisfied at the positions (15, 3), (16, 8) and (17, 9), respectively in population, POP. A solution matrix, SOL is introduced of the identical size as POP. The above three positions in SOL are set to 1 where the remaining positions are set to 0. Therefore, the optimum threshold values are determined as $(15 \times 3) = 45$, $(16 \times 8) = 128$ and $(17 \times 9) = 153$ (in the case of multi-level thresholding) (shown by circle) for that generation. This phenomenon is depicted in Figure 5.2.

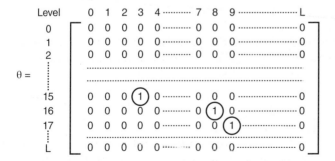

Figure 5.2 The solution matrix for multi-level image thresholding.

5.2.4 Selection

This phase leads to an elitist selection mechanism for each chromosome in *POP* that possesses the best fitness value. At each generation, a tournament selection strategy is invoked to fill the pool of chromosomes in population, *POP*, and the best chromosome is recorded (elitism). The best chromosome is merged at the bottom of the pool in *POP* before the initiation of the next generation.

5.2.5 Quantum Crossover

In this phase, a population diversity may occur due to quantum crossover. Based on a predefined crossover probability, P_{cr}, two random chromosomes are selected for crossing over at an random position. After the occurrence of quantum crossover in each generation, a new pool of chromosomes is created in *POP*. A distinct example of quantum crossover has already been presented in Figure 4.4.

5.2.6 Quantum Mutation

Like quantum crossover, another population diversity may crop up in quantum mutation. In this phase, all positions in each chromosome in *POP* may undergo mutation with another real random number based on a predefined mutation probability, P_m. A typical example of quantum mutation has already been shown in Figure 4.5.

5.2.7 Complexity Analysis

The worst case time complexity of the proposed QIGA is analyzed by describing the following steps:

1. The time complexity to generate the initial chromosomes in population, *POP* in QIGA becomes $O(\mathcal{V} \times \mathcal{L})$, where $\mathcal{V}$ represents the population size in QIGA. Note that, the length of each chromosome is $\mathcal{L} = \sqrt{L}$ where L is the utmost pixel intensity value in a gray-scale image.
2. To assign real value to each pixel encoded in the population of chromosomes in *POP*, the time complexity of QIGA turns into $O(\mathcal{V} \times \mathcal{L})$.
3. The time complexity for updating *POP'* in QIGA is $O(\mathcal{V} \times \mathcal{L})$.
4. The time complexity for *quantum orthogonality* in QIGA is $O(\mathcal{V} \times \mathcal{L})$.
5. The time complexity to create *POP** is $O(\mathcal{V} \times \mathcal{L})$.
6. QIGA employs Otsu's method as an evaluation function to compute the fitness of each chromosome in *POP*. The time complexity to evaluate the fitness values of each chromosome in *POP* is $O(\mathcal{V} \times K)$.
7. The time complexity for performing selection using the tournament selection strategy in *POP* is $O(\mathcal{V})$.
8. Again, the time complexities for mutation and crossover are $O(\mathcal{V} \times \mathcal{L})$ each.
9. So, to execute for a predefined number of generations, the time complexity of QIGA becomes $O(\mathcal{V} \times \mathcal{L} \times MxGen)$. Here, *MxGen* stands for number of generations.

Therefore, the overall worst case time complexity (by summing up all above steps) for the proposed QIGA for multi-level thresholding becomes $O(\mathcal{V} \times \mathcal{L} \times MxGen)$.

5.3 Quantum Inspired Particle Swarm Optimization

The influence of the quantum mechanism over PSO facilitated a new quantum version of the meta-heuristic algorithm, namely, Quantum Inspired Particle Swarm Optimization for multi-level thresholding (QIPSO). The outline of the proposed QIPSO is expressed in Algorithm 3. The step-wise discussion of QIPSO is illustrated in this subsection.

The first step of QIPSO is to produce a population, POP having $\mathcal{V}$ number of initial particles by picking up pixel intensities at random. The length of each particle in POP is $\mathcal{L} = \sqrt{L}$, where L represents the maximum pixel intensity value of a selected image. Afterwards, using the theory of qubits, a random real number between $(0,1)$ is selected and assigned to each pixel encoded in POP to create POP'. Then POP' endures a quantum orthogonality to produce POP''. A predefined number of threshold values as pixel intensities are derived based on some probability criteria. The particles of QIPSO are considered the participating points in a D dimensional space. The jth particle in the swarm is represented as $S = (s_{j1}, s_{j2}, \cdots, s_{jD})$. The best prior position of each particle is documented, which can be symbolized as $P_k = (p_{k1}, p_{k2}, \cdots, p_{kD})^T$. b represents the index of the best particle in the swarm. Here, at tth generation, $v_k^t(D)$ is regarded as the

---

Algorithm 3: Steps of QIPSO for multi-level thresholding

Input: Number of generation: $MxGen$
Size of the population: $\mathcal{V}$
Acceleration coefficients: c_1 and c_2
Inertia weight: ω
No. of thresholds: K
Output: Optimal threshold values: θ

1: Select pixel values randomly to generate $\mathcal{V}$ number of initial particles, POP, where length of each particleis $\mathcal{L} = \sqrt{L}$, where L represents the maximum intensity values of the selected image.
2: Use the notion of qubits to allocate real value between $(0,1)$ to each pixel encoded in POP. Let it create POP'.
3: Update POP' by using quantum rotation gate as given in Eq. (1.10).
4: Each particle in POP' experiences a *quantum orthogonality* to generate POP''.
5: Find K number of thresholds as pixel values from each particle in POP satisfying corresponding value in $POP'' > rand(0,1)$. Let it produce POP^*.
6: Work out fitness of each particle in POP^* using Eq. (2.42).
7: Record the best particle $b \in POP''$ and its threshold values in $T_B \in POP^*$.
8: **For** a predefined number of generations, $MxGen$ **do**
9:    **for** all $k \in POP''$ **do**
10:     The best prior position of each particle and the index of the best particle in POP'' are recorded.
11:     $v_k^{t+1}(D) = \omega * v_k^t(D) + c_1 * rand(0,1) * (p_k^t(D) - s_k^t(D)) + c_2 * rand(0,1) * (p_u^t(D) - s_k^t(D))$.
12:     $s_k^{t+1}(D) = s_k^t(D) + v_k^{t+1}(D)$.
13:     Repeat steps 3 and 5 to update POP' and POP^* respectively.
14:     Evaluate the fitness of particles in POP^* using Eq. (2.42).
15:     The best particle in $c \in POP'$ and its threshold values in $T_B \in POP^*$ are recorded.
16:     Compare the fitness of b and c. Update the best particle in b and its corresponding threshold values in $T_B \in POP^*$.
17:    **end for**
18: **end for**
19: Report the threshold values $\theta = T_B$.

---

current velocity whereas, $s_k^t(D)$ represents the position at the search space of the ith particle of dimension D. $rand$ represents a random real number where $0 \le rand(0, 1) \le 1$. c_1, c_2 are called positive acceleration constants and ω refers to inertia weight.

5.3.1 Complexity Analysis

The following steps are to be followed while computing the worst case time complexity of the proposed QIPSO. The time complexity analysis for the first six steps of QIPSO have already been discussed in Section 5.2.7. The remaining parts of time complexity for the proposed algorithm are described below:

1. The time complexity for manipulation of swarm at each generation is $O(\mathcal{V} \times \mathcal{L})$.
2. So, time complexity to execute QIPSO for a predefined number of generations is $O(\mathcal{V} \times \mathcal{L} \times MxGen)$ where $MxGen$ represents the number of generations.

Therefore, summing up all the steps discussed above, the overall worst case time complexity for the proposed QIPSO for multi-level thresholding becomes $O(\mathcal{V} \times \mathcal{L} \times MxGen)$.

5.4 Quantum Inspired Differential Evolution

In this subsection, a new quantum version of conventional differential evolution for multi-level thresholding, namely, Quantum Inspired Differential Evolution (QIDE) is proposed. The details of QIDE are delineated in Algorithm 4. The step-wise discussion of QIDE is presented in this subsection.

In QIDE, pixel intensities are randomly picked to produce a population POP having $\mathcal{V}$ number of initial vectors. Each vector in POP has a length of $\mathcal{L} = \sqrt{L}$, where L denotes the maximum pixel intensity value of the selected gray-scale image. A real number between $(0,1)$ is randomly chosen and assigned to each pixel encoded in POP to produce POP' using the notion of qubits. POP' goes for a quantum orthogonality to create POP''. QIDE provides a user-defined number of thresholds as pixel intensities based on a probability condition. Three basic operators, namely, mutation, crossover, and selection are consecutively applied in each generation. The applications of mutation and crossover are different in QIDE from those exercised in QIGA. For quantum mutation, three vectors, namely, r_1, r_2 and r_3 are randomly selected from POP'' satisfying r_1, r_2 and $r_3 \in [1, \mathcal{V}], j \ne r_2 \ne r_3 \ne r_1$ where $\mathcal{V}$ and j represent the number of vectors and a vector individual in POP''. The weighted difference between the two population vectors is scaled by F and then added to the third vector to get the new mutant vector solution POP'' where, φ is called the scaling factor. Afterwards, POP'' undergoes the crossover operation to obtain another new vector solution POP''. A random integer, ct is generated from $[1, \mathcal{V}]$ and the mutant vector in POP'' goes for crossover based on the condition given by $j \ne ct$ and $rand(0, 1) > P_c$ where P_c represents a predefined crossover probability and j represents a particular position in a selected mutant vector.

The tournament selection mechanism is applied to get the population of vectors for the next generation. Based on the fitness value, a particular vector solution j is replaced with the new vector solution having better fitness value in POP''. The entire process is repeated for a predefined number of generations, $MxGen$. In QIDE, the values of φ and P_c are in the range $0 < \varphi \le 1.2$ and $[0, 1]$ respectively.

Algorithm 4: Steps of QIDE for multi-level thresholding

Input: Number of generation: $MxGen$
Size of the population: $\mathcal{V}$
Scaling factor: φ
Crossover probability: P_c
No. of thresholds: K
Output: Optimal threshold values: θ

1: Choose the pixel values randomly to produce $\mathcal{V}$ number of initial vectors, POP, where the length of each vector is $\mathcal{L} = \sqrt{L}$, where L is the maximum pixel intensity value of an image.
2: Using the conception of qubits to allocate real value between $(0,1)$ to each pixel encoded in POP. Let it make POP'.
3: Update POP' by using quantum rotation gate as depicted in Eq. (1.10).
4: POP' goes for a *quantum orthogonality* to produce POP''.
5: Locate K number of thresholds as pixel values from each vector in POP satisfying the corresponding value in $POP'' > rand(0,1)$. Let it make POP^*.
6: Calculate the fitness of each vector in POP^* using Eq. (2.42).
7: Save the best vector b from POP^* and its threshold values in $T_B \in POP^*$.
8: $BKPOP = POP''$.
9: **For all** $k \in POP''$ **do**
10:   **for** all j_{th} position in k **do**
11:     Select three random integers r_1, r_2 and r_3 from $[1..\mathcal{V}]$ satisfying $r_1 \neq r_2 \neq r_3 \neq j$.
12:     $POP''(j) = BKPOP(r_1) + \varphi(BKPOP(r_2) - BKPOP(r_3))$
13:   **end for**
14: **end for**
15: **for all** $k \in POP''$ **do**
16:   **for** all j_{th} position in k **do**
17:     Generate a random integer ct from $[1..\mathcal{V}]$.
18:     **if** $(j \neq ct$ and $rand(0,1) > P_c)$ **then**
19:       $POP''_{kj} = BKPOP_{kj}$.
20:     **end if**
21:   **end for**
22: **end for**
23: Follow the steps 5 and 6 to evaluate the fitness of each vector in POP'' using Eq. (2.42).
24: The best vector in $c \in POP''$ and its threshold values in $T_B \in POP^*$ are recorded.
25: Compare the fitness of b and c.
26: Update the best vector in b and its threshold values in $T_B \in POP^*$.
27: **for all** $k \in POP''$ **do**
28:   **if** (Fitness of $BKPOP_k$ is better than the fitness of POP''_k) **then**
29:     $POP''(k) = BKPOP_k$.
30:   **end if**
31: **end for**
32: POP'' undergoes a *quantum orthogonality*.
33: Repeat steps 5 to 31 for a fixed number of generations, $MxGen$.
34: The optimal threshold values, $\theta = T_B$ are reported.

5.4.1 Complexity Analysis

The outline of worst case time complexity analysis of the proposed QIDE is described below. The first six steps of the time complexity analysis for QIDE have already been illustrated in Section 5.2.7. The worst case time complexity analysis for the remaining parts of QIDE is discussed below.

1. Time complexities for mutation, crossover and the selection parts entail $O(\mathcal{V} \times \mathcal{L})$.
2. So, to execute QIDE for a predefined number of generations, the time complexity becomes $O(\mathcal{V} \times \mathcal{L} \times MxGen)$. Here, $MxGen$ signifies the number of generations.

Therefore, aggregating the overall discussion as stated above, it can be concluded that the worst case time complexity for the proposed QIDE for multi-level thresholding is $O(\mathcal{V} \times \mathcal{L} \times MxGen)$.

5.5 Quantum Inspired Ant Colony Optimization

The influence of quantum principles over ACO helped to design a new quantum behaved meta-heuristic algorithm, namely, Quantum Inspired Ant Colony Optimization for multi-level thresholding (QIACO). The details of the proposed QIACO are illustrated in Algorithm 5. QIACO is discussed step by step in this subsection.

In the proposed QIACO, pixel intensities are randomly selected to create a population POP having $\mathcal{V}$ number initial strings. The length of each string in POP is $\mathcal{L} = \sqrt{L}$, where L is the utmost pixel intensity value in a gray-scale image. Using the concept of qubits, a real random number between $(0,1)$ is generated and allocated to each pixel encoded in POP to create POP'. Then POP' undergoes a quantum orthogonality to produce POP''. Based on a probability condition, the algorithm resorts to a user-defined number of thresholds as pixel intensity values. At every generation, the algorithm explores the best search path. At the outset, a pheromone matrix, τ_j is generated for each ant j. For each individual $j \in POP''$, the maximum pheromone integration is deduced as the threshold value in the gray-scale image, if $POP''(j) > q_0$. Here, q_0 is the priori defined number where $0 \leq q_0 \leq 1$. This leads to $POP''(kj) = \arg\max\tau_{kj}$. If $POP''(j) \leq q_0$, $POP''(kj) = rand(0,1)$. The pheromone trial matrix is updated at the end of each generation using $\tau_{kj} = \rho\tau_{kj} + (1-\rho)b$ where b represents the best string of each generation and k represents the corresponding position in a particular string j. In QIACO, $MxGen$ represents the number of generations to be executed, K and $\mathcal{V}$ are the user-defined number of thresholds and population size respectively. ρ is known as persistence of trials, $\rho \in [0,1]$.

5.5.1 Complexity Analysis

The worst case time complexity of the proposed QIACO is analyzed in this section. The time complexities of the first six steps of QIGA and QIACO are identical, already discussed in Section 5.2.7. The time complexity analysis for the remaining parts of proposed QIACO is given below:

1. The time complexity to construct the pheromone matrix, τ is $O(\mathcal{V} \times \mathcal{L})$.
2. The time complexity to determine POP** from τ at each generation is $O(\mathcal{V} \times \mathcal{L})$.
3. The pheromone matrix needs to be updated at each generation. The time complexity for this computation is $O(\mathcal{L})$.
4. Again, time complexity to execute for a predefined number of generations for QIACO is $O(\mathcal{V} \times \mathcal{L} \times MxGen)$, where $MxGen$ is the number of generations.

Therefore, summing up all the steps discussed above, the overall worst case time complexity for the proposed QIACO for multi-level thresholding becomes $O(\mathcal{V} \times \mathcal{L} \times MxGen)$.

Algorithm 5: Steps of QIACO for multi-level thresholding

Input: Number of generation: $MxGen$
Size of the population: $\mathcal{V}$
No. of thresholds: K
Persistence of trials: ρ
Priori defined number: q_0
Output: Optimal threshold values: θ

1: The pixel values are randomly selected to generate $\mathcal{V}$ number of initial strings, POP, where the length of each string is $\mathcal{L} = \sqrt{L}$, where L signifies the greatest pixel value of the gray-scale image.
2: The idea of qubits is employed to assign real value between (0,1) to each pixel encoded in POP. Let it produce POP'.
3: Quantum rotation gate is employed to update POP' using Eq. (1.10).
4: POP' endures a *quantum orthogonality* to create POP''.
5: Find K number of thresholds as pixel values from each string in POP satisfying corresponding value in $POP'' > rand(0, 1)$. Let it create POP^*.
6: Evaluate fitness of each string in POP^* using Eq. (2.42).
7: Save the best string b from POP^*.
8: Repeat step 5 to produce POP^{**}.
9: Construct the pheromone matrix, τ.
10: **For** a fixed number of generations (MxGen) **do**
11:     **for** all $j \in POP''$ **do**
12:         **for** all k_{th} location in j **do**
13:             **if** $(rand(0, 1) > q_0)$ **then**
14:                 $POP''_{jk} = \arg\max \tau_{jk}$.
15:             **else**
16:                 $POP''_{jk} = rand(0, 1)$.
17:             **end if**
18:         **end for**
19:     **end for**
20:     Use Eq. (2.42) to calculate the fitness of POP''.
21:     Save the best string c from POP^{**}.
22:     The fitness of b and c is compared.
23:     Use POP^{**} to update the string with best string along with its corresponding threshold values in T_{BS}.
24:     Save the best string of step 22 in b and its corresponding thresholds $T_{BS} \in POP^{**}$.
25:     **for** all $j \in POP''$ **do**
26:         **for** all k_{th} location in j **do**
27:             $\tau_{jk} = \rho\tau_{jk} + (1 - \rho)b$.
28:         **end for**
29:     **end for**
30: **end for**
31: Report the threshold values $\theta = T_{BS}$.

5.6 Quantum Inspired Simulated Annealing

A simulated annealing inspired by the inherent principles of quantum mechanical systems, called Quantum Inspired Simulated Annealing for multi-level thresholding (QISA), is presented in this subsection. The working procedure of QISA is depicted in Algorithm 6. We present the step-by-step illustration of the proposed QISA below.

Initially, an initial configuration, P is created by choosing the pixel intensities randomly of length $\mathcal{L} = \sqrt{L}$, where L represents the highest pixel intensity value in

---

Algorithm 6: Steps of QISA for multi-level thresholding

Input: Initial temperature: $\mathcal{T}_{max}$
Final temperature: $\mathcal{T}_{min}$
Reduction factor: υ
No. of thresholds: K
Number of iterations: $\mathcal{I}$
Output: Optimal threshold values: θ

1: Randomly select pixel intensities to create an initial configuration, P, where length of the configuration is denoted by $\mathcal{L} = \sqrt{L}$, where L is taken as the maximum intensity value of a gray-scale image.
2: Apply the idea of qubits to assign real value between $(0,1)$ to each pixel encoded in P. Let it generate P'.
3: Quantum rotation gate is used to update P' using Eq. (1.10).
4: The configuration in P' passes through a *quantum orthogonality* to generate P''.
5: Discover K number of thresholds as pixel intensities from the configuration in P. It should hold corresponding value in $P'' > rand(0,1)$. Let it generate P^*.
6: Evaluate fitness of the configuration in P^* using Eq. (2.42). Let it be symbolized by $\mathcal{F}(P^*)$.
7: $\mathcal{T} = \mathcal{T}_{max}$.
8: **repeat**
9:   **for** $j = 1$ to $\mathcal{I}$ **do**
10:     Perturb P. Let it create V.
11:     Repeat steps 2–4 to create V^*.
12:     Use Eq. (2.42) to evaluate fitness $E(V^*, T)$ of the configuration V^*.
13:     **if** $(\mathcal{F}(S^*) - \mathcal{F}(P^*) > 0)$ **then**
14:       Set $P = S$, $P^* = S^*$ and $\mathcal{F}(P^*) = \mathcal{F}(S^*)$.
15:     **else**
16:       Set $P = S$, $P^* = S^*$ and $\mathcal{F}(P^*) = \mathcal{F}(S^*)$ with probability $\exp(-(\mathcal{F}(P^*) - \mathcal{F}(S^*)))/\mathcal{T}$.
17:     **end if**
18:   **end for**
19:   $\mathcal{T} = \mathcal{T} \times \upsilon$.
20: **until** $\mathcal{T} >= \mathcal{T}_{min}$
21: Report the optimal threshold values, $\theta = P^*$.

---

a gray-scale image. The theory of qubits is applied to a newly introduced encoded scheme to allocate real random value between $(0,1)$ to each pixel encoded in P and it is named P'. Afterwards, P' goes for quantum orthogonality to create P''. QISA finds a user-defined number of thresholds as pixel intensities based on a defined probability condition. First, a very high temperature, $\mathcal{T}_{max}$ is assigned to a newly invoked temperature parameter, $\mathcal{T}$. QISA is allowed to execute for $\mathcal{I}$ number of iterations for every temperature and after the completion of one generation, $\mathcal{T}$ is reduced by a reduction factor, υ. The execution continues until $\mathcal{T}$ reaches the predefined final temperature, $\mathcal{T}_{min}$. At each iteration of $\mathcal{I}$, a better configuration, (S) is expected because the old configuration is perturbed randomly at multiple points. Subsequently, a new configuration, S^* is generated from S by enduring the similar process which was acclimatized before to create P^* from P. The acceptance of the configurations S and S^* depends on the condition given by $\mathcal{F}(S^*) > \mathcal{F}(P^*)$, by revising the previous configurations P and P^*, respectively; otherwise, the newly generated configuration may be admitted with a probability $\exp(-(\mathcal{F}(P^*) - \mathcal{F}(S^*)))/\mathcal{T}$. In general, the probability is determined by the Boltzmann distribution. υ is chosen within the range of $[0.5, 0.99]$ whereas K signifies the user-defined number of thresholds. θ represents the output as thresholds.

5.6.1 Complexity Analysis

The following steps describe the worst case time complexity for the proposed QISA for multi-level thresholding.

1. For initial configuration in QISA, the time complexity is $O(\mathcal{L})$, where the length of configuration is $\mathcal{L} = \sqrt{L}$ where L signifies the maximum intensity value of a gray-scale image.
2. For the assignment of real value to each pixel encoded in the population of configuration, the time complexity is $O(\mathcal{L})$.
3. The time complexity for updating P' turns into $O(\mathcal{L})$.
4. The time complexity to perform quantum orthogonality is $O(\mathcal{L})$.
5. The time complexity to create P^* is $O(\mathcal{L})$.
6. The time complexity for fitness computation using Eq. (2.42) turns into $O(\mathcal{K})$.
7. In a similar way, for the fitness computation of the configuration after perturbation, the time complexity is $O(\mathcal{K})$.
8. Let the outer loop and inner loop be executed $MxIn$ and $\mathcal{I}$ times respectively in QISA. Therefore, the time complexity to execute this step in QISA happens to be $O(\mathcal{I} \times MxIn)$.

Therefore, aggregating the steps discussed above, the proposed QISA for multi-level thresholding possesses the worst case time complexity $O(\mathcal{L} \times \mathcal{I} \times MxIn)$.

5.7 Quantum Inspired Tabu Search

A Tabu search inspired by the inherent features of quantum mechanical systems, referred to Quantum Inspired Tabu Search for multi-level thresholding (QITS), is developed in this subsection. The working steps of QITS are portrayed in Algorithm 7. The step-wise illustration of the proposed QITS is given below.

In QITS, at the beginning, pixel intensities are randomly selected to produce a string, P. The length of P is taken as $\mathcal{L} = \sqrt{L}$, where L represents the maximum gray-scale intensity value of an image. A new encoded scheme has been introduced to assign a real random value between $(0,1)$ to each pixel encoded in P to produce P'. For this purpose the concept of qubits was applied along with the encoded technique. P' again passes through a quantum orthogonality to produce P''. A user-defined number of thresholds as pixel intensities is computed based on a defined probability condition. The tabu memory, *mem* is introduced and assigned to null prior to its execution. QITS starts with a single vector, v_{best} and stops executing when it reaches the predefined number of iterations $\mathcal{I}$. For each iteration of $\mathcal{I}$, a new set of vectors, $V(BS)$ is created in the neighborhood areas of v_{best}. For each vector in $V(BS)$, if it is not in *mem* and possesses more fitness value than v_{best}, v_{best} is updated with the new vector. The vector is pushed in *mem*. When the tabu memory is full, it follows FIFO to eliminate a vector from the list.

5.7.1 Complexity Analysis

The worst case time complexity of the proposed QITS is analyzed in thi subsection. The worst case time complexity for the first six steps have already been discussed in

---

Algorithm 7: Steps of QITS for multi-level thresholding

Input: Number of generation: *MxGen*
Number of thresholds: *K*
Output: Optimal threshold values: θ

1: Select the pixel values randomly to create an initial string, P, where length of the string is $\mathcal{L} = \sqrt{L_1 \times L_2}$, where L is the greatest pixel intensity value of a gray-scale image.
2: Apply the concept of qubits to assign a real value between (0,1) to each pixel encoded in P. Let it create P'.
3: Quantum rotation gate is utilized to update P' using Eq. (1.10).
4: The string in P' goes for a *quantum orthogonality* to create P''.
5: Find K number of thresholds as pixel values from the string in P satisfying the corresponding value in $P'' > rand(0,1)$. Let it create P^*.
6: Compute fitness of the string in P^* using Eq. (2.42).
7: Record the best string b from P^*.
8: Initialize the tabu memory, $mem = \phi$.
9: **For** $j = 1$ to *MxGen* **do**
10:   $v_{best} = b$.
11:   Create a set of neighborhood, $V(BS)$ of vector v_{best}
12:   **for** each $v \in V(BS)$ **do**
13:     **if** $v \notin mem$ and $(Fitness(v) > Fitness(v_{best}))$ **then**
14:       $v_{best} = v$.
15:     **end if**
16:   **end for**
17:   Set $c = v_{best}$.
18:   $mem = mem \cup v_{best}$.
19:   Compare the fitness of b and c.
20:   Store the best individual in $b \in POP''$ and its corresponding threshold values in $T_B \in POP^*$.
21: **end for**
22: Report the optimal threshold values, $\theta = T_B$.

Section 5.6.1. The time complexity analysis for the remaining parts of QITS is discussed below.

1. For each generation, the time complexity to create a set of neighborhood of the best vector turns into $O(\mathcal{W})$, where $\mathcal{W}$ is the number of neighborhoods.
2. To assign the best vector at each generation, the time complexity happens to be $O(\mathcal{W} \times \mathcal{L})$, where the length of string is $\mathcal{L} = \sqrt{L}$ where L signifies the maximum pixel intensity value of a gray-scale image.
3. Hence, the time complexity to execute for a predefined number of generations for QITS happens to be $O(\mathcal{W} \times \mathcal{L} \times MxGen)$, where *MxGen* is the number of generations.

Therefore, summing up all the above steps, it can be concluded that the overall worst case time complexity for the proposed QITS for multi-level thresholding is $O(\mathcal{W} \times \mathcal{L} \times MxGen)$.

5.8 Implementation Results

Application of each technique is exhibited through the thresholding of multi-level real/synthetic gray-scale images. The optimal threshold values are reported for ten gray-scale images and five synthetic images, each of size 256×256 at different levels. The synthetic images with a complex background and low contrast are depicted in Figures 5.3 (a) and (b), respectively. The original test images, namely, Lena, Woman and Barbara, are shown in Figures 4.7 (a), (c) and (f) and the images, namely, B2, Boat, Cameraman, Jetplane, Pinecone, Car, and House are presented in Figures 5.5(a)–(g). Four new synthetic images have been designed by adding different noise and contrast levels on the synthetic image of Figure 5.3, which are portrayed in Figures 5.4(a)–(d). These noisy images are named ImageN30C30 (noise-30, contrast-30), ImageN30C80 (noise-30, contrast-80), ImageN80C30 (noise-80, contrast-30), and ImageN80C80 (noise-80, contrast-80). The first synthetic image is shown in Figure 5.3(a) and these four modified versions of synthetic images have been used as the test images.

In Sections 5.2–5.7, the proposed quantum inspired meta-heuristic techniques are described in detail. The selection of the best combination of parameters in each algorithm can accelerate its performance. The parameters set for the proposed algorithms are listed in Table 5.2. The worst case time complexity for each algorithm is analyzed in detail in the respective section. The list of worst case time complexities of each algorithm is reported in Table 5.1. It should be noted that the worst case time complexities for QIGA, QIPSO, QIDE, and QIACO are the same. These four algorithms possess lower time complexities than QITS if $\mathcal{V} < \mathcal{W}$. In the case of QISA, the worst case time complexity depends on the parameters selected for execution. Each algorithm has been

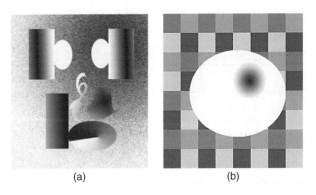

(a) (b)

Figure 5.3 Original synthetic images with (a) complex backgrounds and (b) low contrast.

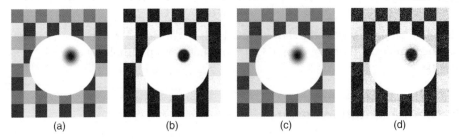

(a) (b) (c) (d)

Figure 5.4 Synthetic images: (a) Noise-30,Contrast-30, (b) Noise-30,Contrast-80, (c) Noise-80,Contrast-30, and (d) Noise-80,Contrast-80.

Figure 5.5 Original test images (a) B2, (b) Boat, (c) Cameraman, (d) Jetplane, (e) Pinecone, (f) Car, and (g) House.

Table 5.1 List of worst case time complexity for the proposed algorithms.

Proposed algorithm	Time complexity
QIGA	$O(\mathcal{V} \times \mathcal{L} \times MxGen)$
QIPSO	$O(\mathcal{V} \times \mathcal{L} \times MxGen)$
QIDE	$O(\mathcal{V} \times \mathcal{L} \times MxGen)$
QIACO	$O(\mathcal{V} \times \mathcal{L} \times MxGen)$
QISA	$O(\mathcal{L} \times \mathcal{I} \times MxIn)$
QITS	$O(\mathcal{W} \times \mathcal{L} \times MxGen)$

executed 20 times for the first four levels of thresholds. These algorithms are executed for one to four optimal threshold values depending on the level chosen for optimization. The results of the proposed algorithms are reported in Tables 5.3–5.6. These tables encompass the number of threshold (K), optimum threshold (θ) at each level, the corresponding fitness value ($\mathcal{U}_{best}$), and also the execution time (t) (in seconds). Later, the mean fitness ($\mathcal{U}_{avg}$) and standard deviation (σ) of the mean fitness over 20 individual runs are reported in Tables 5.7–5.9.

Each algorithm sounds a good result for $K = 2$ and 3. Each algorithm reports the same mean fitness ($\mathcal{U}_{avg}$) value and the same standard deviation (σ) in all test images for $K = 2$. It has been noted that the average fitness ($\mathcal{U}_{avg}$) values for all test images remain equal for each algorithm when $K = 3$. Unlike the mean fitness values, the standard deviations (σ) vary for QIACO, QISA, and QITS by a small amount for this level of computation. Each method reports unequal values of optimal thresholds, average fitness values, and the standard deviations for all test images for $K = 4$ and 5.

For $K = 4$ and 5, the Friedman test [94, 95] was conducted for the proposed algorithms using 15 different data sets for both real and synthetic test images. The result of

Table 5.2 Parameter set for the proposed QIGA, QIPSO, QIDE, QIACO, QISA, and QITS for multi-level thresholding.

QIGA	QIPSO
Population size: $\mathcal{V} = 50$	Population size: $\mathcal{V} = 50$
Number of Generation: $MxGen = 1000$	Number of Generation: $MxGen = 1000$
Crossover probability: $P_{cr} = 0.9$	Inertia weight: $\omega = 0.4$
Mutation probability: $P_m = 0.1$	Acceleration coefficients: $c_1, c_2 = 0.5$
No. of thresholds: $K = 2, 3, 4, 5$	No. of thresholds: $K = 2, 3, 4, 5$
QIDE	QIACO
Population size: $\mathcal{V} = 50$	Population size: $\mathcal{V} = 50$
Number of Generation: $MxGen = 1000$	Number of Generation: $MxGen = 1000$
Scaling factor: $F = 0.5$	Priori defined number: $q_0 = 0.5$
Crossover constant: $P_c = 0.9$	Persistence of trials: $\rho = 0.2$
No. of thresholds: $K = 2, 3, 4, 5$	No. of thresholds: $K = 2, 3, 4, 5$
QISA	QITS
Initial temperature: $\mathcal{T}_{max} = 100$	Number of Generation: $MxGen = 1000$
Final temperature: $\mathcal{T}_{min} = 0.5$	No. of thresholds: $K = 2, 3, 4, 5$
Number of Iterations: $\mathcal{I} = 50$	
Reduction factor: $\upsilon = 0.9$	
No. of thresholds: $K = 2, 3, 4, 5$	

this test is reported in Table 5.10. In general, the Friedman test is introduced to compare the performances of multiple methods using multiple data sets. This test exposes the average rank of each individual method as the output. In this statistical test, the null hypothesis, H_0 affirms the equal behavior of the participating methods. Hence, under H_0, each algorithm possesses equal rank, which confirms that each method is equally efficient as the rest. The alternative hypothesis, H_1 endorses the difference in performances among the participating methods. Since each method reports equal fitness values for all test images for $K = 2$ and 3, the Friedman test [94, 95] finds equal average rank for all of them. Hence, the null hypothesis is accepted, which proves that the proposed methods behave equally for $K = 2$ and $K = 3$. Table 5.10 shows that average ranks for QIGA, QIPSO, QIDE, QIACO, QISA, and QITS are found to be 2.46, 1.73, 2.86, 3.23, 5.16, and 5.53, respectively for $K = 4$, whereas for $K = 5$, the corresponding average ranks are 2.73, 1.30, 2.86, 3.50, 5.20 and 5.40, respectively. Note that each method acquires an unequal rank for $K = 4$ and 5. Moreover, the Friedman test finds the chi-square ($\mathcal{X}^2$) value of 48.854 for $K = 4$ and 50.344 for $K = 5$. This statistical test also determines the p-values for $K = 4$ and 5 as 2.1×10^{-4} and 3.5×10^{-4}, respectively. From the chi-square ($\mathcal{X}^2$) distribution table, we find that the critical value for $(6 - 1) = 5$ degrees of freedom with 0.05 significance level is 11.070. Since both these chi-square values for $K = 4$ and 5 are greater than the critical value, H_0 is rejected and H_1 is accepted. Furthermore, the p-values for $K = 4$ and 5 are found to be very small (very close to zero), which confirms the rejection of H_0 and find some significant difference in behavior among the proposed methods. Furthermore, for $K = 4$ and 5, QIPSO possesses the lowest rank and QISA has the highest rank among the six different methods. Hence, QIPSO can be declared the best performing method, whereas QISA is the worst performing method among others.

Table 5.3 Best results of QIGA, QIPSO, and QIDE for multi-level thresholding for Image1, ImageN30C30, ImageN30C80, ImageN80C30 and ImageN80C80.

Image1									
K	QIGA			QIPSO			QIDE		
	θ	$\mathcal{U}_{best}$	t	θ	$\mathcal{U}_{best}$	t	θ	$\mathcal{U}_{best}$	t
2	149	4970.47	4.37	149	4970.47	0.02	149	4970.47	4.06
3	95,183	5624.56	10.36	95,183	5624.56	2.13	95,183	5624.56	10.12
4	64,127,199	5850.34	16.22	64,128,197	5850.62	2.60	65,130,199	5850.53	15.22
5	52,105,163,212	5962.32	21.28	51,105,162,213	5962.95	4.44	55,106,160,212	5962.94	23.21
ImageN30C30									
K	QIGA			QIPSO			QIDE		
	θ	$\mathcal{U}_{best}$	t	θ	$\mathcal{U}_{best}$	t	θ	$\mathcal{U}_{best}$	t
2	138	5710.32	4.41	138	5710.32	0.03	138	5710.32	4.13
3	107,199	6677.16	10.23	107,199	6677.16	2.31	107,199	6677.16	10.58
4	64,123,199	7027.37	18.15	57,123,199	7027.44	3.12	59,123,201	7027.30	17.54
5	60,122,191,241	7131.23	23.24	60,123,190,237	7131.38	4.39	62,120,191,241	7131.13	23.18
ImageN30C80									
K	QIGA			QIPSO			QIDE		
	θ	$\mathcal{U}_{best}$	t	θ	$\mathcal{U}_{best}$	t	θ	$\mathcal{U}_{best}$	t
2	133	7340.33	4.45	133	7340.33	0.02	133	7340.33	4.13
3	124,240	7466.66	11.10	124,240	7466.66	2.10	124,240	7466.66	10.22
4	105,215,245	7510.31	22.12	108,216,243	7510.42	3.12	107,213,244	7510.18	21.40
5	33,129,220,245	7541.28	24.36	27,123,219,244	7541.84	4.09	28,123,221,245	7541.46	22.38
ImageN80C30									
K	QIGA			QIPSO			QIDE		
	θ	$\mathcal{U}_{best}$	t	θ	$\mathcal{U}_{best}$	t	θ	$\mathcal{U}_{best}$	t
2	138	5528.15	4.16	138	5528.15	0.02	138	5528.15	4.10
3	110,201	6488.54	10.18	110,201	6488.54	2.12	110,201	6488.54	10.13
4	67,127,201	6822.28	20.40	62,127,202	6822.51	3.24	58,127,202	6822.44	20.47
5	61,127,189,239	6936.35	24.18	61,125,187,237	6936.49	4.20	64,119,186,238	6936.22	21.17
ImageN80C80									
K	QIGA			QIPSO			QIDE		
	θ	$\mathcal{U}_{best}$	t	θ	$\mathcal{U}_{best}$	t	θ	$\mathcal{U}_{best}$	t
2	142	5678.56	4.22	142	5678.56	0.02	142	5678.56	4.07
3	114,218	6075.08	10.18	114,218	6075.08	2.12	114,218	6075.08	10.13
4	79,158,226	6186.01	20.28	75,159,227	6186.48	3.15	76,162,228	6186.31	23.42
5	58,134,181,229	6244.34	24.16	58,124,182,228	6244.68	4.25	59,120,185,227	6244.03	21.31

We performed our experiments on Toshiba Intel® Core™ i3, 2.53GHz PC with 2GB RAM. The computational times at different levels for each of the proposed algorithm are reported in Tables 5.3–5.6. It can be noted from these tables that the computational time for QIPSO is the shortest of them all. As QIPSO outperforms the other algorithms and the threshold values of all the proposed algorithms are identical or very close to each other, only the set of images for QIPSO after thresholding are reported as visual representation of the test results. The thresholded images of QIPSO are presented for the synthetic test images in Figure 5.6 and for the real test images in Figures 5.7 and 5.8, respectively for different levels. Since the experimental results of the proposed algorithms show no or very little significant differences for $K = 2, 3$, only the convergence curves for $K = 4, 5$ are reported for each test image. The convergence curves

Table 5.4 Best results of QIACO, QISA, and QITS for multi-level thresholding for Image1, ImageN30C30, ImageN30C80, ImageN80C30, and ImageN80C80.

	Image1								
K	QIACO			QISA			QITS		
	θ	$\mathcal{V}_{best}$	t	θ	$\mathcal{V}_{best}$	t	θ	$\mathcal{V}_{best}$	t
2	149	4970.47	4.29	149	4970.47	5.56	149	4970.47	6.07
3	95,183	5624.56	12.40	95,183	5624.56	17.22	95,183	5624.56	16.21
4	64,127,199	5850.34	16.22	64,128,197	5850.62	2.60	65,130,199	5850.53	15.22
5	50,101,160,214	5962.06	25.19	46,98,157,211	5959.67	54.11	48,99,150,210	5959.12	53.17
	ImageN30C30								
K	QIACO			QISA			QITS		
	θ	$\mathcal{V}_{best}$	t	θ	$\mathcal{V}_{best}$	t	θ	$\mathcal{V}_{best}$	t
2	138	5710.32	5.17	138	5710.32	6.31	138	5710.32	6.01
3	107,199	6677.16	12.12	107,199	6677.16	12.48	107,199	6677.16	10.32
4	63,124,202	7027.04	18.33	62,133,201	7026.67	44.64	72,125,200	7026.57	41.14
5	59,120,194,237	7131.08	24.18	59,129,182,239	7130.65	53.18	75,129,193,237	7130.06	52.14
	ImageN30C80								
K	QIACO			QISA			QITS		
	θ	$\mathcal{V}_{best}$	t	θ	$\mathcal{V}_{best}$	t	θ	$\mathcal{V}_{best}$	t
2	133	7340.33	5.25	133	7340.33	18.35	133	7340.33	21.36
3	124,240	7466.66	11.19	124,240	7466.66	32.12	124,240	7466.66	34.23
4	108,218,243	7510.19	19.45	115,221,245	7509.39	42.20	111,221,247	7509.12	41.27
5	30,122,217,240	7541.02	22.47	34,112,212,245	7540.33	54.34	31,122,223,245	7540.58	55.15
	ImageN80C30								
K	QIACO			QISA			QITS		
	θ	$\mathcal{V}_{best}$	t	θ	$\mathcal{V}_{best}$	t	θ	$\mathcal{V}_{best}$	t
2	138	5528.15	5.21	138	5528.15	18.34	138	5528.15	21.36
3	110,201	6488.54	11.49	110,201	6488.54	33.18	110,201	6488.54	35.53
4	55,129,202	6822.24	21.31	57,131,202	6822.28	44.28	58,120,200	6821.89	43.12
5	57,126,189,232	6936.09	26.01	71,113,189,237	6935.28	55.17	57,133,183,232	6935.15	54.24
	ImageN80C80								
K	QIACO			QISA			QITS		
	θ	$\mathcal{V}_{best}$	t	θ	$\mathcal{V}_{best}$	t	θ	$\mathcal{V}_{best}$	t
2	142	5678.56	5.08	142	5678.56	18.36	142	5678.56	21.35
3	114,218	6075.08	11.49	114,218	6075.08	33.18	114,218	6075.08	35.53
4	71,160,226	6185.93	20.01	71,156,227	6185.43	45.20	80,157,226	6185.51	43.27
5	59,124,180,226	6243.96	24.17	54,113,180,227	6243.43	56.24	48,112,182,226	6243.52	55.20

of the proposed algorithms for $K = 4, 5$ are shown in Figures 5.9–5.14. It is clearly noticeable from Figures 5.9–5.14 that QIPSO is the fastest converging algorithm of them all.

5.8.1 Consensus Results of the Quantum Algorithms

A consensus of the six inspired quantum algorithms is given in this subsection. The optimum set of thresholds has been reported for $K = 2, 3, 4$ and 5 in Table 5.11 for synthetic test images and in Tables 5.12 and 5.13 for real test images, respectively. Here, 1∗ signifies that a particular set of threshold values is reported by a single algorithm at

Table 5.5 Best results of QIGA, QIPSO, and QIDE for multi-level thresholding for Lena, B2, Barbara, Boat, Cameraman, Jetplane, Pinecone, Car, Woman, and House.

	Lena								
K	QIGA			QIPSO			QIDE		
	θ	v_{best}	t	θ	v_{best}	t	θ	v_{best}	t
2	118	1424.30	4.48	118	1424.30	0.02	118	1424.30	4.12
3	99,157	1763.29	10.47	99,157	1763.29	2.13	99,157	1763.29	10.39
4	84,127,169	1913.75	17.12	84,126,169	1913.79	3.32	83,125,168	1913.74	20.70
5	80,116,146,180	1970.44	23.10	79,115,146,180	1970.47	4.46	79,115,146,180	1970.47	21.20

	B2								
K	QIGA			QIPSO			QIDE		
	θ	v_{best}	t	θ	v_{best}	t	θ	v_{best}	t
2	128	3108.56	4.49	128	3108.56	0.03	128	3108.56	4.28
3	93,175	3612.92	10.44	93,175	3612.92	2.37	93,175	3612.92	10.12
4	73,139,186	3756.66	17.35	75,138,185	3756.72	3.00	73,139,186	3756.66	17.34
5	75,135,178,216	3838.73	22.18	75,135,179,215	3839.29	4.34	75,139,180,214	3839.23	20.55

	Barbara								
K	QIGA			QIPSO			QIDE		
	θ	v_{best}	t	θ	v_{best}	t	θ	v_{best}	t
2	115	1550.91	4.22	115	1550.91	0.02	115	1550.91	4.04
3	79,135	1912.75	10.05	79,135	1912.75	2.10	79,135	1912.75	10.10
4	73,121,165	2046.14	21.12	73,121,164	2046.10	3.01	73,121,165	2046.14	18.51
5	61,97,130,172	2111.66	22.46	61,97,131,170	2112.66	4.01	60,96,132,172	2111.90	20.28

	Boat								
K	QIGA			QIPSO			QIDE		
	θ	v_{best}	t	θ	v_{best}	t	θ	v_{best}	t
2	91	2270.87	5.15	91	2270.87	0.06	91	2270.87	4.14
3	73,135	2538.32	9.10	73,135	2538.32	2.40	73,135	2538.32	8.07
4	58,109,147	2631.52	17.40	57,109,147	2631.60	3.27	56,109,147	2631.54	16.52
5	51,98,135,160	2683.65	23.15	48,94,132,160	2683.90	4.16	47,95,134,161	2683.90	20.10

	Cameraman								
K	QIGA			QIPSO			QIDE		
	θ	v_{best}	t	θ	v_{best}	t	θ	v_{best}	t
2	88	3147.03	5.10	88	3147.03	0.08	88	3147.03	4.03
3	70,143	3515.77	8.24	70,143	3515.77	2.32	70,143	3515.77	8.05
4	45,99,147	3585.38	16.55	45,100,147	3585.41	3.46	45,100,147	3585.41	17.00
5	40,93,137,166	3642.19	23.55	39,90,137,167	3642.45	4.30	39,90,139,166	3641.76	20.40

	Jetplane								
K	QIGA			QIPSO			QIDE		
	θ	v_{best}	t	θ	v_{best}	t	θ	v_{best}	t
2	151	1764.79	4.55	151	1764.79	0.05	151	1764.79	5.10
3	111,172	1929.18	12.10	111,172	1929.18	3.21	111,172	1929.18	11.11
4	88,141,187	2007.81	17.36	88,140,187	2007.81	3.27	88,140,187	2007.81	16.40
5	82,128,171,202	2054.57	22.47	83,128,172,201	2054.58	4.30	86,132,174,201	2054.03	20.57

	Pinecone								
K	QIGA			QIPSO			QIDE		
	θ	v_{best}	t	θ	v_{best}	t	θ	v_{best}	t
2	135	752.67	5.12	135	752.67	0.04	135	752.67	5.10
3	117,160	945.81	13.38	117,160	945.81	2.01	117,160	945.81	13.40
4	108,139,177	1027.78	18.05	107,139,178	1027.76	2.34	107,139,177	1027.77	16.30
5	99,123,149,186	1068.08	23.08	102,126,152,188	1068.08	4.42	101,125,153,187	1068.09	20.11

	Car								
K	QIGA			QIPSO			QIDE		
	θ	v_{best}	t	θ	v_{best}	t	θ	v_{best}	t
2	152	4260.51	4.44	152	4260.51	0.06	152	4260.51	4.05
3	107,186	4666.19	12.30	107,186	4666.19	2.12	107,186	4666.19	11.22
4	82,143,213	4901.29	17.24	81,143,214	4901.31	3.06	82,142,213	4901.18	16.00
5	79,121,170,224	4995.80	24.18	78,121,173,225	4996.49	4.41	78,120,174,224	4996.26	21.25

Table 5.5 (Continued)

	Woman								
K	QIGA			QIPSO			QIDE		
	θ	v_{best}	t	θ	v_{best}	t	θ	v_{best}	t
2	124	1298.23	4.25	124	1298.23	0.04	124	1298.23	4.15
3	108,157	1529.55	11.11	108,157	1529.55	2.18	108,157	1529.55	10.01
4	96,134,167	1606.12	17.23	95,134,167	1606.17	2.32	97,135,168	1605.93	17.09
5	56,103,138,169	1666.02	25.47	54,102,139,168	1666.05	4.27	57,103,140,170	1666.01	23.40
	House								
K	QIGA			QIPSO			QIDE		
	θ	v_{best}	t	θ	v_{best}	t	θ	v_{best}	t
2	146	1653.59	5.10	146	1653.59	0.03	146	1653.59	5.02
3	95,154	1962.94	13.10	95,154	1962.94	2.13	95,154	1962.94	12.54
4	88,111,157	2005.45	17.18	81,112,157	2005.43	3.50	81,112,157	2005.43	16.42
5	80,110,154,205	2042.14	23.57	80,110,155,202	2042.11	4.42	84,112,154,201	2041.41	22.18

Table 5.6 Best results of QIACO, QISA, and QITS for multi-level thresholding for Lena, B2, Barbara, Boat, Cameraman, Jetplane, Pinecone, Car, Woman, and House.

	Lena								
K	QIACO			QISA			QITS		
	θ	v_{best}	t	θ	v_{best}	t	θ	v_{best}	t
2	118	1424.30	4.41	118	1424.30	6.30	118	1424.30	6.18
3	99,157	1763.29	13.07	99,157	1763.29	15.50	99,157	1763.29	16.41
4	84,127,169	1913.75	18.38	85,125,170	1912.85	45.19	84,128,168	1913.14	42.00
5	80,116,145,180	1970.31	23.24	81,118,147,179	1969.87	56.14	81,120,147,182	1968.91	53.29
	B2								
K	QIACO			QISA			QITS		
	θ	v_{best}	t	θ	v_{best}	t	θ	v_{best}	t
2	128	3108.56	5.07	128	3108.56	6.51	128	3108.56	5.55
3	93,175	3612.92	11.01	93,175	3612.92	12.35	93,175	3612.92	11.11
4	75,139,185	3756.69	17.34	75,138,186	3756.74	43.34	77,139,186	3756.56	42.16
5	73,138,181,218	3839.03	22.23	76,135,179,217	3838.77	53.28	76,136,180,220	3837.72	48.04
	Barbara								
K	QIACO			QISA			QITS		
	θ	v_{best}	t	θ	v_{best}	t	θ	v_{best}	t
2	115	1550.91	5.01	115	1550.91	18.15	115	1550.91	21.35
3	79,135	1912.75	11.29	79,135	1912.75	32.10	79,135	1912.75	34.23
4	73,121,165	2046.14	21.20	73,121,164	2046.10	42.20	72,122,168	2045.14	41.20
5	60,96,132,172	2111.90	23.28	65,102,133,173	2111.28	55.51	60,97,132,166	2111.49	52.16
	Boat								
K	QIACO			QISA			QITS		
	θ	v_{best}	t	θ	v_{best}	t	θ	v_{best}	t
2	91	2270.87	5.40	91	2270.87	6.25	91	2270.87	6.30
3	73,135	2538.32	10.34	73,135	2538.32	12.40	73,135	2538.32	11.26
4	55,108,147	2631.43	18.30	51,104,144	2630.73	45.32	52,103,144	2631.00	43.00
5	49,98,134,160	2683.80	23.22	52,102,133,162	2681.97	55.14	54,102,139,165	2681.63	51.25
	Cameraman								
K	QIACO			QISA			QITS		
	θ	v_{best}	t	θ	v_{best}	t	θ	v_{best}	t
2	88	3147.03	5.13	88	3147.03	8.30	88	3147.03	8.11
3	70,143	3515.77	9.16	70,143	3515.77	10.47	70,143	3515.77	10.36
4	43,97,146	3585.35	17.18	40,97,145	3584.83	40.22	45,97,145	3584.83	40.01
5	39,90,138,167	3642.45	25.45	35,86,138,165	3640.83	51.04	30,87,137,167	3640.21	49.16
	Jetplane								
K	QIACO			QISA			QITS		
	θ	v_{best}	t	θ	v_{best}	t	θ	v_{best}	t
2	151	1764.79	5.03	151	1764.79	9.38	151	1764.79	8.20
3	111,172	1929.18	13.42	111,172	1929.18	14.17	111,172	1929.18	14.23
4	89,141,188	2007.80	18.04	89,140,186	2007.74	45.12	88,143,190	2007.38	40.36
5	82,130,172,202	2054.49	23.31	80,122,171,201	2052.85	55.10	81,124,174,203	2052.32	50.08

Table 5.6 (Continued)

Pinecone										
K	QIACO			QISA			QITS			
	θ	$\mathcal{V}_{best}$	t	θ	$\mathcal{V}_{best}$	t	θ	$\mathcal{V}_{best}$	t	
2	135	752.67	5.47	135	752.67	9.48	135	752.67	8.40	
3	117,160	945.81	14.50	117,160	945.81	23.40	117,160	945.81	21.34	
4	108,140,179	1027.69	18.50	111,142,180	1026.45	45.18	108,141,180	1027.38	44.16	
5	101,122,150,187	1067.15	23.42	101,125,153,191	1067.75	51.39	101,126,151,189	1067.91	51.04	

Car										
K	QIACO			QISA			QITS			
	θ	$\mathcal{V}_{best}$	t	θ	$\mathcal{V}_{best}$	t	θ	$\mathcal{V}_{best}$	t	
2	152	4260.51	5.04	152	4260.51	9.42	152	4260.51	8.27	
3	107,186	4666.19	13.27	107,186	4666.19	17.45	107,186	4666.19	15.17	
4	82,143,214	4901.33	18.16	83,144,216	4900.80	40.27	80,141,213	4900.75	40.01	
5	77,121,172,223	4996.47	26.27	78,122,176,223	4996.33	51.40	77,121,171,229	4994.41	50.10	

Woman										
K	QIACO			QISA			QITS			
	θ	$\mathcal{V}_{best}$	t	θ	$\mathcal{V}_{best}$	t	θ	$\mathcal{V}_{best}$	t	
2	124	1298.23	5.02	124	1298.23	10.40	124	1298.23	9.10	
3	108,157	1529.55	11.37	108,157	1529.55	13.27	108,157	1529.55	12.00	
4	95,133,167	1606.01	18.12	96,133,168	1605.54	40.56	94,132,165	1605.67	41.08	
5	56,101,140,168	1665.41	26.18	59,104,141,168	1665.16	52.44	49,100,138,168	1665.30	51.10	

House										
K	QIACO			QISA			QITS			
	θ	$\mathcal{V}_{best}$	t	θ	$\mathcal{V}_{best}$	t	θ	$\mathcal{V}_{best}$	t	
2	146	1653.59	5.20	146	1653.59	10.28	146	1653.59	9.30	
3	95,154	1962.94	13.20	95,154	1962.94	18.27	95,154	1962.94	16.43	
4	82,112,157	2005.44	18.11	77,108,155	2004.65	42.50	80,111,156	2005.38	42.01	
5	78,111,154,207	2041.51	24.40	81,113,156,213	2040.84	51.51	78,111,156,209	2041.33	51.04	

a particular level, 2 * signifies that two out of six quantum algorithms report another set of thresholds for different levels, and so on. The name of the algorithms and the corresponding threshold values are also reported in the abovementioned tables. It is clearly visible from Tables 5.11–5.13 that all the algorithms report the same threshold value/values for all the test images when $K = 2$ and 3. The results vary for $K \geq 3$. A typical example can be outlined for $K = 4$ where each algorithm reports 73 and 121 as the optimum threshold value for the Barbara image except QITS. In some cases it may be observed that most of the proposed algorithms report the maximum number of threshold values for some images for any level of thresholding. In another observation, it is seen that very few algorithms report the maximum number of threshold values for another set of images at other levels as shown in Tables 5.11–5.13, respectively.

5.9 Comparison of QIPSO with Other Existing Algorithms

The experimental results prove that QIPSO is the best performing algorithm of the six different proposed algorithms. Hence, QIPSO is compared with five other algorithms, namely, TSMO [129], MTT [288], MBF [222], conventional PSO [144, 167], and conventional GA [127, 210] to find the effectiveness of this proposed algorithms. These five other algorithms have been executed for the same number of runs as the proposed algorithms on the similar images. The best results for all comparable algorithms are reported in Tables 5.14–5.16, respectively with the same format as used for the proposed algorithms. Later, the mean fitness ($\mathcal{V}_{avg}$) and standard deviation (σ) for different runs are reported in Tables 5.17–5.19, respectively for different levels of thresholding. These five

Table 5.7 Average values of fitness ($\mathcal{U}_{avg}$) and standard deviation (σ) of QIGA, QIPSO, QIDE, QIACO, QISA, and QITS for multi-level thresholding for Image1, ImageN30C30, ImageN30C80, ImageN80C30, and ImageN80C80.

	Image1											
K	QIGA		QIPSO		QIDE		QIACO		QISA		QITS	
	$\mathcal{U}_{avg}$	σ	$\mathcal{U}_{avg}$	σ	$\mathcal{U}_{avg}$	σ	$\mathcal{U}_{avg}$	σ	$\mathcal{U}_{avg}$	σ	$\mathcal{U}_{avg}$	σ
2	4970.47	0	4970.47	0	4970.47	0	4970.47	0	4970.47	0	4970.47	0
3	5624.56	0	5624.56	0	5624.56	0	5624.56	0.011	5624.54	0.039	5624.53	0.042
4	5850.30	0.050	5850.58	0.047	5850.44	0.073	5850.17	0.096	5849.86	0.097	5849.63	0.084
5	5962.14	0.156	5962.89	0.150	5962.80	0.184	5961.92	0.195	5959.48	0.188	5958.86	0.188
	ImageN30C30											
K	QIGA		QIPSO		QIDE		QIACO		QISA		QITS	
	$\mathcal{U}_{avg}$	σ	$\mathcal{U}_{avg}$	σ	$\mathcal{U}_{avg}$	σ	$\mathcal{U}_{avg}$	σ	$\mathcal{U}_{avg}$	σ	$\mathcal{U}_{avg}$	σ
2	5710.32	0	5710.32	0	5710.32	0	5710.32	0	5710.32	0	5710.32	0
3	6677.16	0	6677.16	0	6677.15	0.019	6677.16	0	6677.14	0.024	6677.14	0.024
4	7027.27	0.099	7027.27	0.081	7027.21	0.083	7026.96	0.097	7026.59	0.091	7026.45	0.093
5	7131.10	0.190	7131.24	0.116	7131.05	0.157	7130.83	0.179	7130.46	0.182	7129.93	0.146
	ImageN30C80											
K	QIGA		QIPSO		QIDE		QIACO		QISA		QITS	
	$\mathcal{U}_{avg}$	σ	$\mathcal{U}_{avg}$	σ	$\mathcal{U}_{avg}$	σ	$\mathcal{U}_{avg}$	σ	$\mathcal{U}_{avg}$	σ	$\mathcal{U}_{avg}$	σ
2	7340.33	0	7340.33	0	7340.33	0	7340.33	0	7340.33	0	7340.33	0
3	7466.66	0	7466.66	0	7466.66	0	7466.66	0	7466.66	0	7466.65	0.017
4	7510.20	0.096	7510.27	0.092	7510.10	0.089	7510.09	0.095	7509.30	0.091	7509.01	0.094
5	7541.08	0.177	7541.73	0.152	7541.29	0.168	7541.28	0.118	7540.61	0.188	7540.00	0.199
	ImageN80C30											
K	QIGA		QIPSO		QIDE		QIACO		QISA		QITS	
	$\mathcal{U}_{avg}$	σ	$\mathcal{U}_{avg}$	σ	$\mathcal{U}_{avg}$	σ	$\mathcal{U}_{avg}$	σ	$\mathcal{U}_{avg}$	σ	$\mathcal{U}_{avg}$	σ
2	5528.15	0	5528.15	0	5528.15	0	5528.15	0	5528.15	0	5528.15	0
3	6488.54	0	6488.54	0	6488.54	0	6488.54	0	6488.51	0.042	6488.52	0.015
4	6822.23	0.094	6822.43	0.094	6822.36	0.072	6822.16	0.107	6822.18	0.095	6821.82	0.114
5	6936.12	0.192	6936.25	0.165	6936.14	0.168	6935.98	0.190	6935.11	0.189	6935.05	0.146
	ImageN80C80											
K	QIGA		QIPSO		QIDE		QIACO		QISA		QITS	
	$\mathcal{U}_{avg}$	σ	$\mathcal{U}_{avg}$	σ	$\mathcal{U}_{avg}$	σ	$\mathcal{U}_{avg}$	σ	$\mathcal{U}_{avg}$	σ	$\mathcal{U}_{avg}$	σ
2	5678.56	0	5678.56	0	5678.56	0	5678.56	0	5678.56	0	5678.56	0
3	6075.08	0	6075.08	0	6075.08	0	6075.08	0.020	6075.06	0.029	6075.05	0.033
4	6185.86	0.106	6186.37	0.083	6186.24	0.115	6185.83	0.098	6185.25	0.124	6185.23	0.125
5	6244.20	0.198	6244.37	0.167	6243.91	0.172	6243.80	0.189	6242.91	0.192	6242.97	0.199

algorithms have the identical threshold value as QIPSO for all images for $K = 2$. The mean fitness ($\mathcal{U}_{avg}$) and standard deviation (σ) obtained for each test image are found to be identical for the lowest level of thresholding. The results vary when the upper level of thresholding is considered. There are some significant changes found in the mean fitness values and standard deviations for these five algorithms for $K > 2$. Hence, QIPSO outperforms the comparable algorithms in terms of accuracy and effectiveness. As the level of thresholding increases, the effectiveness of QIPSO increases as well. In addition, QIPSO takes the least time to execute as compared to the others.

For $K = 4$ and 5, the Friedman test [94, 95] was again conducted for these five identical algorithms along with QIPSO, using the same data set as earlier. Since there are significant changes in mean fitness values and standard deviations for $K \geq 3$, the efficiency of QIPSO can easily be determined by calculating the average rank among these six comparable algorithms. The average ranks for the comparable algorithms are reported

Table 5.8 Average values of fitness ($\mathcal{U}_{avg}$) and standard deviation (σ) of QIGA, QIPSO, QIDE, QIACO, QISA, and QITS for multi-level thresholding for Lena, B2, Barbara, Boat, and Cameraman.

Lena

K	QIGA $\mathcal{U}_{avg}$	σ	QIPSO $\mathcal{U}_{avg}$	σ	QIDE $\mathcal{U}_{avg}$	σ	QIACO $\mathcal{U}_{avg}$	σ	QISA $\mathcal{U}_{avg}$	σ	QITS $\mathcal{U}_{avg}$	σ
2	1424.30	0	1424.30	0	1424.30	0	1424.30	0	1424.30	0	1424.30	0
3	1763.29	0	1763.29	0	1763.29	0	1763.28	0.026	1763.27	0.034	1763.26	0.041
4	1913.69	0.089	1913.72	0.040	1913.68	0.088	1913.68	0.086	1912.76	0.094	1912.90	0.115
5	1969.92	0.215	1969.97	0.213	1969.91	0.290	1969.94	0.289	1969.63	0.297	1968.64	0.290

B2

K	QIGA $\mathcal{U}_{avg}$	σ	QIPSO $\mathcal{U}_{avg}$	σ	QIDE $\mathcal{U}_{avg}$	σ	QIACO $\mathcal{U}_{avg}$	σ	QISA $\mathcal{U}_{avg}$	σ	QITS $\mathcal{U}_{avg}$	σ
2	3108.56	0	3108.56	0	3108.56	0	3108.56	0	3108.56	0	3108.56	0
3	3612.92	0	3612.92	0	3612.92	0	3612.91	0.023	3612.91	0.032	3612.90	0.034
4	3756.56	0.104	3756.59	0.068	3756.57	0.079	3756.56	0.109	3756.58	0.125	3756.45	0.112
5	3838.41	0.253	3839.06	0.153	3839.03	0.205	3838.87	0.158	3838.58	0.217	3837.46	0.214

Barbara

K	QIGA $\mathcal{U}_{avg}$	σ	QIPSO $\mathcal{U}_{avg}$	σ	QIDE $\mathcal{U}_{avg}$	σ	QIACO $\mathcal{U}_{avg}$	σ	QISA $\mathcal{U}_{avg}$	σ	QITS $\mathcal{U}_{avg}$	σ
2	1550.91	0	1550.91	0	1550.91	0	1550.91	0	1550.91	0	1550.91	0
3	1912.75	0	1912.75	0	1912.75	0	1912.71	0.057	1912.70	0.050	1912.72	0.059
4	2046.01	0.085	2046.03	0.075	2046.02	0.082	2046.01	0.088	2046.01	0.092	2045.04	0.077
5	2111.03	0.240	2112.27	0.223	2111.07	0.234	2111.09	0.224	2110.98	0.158	2110.97	0.160

Boat

K	QIGA $\mathcal{U}_{avg}$	σ	QIPSO $\mathcal{U}_{avg}$	σ	QIDE $\mathcal{U}_{avg}$	σ	QIACO $\mathcal{U}_{avg}$	σ	QISA $\mathcal{U}_{avg}$	σ	QITS $\mathcal{U}_{avg}$	σ
2	2270.87	0	2270.87	0	2270.87	0	2270.87	0	2270.87	0	2270.87	0
3	2538.32	0	2538.32	0	2538.32	0	2538.31	0.019	2538.30	0.023	2538.29	0.025
4	2631.35	0.099	2631.45	0.098	2631.36	0.099	2631.35	0.077	2630.64	0.115	2630.93	0.111
5	2683.39	0.239	2683.49	0.219	2683.42	0.278	2683.43	0.284	2681.81	0.264	2681.42	0.252

Cameraman

K	QIGA $\mathcal{U}_{avg}$	σ	QIPSO $\mathcal{U}_{avg}$	σ	QIDE $\mathcal{U}_{avg}$	σ	QIACO $\mathcal{U}_{avg}$	σ	QISA $\mathcal{U}_{avg}$	σ	QITS $\mathcal{U}_{avg}$	σ
2	3147.03	0	3147.03	0	3147.03	0	3147.03	0	3147.03	0	3147.03	0
3	3515.77	0	3515.77	0	3515.77	0	3515.75	0.025	3515.72	0.061	3515.70	0.063
4	3585.28	0.089	3585.29	0.062	3585.27	0.094	3585.26	0.085	3584.76	0.089	3584.72	0.095
5	3641.68	0.266	3642.35	0.154	3641.56	0.220	3642.30	0.155	3640.54	0.264	3639.73	0.263

in Table 5.20 for $K = 4$ and 5. This test determines the average ranks for QIPSO, TSMO, MTT, MBF, PSO, and GA as 1.00, 3.13, 3.73, 2.13, 5.20, and 5.80, respectively for $K = 4$ and that of 1.00, 3.23, 3.60, 2.20, 4.90, and 5.86, respectively for $K = 5$. For both $K = 4$ and 5, QIPSO possesses the lowest average rank of these six algorithms, which proves its superiority compared to the others. Furthermore, this test determines the chi-square (χ^2) value and p-values for $K = 4$ as 69.847 and a very small value (close to zero), whereas these two measures for $K = 5$ are 69.847 and another very small number (close to zero). Furthermore, at 0.05 significance level and ($6-1 = 5$) degree of freedom, the chi-square values are larger than the predetermined critical values for $K = 4$ and 5. Hence, these two measures validate the rejection of H_0 and confirm the acceptance of H_1, which substantiates the behavioral disparity among the comparable algorithms. The convergence curves for QIPSO and other comparable algorithms are presented in Figures 5.15–5.20. It is easily noted from these figures that QIPSO is the

Table 5.9 Average values of fitness ($\mathcal{V}_{avg}$) and standard deviation (σ) of QIGA, QIPSO, QIDE, QIACO, QISA, and QITS for multi-level thresholding for Jetplane, Pinecone, Car, Woman, and House.

	Jetplane											
K	QIGA		QIPSO		QIDE		QIACO		QISA		QITS	
	$\mathcal{V}_{avg}$	σ	$\mathcal{V}_{avg}$	σ	$\mathcal{V}_{avg}$	σ	$\mathcal{V}_{avg}$	σ	$\mathcal{V}_{avg}$	σ	$\mathcal{V}_{avg}$	σ
2	1764.79	0	1764.79	0	1764.79	0	1764.79	0	1764.79	0	1764.79	0
3	1929.18	0	1929.18	0	1929.18	0	1929.16	0.032	1929.15	0.045	1929.14	0.052
4	2007.73	0.103	2007.71	0.097	2007.69	0.112	2007.69	0.107	2007.66	0.082	2007.24	0.096
5	2054.37	0.291	2054.31	0.289	2053.80	0.263	2054.31	0.295	2052.51	0.256	2052.08	0.276

	Pinecone											
K	QIGA		QIPSO		QIDE		QIACO		QISA		QITS	
	$\mathcal{V}_{avg}$	σ	$\mathcal{V}_{avg}$	σ	$\mathcal{V}_{avg}$	σ	$\mathcal{V}_{avg}$	σ	$\mathcal{V}_{avg}$	σ	$\mathcal{V}_{avg}$	σ
2	752.67	0	752.67	0	752.67	0	752.67	0	752.67	0	752.67	0
3	945.81	0	945.81	0	945.81	0	945.78	0.052	945.74	0.094	945.76	0.089
4	1027.64	0.132	1027.67	0.118	1027.65	0.114	1027.62	0.083	1026.39	0.118	1027.29	0.102
5	1067.68	0.280	1067.62	0.248	1067.69	0.274	1066.75	0.237	1067.46	0.295	1067.55	0.239

	Car											
K	QIGA		QIPSO		QIDE		QIACO		QISA		QITS	
	$\mathcal{V}_{avg}$	σ	$\mathcal{V}_{avg}$	σ	$\mathcal{V}_{avg}$	σ	$\mathcal{V}_{avg}$	σ	$\mathcal{V}_{avg}$	σ	$\mathcal{V}_{avg}$	σ
2	4260.51	0	4260.51	0	4260.51	0	4260.51	0	4260.51	0	4260.51	0
3	4666.19	0	4666.19	0	4666.19	0	4666.19	0	4666.17	0.030	4666.16	0.042
4	4901.21	0.124	4901.25	0.119	4901.01	0.129	4901.03	0.063	4900.71	0.122	4900.70	0.113
5	4995.63	0.265	4996.34	0.215	4996.01	0.273	4996.14	0.267	4996.11	0.233	4994.17	0.256

	Woman											
K	QIGA		QIPSO		QIDE		QIACO		QISA		QITS	
	$\mathcal{V}_{avg}$	σ	$\mathcal{V}_{avg}$	σ	$\mathcal{V}_{avg}$	σ	$\mathcal{V}_{avg}$	σ	$\mathcal{V}_{avg}$	σ	$\mathcal{V}_{avg}$	σ
2	1298.23	0	1298.23	0	1298.23	0	1298.23	0	1298.23	0	1298.23	0
3	1529.55	0	1529.55	0	1529.55	0	1529.55	0	1529.53	0.054	1529.52	0.056
4	1606.03	0.085	1606.01	0.083	1605.84	0.092	1605.86	0.112	1605.41	0.108	1605.43	0.130
5	1665.83	0.185	1665.80	0.179	1665.71	0.236	1664.86	0.242	1664.87	0.210	1664.89	0.234

	House											
K	QIGA		QIPSO		QIDE		QIACO		QISA		QITS	
	$\mathcal{V}_{avg}$	σ	$\mathcal{V}_{avg}$	σ	$\mathcal{V}_{avg}$	σ	$\mathcal{V}_{avg}$	σ	$\mathcal{V}_{avg}$	σ	$\mathcal{V}_{avg}$	σ
2	1653.59	0	1653.59	0	1653.59	0	1653.59	0	1653.59	0	1653.59	0
3	1962.94	0	1962.94	0	1962.94	0	1962.92	0.059	1962.90	0.078	1962.89	0.098
4	2005.26	0.122	2005.25	0.116	2005.24	0.133	2005.23	0.151	2004.37	0.152	2005.28	0.142
5	2041.96	0.239	2041.95	0.230	2041.14	0.256	2041.15	0.269	2040.64	0.224	2041.10	0.259

fastest converging algorithm of them all. Summarizing the facts discussed above, it can be concluded that QIPSO is the best performing algorithm of them all.

In an another approach, the proposed Quantum Inspired Ant Colony Optimization and Quantum Inspired Simulated Annealing for multi-level thresholding have been tested on two gray-scale images by maximizing an objective function ($\mathcal{F}$) given by Eq. (2.42). The selected gray-scale images are B2 and Barbara, which are presented in Figures 5.5(a) and 4.7(f), respectively. Experiments have been conducted for the proposed algorithms and their respective classical counterparts for 20 different runs at different levels of thresholding. The best results of each algorithm are reported in Table 5.21, which contains the number of classes (K), optimal threshold values (θ), the fitness value ($\mathcal{F}_{best}$), and the corresponding execution time (t) (in seconds). Thereafter, the average fitness ($\mathcal{F}_{avg}$) and standard deviation (σ) over 20 runs are reported in Table 5.22. A two-tailed t-test has been conducted at 5% confidence level to judge the

Table 5.10 Data sets for QIGA, QIPSO, QIDE, QIACO, QISA, and QITS used in the Friedman test for K = 4 and 5, respectively. The rank for each method is shown in parentheses in this statistical test.

SN	Image	QIGA	QIPSO	QIDE	QIACO	QISA	QITS
				For K=4			
1	Image1	5850.348(3)	5850.622(1)	5850.539(2)	5850.247(4)	5849.940(5)	5849.709(6)
2	ImageN30C30	7027.372(2)	7027.449(1)	7027.309(3)	7027.044(4)	7026.673(5)	7026.091(6)
3	ImageN30C80	7510.318(2)	7510.425(1)	7510.186(4)	7510.192(3)	7509.397(5)	7509.129(6)
4	ImageN80C30	6822.284(2)	6822.516(1)	6822.444(3)	6822.248(4)	6821.034(6)	6821.149(5)
5	ImageN80C80	6186.015(3)	6186.481(1)	6186.311(2)	6185.935(4)	6185.431(6)	6185.511(5)
6	Lena	1913.752(2.5)	1913.793(1)	1913.745(4)	1913.752(2.5)	1912.853(6)	1913.142(5)
7	B2	3756.667(4.5)	3756.729(2)	3756.667(4.5)	3756.697(3)	3756.749(1)	3756.566(6)
8	Barbara	2046.140(2)	2046.106(4.5)	2046.140(2)	2046.140(2)	2046.106(4.5)	2045.145(6)
9	Boat	2631.525(3)	2631.609(1)	2631.545(2)	2631.430(4)	2630.732(6)	2631.006(5)
10	Cameraman	3585.385(3)	3585.419(1.5)	3585.419(1.5)	3585.359(4)	3584.838(5)	3584.830(6)
11	Jetplane	2007.812(3)	2007.816(1.5)	2007.816(1.5)	2007.801(4)	2007.740(5)	2007.389(6)
12	Pinecone	1027.781(1)	1027.767(3)	1027.771(2)	1027.699(4)	1026.450(6)	1027.384(5)
13	Car	4901.299(3)	4901.312(2)	4901.187(4)	4901.333(1)	4900.805(5)	4900.759(6)
14	Woman	1606.126(2)	1606.179(1)	1605.931(4)	1606.011(3)	1605.540(6)	1605.670(5)
15	House	2005.453(1)	2005.433(3.5)	2005.433(3.5)	2005.442(2)	2004.653(6)	2005.381(5)
	Average rank	2.46	1.73	2.86	3.23	5.16	5.53
				For K=5			
SN	Image	QIGA	QIPSO	QIDE	QIACO	QISA	QITS
1	Image1	5962.326(3)	5962.951(1)	5962.945(2)	5962.067(4)	5959.670(5)	5959.121(6)
2	ImageN30C30	7131.235(2)	7131.386(1)	7131.136(3)	7131.089(4)	7130.657(5)	7130.061(6)
3	ImageN30C80	7541.285(3)	7541.848(1)	7541.469(2)	7541.020(4)	7540.338(6)	7540.589(5)
4	ImageN80C30	6936.352(2)	6936.493(1)	6936.221(3)	6936.093(4)	6935.283(5)	6935.152(6)
5	ImageN80C80	6244.345(2)	6244.686(1)	6244.032(3)	6243.963(4)	6243.436(6)	6243.525(5)
6	Lena	1970.441(3)	1970.475(1.5)	1970.475(1.5)	1970.313(4)	1969.870(5)	1968.917(6)
7	B2	3838.730(5)	3839.299(1)	3839.235(2)	3839.032(3)	3838.779(4)	3837.724(6)
8	Barbara	2111.669(4)	2112.660(1)	2111.902(2.5)	2111.902(2.5)	2111.287(5)	2111.249(6)
9	Boat	2683.653(4)	2683.900(2)	2683.907(1)	2683.803(3)	2681.970(5)	2681.635(6)
10	Cameraman	3642.194(3)	3642.459(1)	3641.766(4)	3642.450(2)	3640.830(5)	3640.218(6)
11	Jetplane	2054.578(2)	2054.581(1)	2054.030(4)	2054.497(3)	2052.853(5)	2052.320(6)
12	Pinecone	1068.089(2)	1068.083(3)	1067.913(4)	1067.153(6)	1067.750(5)	1068.093(1)
13	Car	4996.339(3)	4996.493(1)	4996.267(4)	4996.478(2)	4995.809(5)	4994.418(6)
14	Woman	1666.022(2)	1666.051(1)	1666.017(3)	1665.411(4)	1665.160(6)	1665.302(5)
15	House	2042.143(1)	2042.110(2)	2041.417(4)	2041.516(3)	2040.843(6)	2041.334(5)
	Average rank	2.73	1.30	2.86	3.50	5.20	5.40

statistical superiority for all of the participating algorithms. The results of such test are reported in Table 5.23.

For $K = 2$, each algorithm finds an identical threshold value for all runs. Hence, the average fitness value ($\mathcal{F}_{avg}$) and standard deviation for each algorithm remain the same. So, the t-test finds no conclusive results in this case. The test results vary for $K \geq 3$. The QIACO is found to be superior in terms of accuracy and stability. The statistical superiority of QIACO is also established by means of the t-test. Moreover, the executional time for QIACO is the shortest of them all.

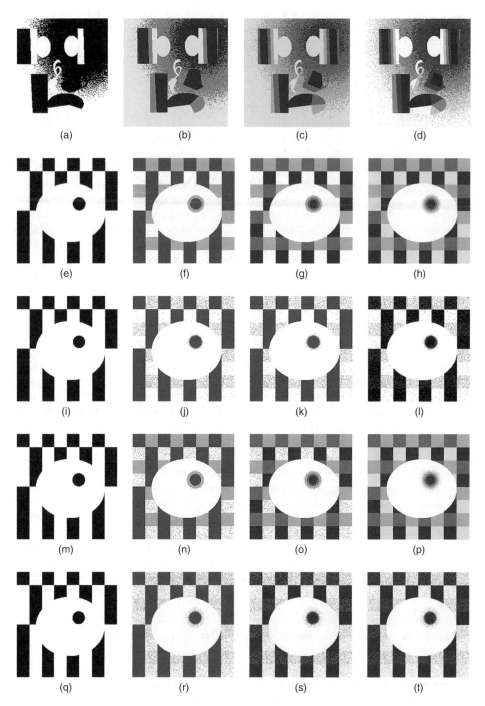

Figure 5.6 For $K = 2, 3, 4, 5$, images (a)–(d), for Image1, (e)–(h), for ImageN30C30, (i)–(l), for ImageN30C80, (m)–(p), for ImageN80C30, and, (q)–(t), for ImageN80C80, after using QIPSO for multi-level thresholding.

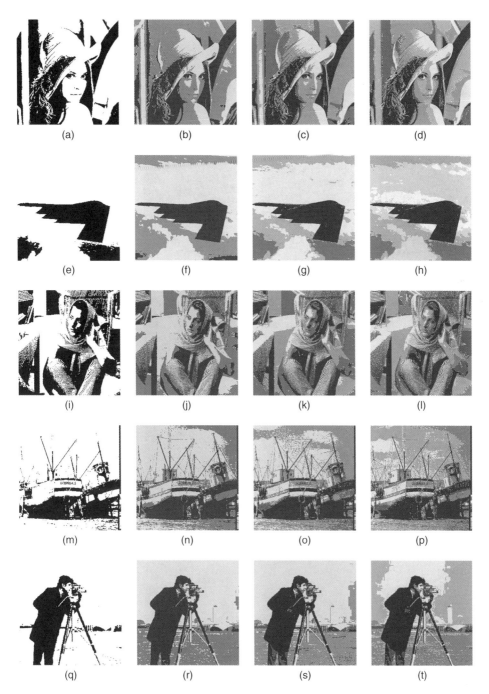

Figure 5.7 For $K = 2, 3, 4, 5$, images (a)–(d), for Lena, (e)–(h), for B2, (i)–(l), for Barbara, (m)–(p), for Boat, and, (q)–(t), for Cameraman, after using QIPSO for multi-level thresholding.

Figure 5.8 For $K = 2, 3, 4, 5$, images (a)–(d), for Jetplane, (e)–(h), for Pinecone, (i)–(l), for Car, (m)–(p), for Woman, and, (q)–(t), for House, after using QIPSO for multi-level thresholding.

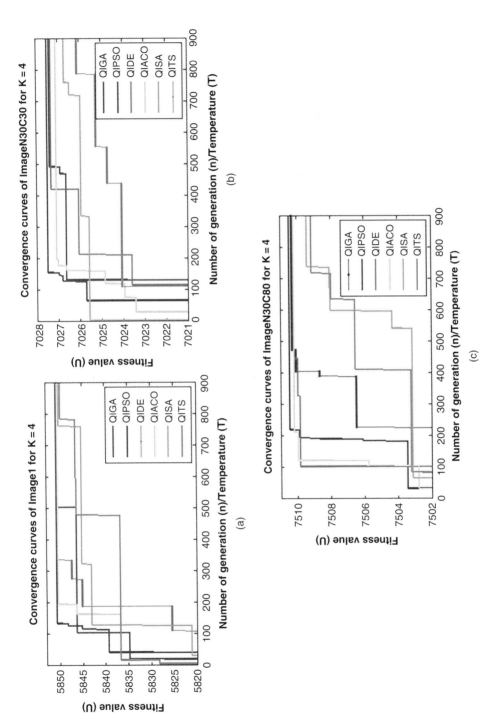

Figure 5.9 For $K = 4$, convergence curves of the proposed algorithms (a) for Image1, (b) for ImageN30C30, (c) for ImageN30C80, (d) for ImageN80C30, and (e) for ImageN80C80.

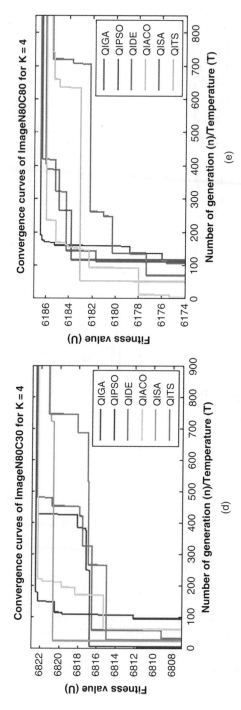

Figure 5.9 (*Continued*)

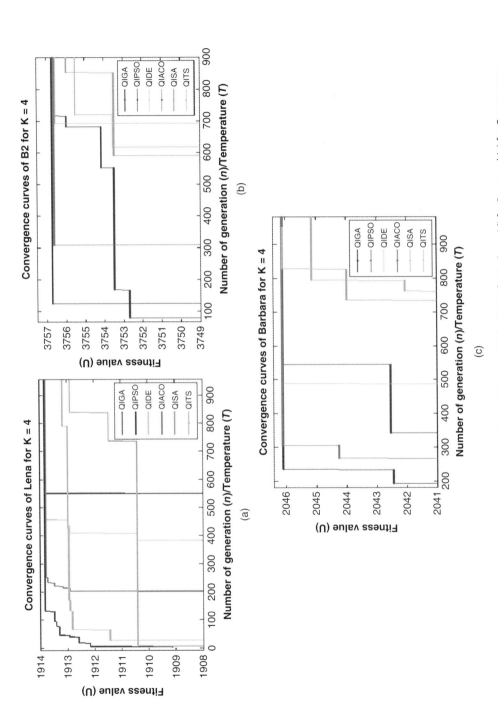

Figure 5.10 For $K = 4$, convergence curves of the proposed algorithms (a) for Lena, (b) for B2, (c) for Barbara, (d) for Boat, and (e) for Cameraman.

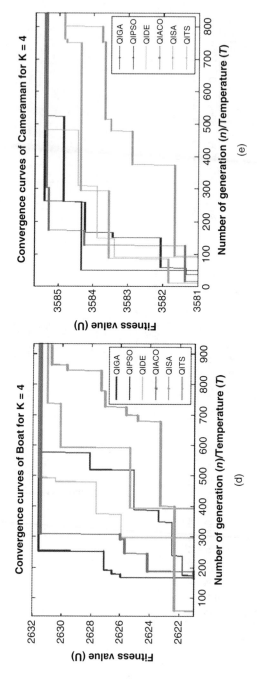

Figure 5.10 (*Continued*)

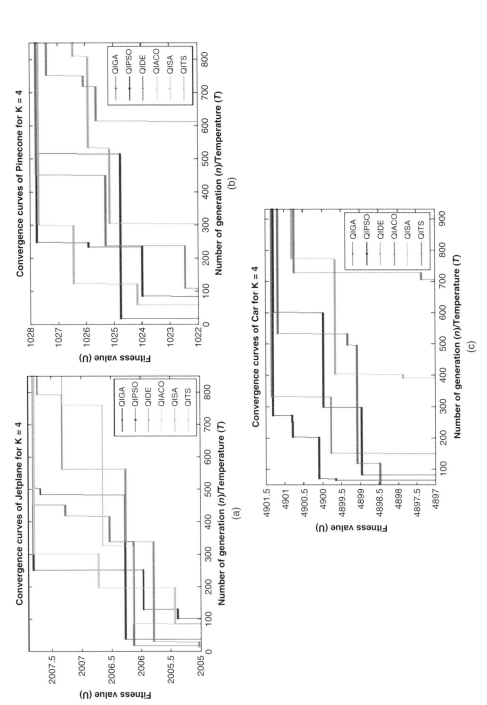

Figure 5.11 For $K = 4$, convergence curves of the proposed algorithms (a) for Jetplane, (b) for Pinecone, (c) for Car, (d) for Woman, and (e), for House.

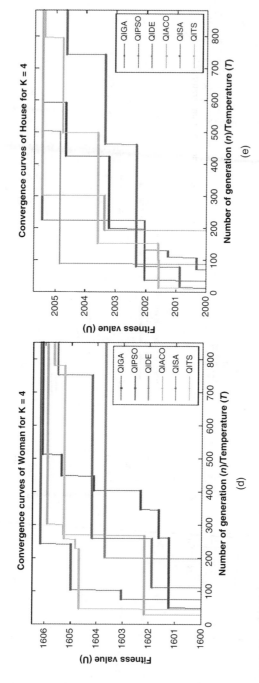

Figure 5.11 (*Continued*)

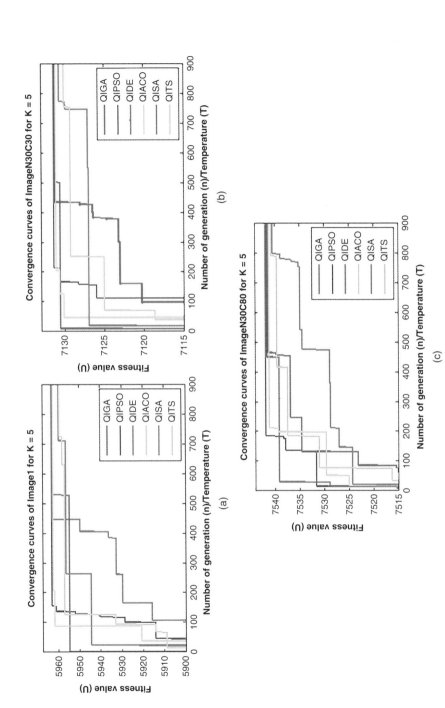

Figure 5.12 For $K = 5$, convergence curves of the proposed algorithms (a) for Image1, (b) for ImageN30C30, (c) for ImageN30C80, (d) for ImageN80C30, and (e) for ImageN80C80.

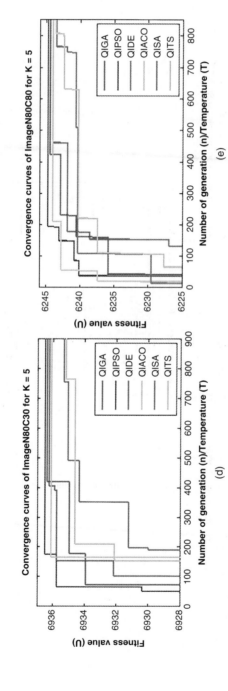

Figure 5.12 (*Continued*)

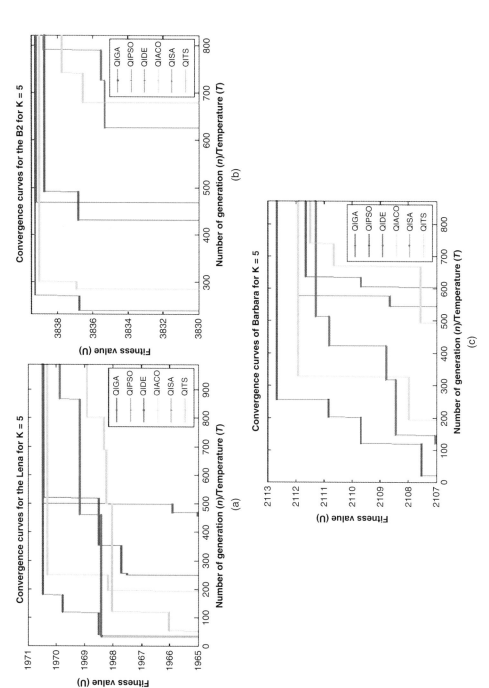

Figure 5.13 For $K = 5$, convergence curves of the proposed algorithms (a) for Lena, (b) for B2, (c) for Barbara, (d) for Boat, and (e) for Cameraman.

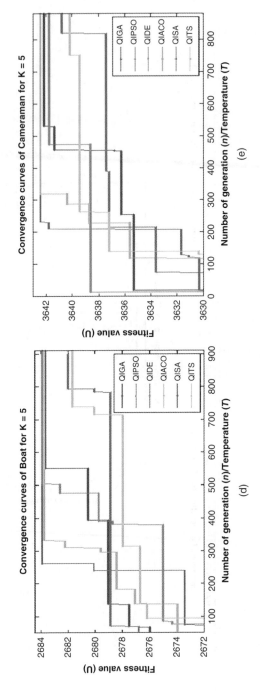

Figure 5.13 (Continued)

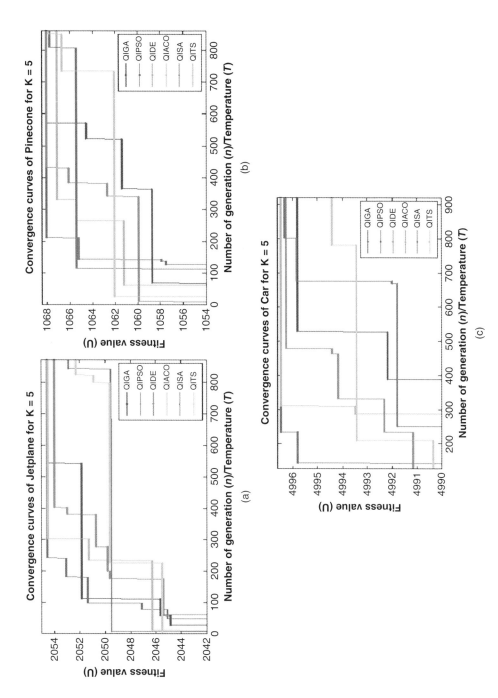

Figure 5.14 For $K = 5$, convergence curves of the proposed algorithms (a) for Jetplane, (b) for Pinecone, (c) for Car, (d) for Woman, and (e) for House.

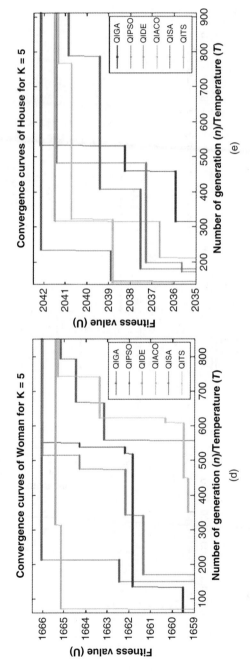

Figure 5.14 (*Continued*)

Table 5.11 Consensus results of multi-level thresholding for Image1, ImageN30C30, ImageN30C80, ImageN80C30, and ImageN80C80.

Image	K	(1*)	(2*)	(3*)	(4*)	(5*)	(6*)
Image1	2	-	-	-	-	-	149
							{g}
	3	-	-	-	-	-	95,183
							{g},{g}
	4	65,66,67,125,127,128,129,130,132,195,197,198,200	199	64	-	-	-
		{c},{d},{f},{e},{a},{b},{d},{c},{f},{e},{b},{f},{d}	{a,c}	{a,b,e}			
	5	46,48,50,51,52,55,98,99,101,106,150,157,162,163,210,211,213,214	105,160,212	-	-	-	-
		{e},{f},{d},{f},{a},{c},{e},{f},{d},{c},{f},{e},{b},{a},{f},{e},{b},{d}	{a,b},{c,d},{a,c}				
Image-N30C30	2	-	-	-	-	-	138
							{g}
	3	-	-	-	-	-	107,199
							{g},{g}
	4	57,59,62,63,64,72,124,125,133,200,202	199,201	123	-	-	-
		{b},{c},{e},{d},{a},{f},{d},{f},{e},{f},{d}	{a,b},{c,e}	{a,b,c}			
	5	62,75,122,123,182,190,193,194,239	59,60,120,129,191,241	237	-	-	-
		{c},{f},{a},{b},{e},{b},{f},{d},{e}	{d,e},{a,b},{c,d},{e,f},{a,c},{a,c}	{b,d,f}			
Image-N30C80	2	-	-	-	-	-	133
							{g}
	3	-	-	-	-	-	124,240
							{g},{g}
	4	105,107,111,115,213,215,216,218,244,247	108,221,243,245	-	-	-	-
		{a},{c},{f},{e},{c},{a},{b},{d},{c},{f}	{b,d},{c,f},{b,d},{a,e}	{a,b,c},{a,b,c}			
	5	27,28,30,31,33,34,112,129,212,217,219,220,221,223,240,244	122,123	-	245	-	-
		{b},{c},{d},{f},{a},{e},{e},{a},{e},{d},{b},{a},{c},{f},{d},{b}	{d,f},{b,c}		{a,c,e,f}		
Image-N80C30	2	-	-	-	-	-	138
							{g}
	3	-	-	-	-	-	110,201
							{g},{g}
	4	55,57,62,67,120,129,131,200,201	58	127	202	-	-
		{d},{e},{b},{a},{f},{d},{e},{f},{a}	{c,f}	{a,b,c}	{b,c,d,e}		
	5	64,71,113,119,125,126,127,133,183,186,187,238,239	57,61,232,237	189	-	-	-
		{c},{e},{e},{c},{b},{d},{a},{f},{f},{c},{b},{c},{a}	{d,f},{a,b},{d,f},{b,e}	{a,d,e}			
Image-N80C80	2	-	-	-	-	-	142
							{g}
	3	-	-	-	-	-	114,218
							{g},{g}
	4	75,76,79,80,156,157,158,159,160,162,228	71,227	226	-	-	-
		{b},{c},{a},{f},{e},{f},{a},{b},{d},{c},{c}	{d,e},{b,e}	{a,d,f}			
	5	48,54,112,113,120,134,181,185,228,229	58,59,124,180,182,226,227	-	-	-	-
		{f},{e},{f},{e},{c},{a},{c},{f},{b},{a}	{a,b},{c,d},{b,d},{d,e},{b,f},{d,f},{c,e}				

a:−Signifies that only QIGA reports   b:−Signifies that only QIPSO reports   c:−Signifies that only QIDE reports   d:−Signifies that only QIACO reports
e:−Signifies that only QISA reports   f:−Signifies that only QITS reports   g:−Signifies that all algorithms report

5.10 Conclusion

In this chapter, the functionality of bi-level gray-scale image thresholding has been extended to its multi-level domain, hence, six different quantum inspired meta-heuristic techniques for multi-level image thresholding are introduced and discussed. The quantum version of meta-heuristic algorithms is developed to determine the predefined number of optimal threshold values from five synthetic images and ten gray-level test images. These proposed algorithms used Otsu's method as an objective function. The effectiveness of the proposed algorithms has been shown at different levels of thresholding in terms of optimal threshold values with fitness measure, average fitness measure, standard deviation of the fitness measure, computational time and the convergence plot. Finally, the Friedman test was used to judge the superiority of these algorithms among themselves. It has been noted that each algorithm performs almost equally at a lower level of thresholding, specially for $K = 2$ and 3. While $K \geq 4$, the efficiency varies among different algorithms. As a result, the Quantum Inspired Particle Swarm Optimization is found to be superior to the other algorithms as far as computational time is concerned. Moreover, it has also been noted that the Quantum Inspired Particle Swarm Optimization is the best performing method while the proposed Quantum Inspired Tabu Search is the worst of them all at the higher level of thresholding. A comparative study of the proposed methods has been carried out with

Table 5.12 Consensus results of multi-level thresholding for Lena, B2, Barbara, Boat, and Cameraman.

Image	K	(1*)	(2*)	(3*)	(4*)	(5*)	(6*)
Lena	2	-	-	-	-	-	118
		-	-	-	-	-	{g}
	3	-	-	-	-	-	99,157
		-	-	-	-	-	{g},{g}
	4	83,85,126,128,170	125,127,168	169	84	-	-
		{c},{e},{b},{f},{e}	{c,e},{a,d},{c,f}	{a,b,d}	{a,b,d,f}	-	-
	5	118,120,145,179,182	79,80,81,115,116,147	146	180	-	-
		{e},{f},{d},{e},{f}	{b,c},{a,d},{e,f},{b,c},{a,d},{e,f}	{a,b,c}	{a,b,c,d}		
B2	2	-	-	-	-	-	128
		-	-	-	-	-	{g}
	3	-	-	-	-	-	93,175
		-	-	-	-	-	{g},{g}
	4	77	73,138,185	75	139,186	-	-
		{f}	{a,c},{b,e},{b,d}	{b,d,e}	{a,c,d,f},{a,c,e,f}	-	-
	5	73,136,138,139,178,181,214,215,216,217,218,220	76,179,180	75,135			
		{d},{f},{d},{c},{a},{d},{c},{b},{a},{e},{d},{f}	{e,f},{b,e},{c,f}	{a,b,c},{a,b,e}	-	-	-
Barbara	2	-	-	-	-	-	115
		-	-	-	-	-	{g}
	3	-	-	-	-	-	79,135
		-	-	-	-	-	{g},{g}
	4	72,122,168	164	165	-	73,121	-
		{f},{f},{f}	{b,e}	{a,c,d}	-	{a,b,c,d,e},{a,b,c,d,e}	-
	5	65,102,130,131,133,166,170,173	61,96	60,97,132,172	-	-	-
		{e},{e},{a},{b},{e},{f},{b},{e}	{a,b},{c,d}	{c,d,f},{a,b,f},{c,d,f},{a,c,d}			
Boat	2	-	-	-	-	-	91
		-	-	-	-	-	{g}
	3	-	-	-	-	-	73,135
		-	-	-	-	-	{g},{g}
	4	51,52,55,56,57,58,103,104,108	144	109	147	-	-
		{e},{f},{d},{c},{b},{a},{f},{e},{d}	{e,f}	{a,b,c}	{a,b,c,d}	-	-
	5	47,48,49,51,52,54,94,95,132,133,135,139,161,162,165	98,102,134	160	-	-	-
		{c},{b},{d},{a},{e},{f},{b},{c},{b},{e},{a},{f},{c},{e},{f}	{a,d},{e,f},{c,d}	{a,b,d}			
Camera-man	2	-	-	-	-	-	88
		-	-	-	-	-	{g}
	3	-	-	-	-	-	70,143
		-	-	-	-	-	{g},{g}
	4	40,43,99,146	100,145	97,147	45	-	-
		{e},{d},{a},{d}	{b,c},{e,f}	{d,e,f},{a,b,c}	{a,b,c,e}	-	-
	5	30,35,40,86,87,93,139,165	138,166	39,90,137,167	-	-	-
		{f},{e},{a},{e},{f},{a},{c},{e}	{d,e},{a,c}	{b,c,d},{b,c,d},{a,b,f},{b,d,f}	-		

a:–Signifies that only QIGA reports   b:–Signifies that only QIPSO reports   c:–Signifies that only QIDE reports   d:–Signifies that only QIACO reports
e:–Signifies that only QISA reports   f:–Signifies that only QITS reports   g:–Signifies that all algorithms report

the two-stage multithreshold Otsu method, the Maximum Tsallis entropy thresholding, the modified bacterial foraging algorithm, the classical Particle Swarm Optimization and the classical genetic algorithm. It has been found from empirical observation that the Quantum Inspired Particle Swarm Optimization outperforms these algorithms in terms of average ranking, accuracy, and robustness.

5.11 Summary

- Six different quantum inspired meta-heuristics were introduced in this chapter.
- The functionality of the quantum inspired meta-heuristic techniques has been extended to the gray-scale domain so as to enable them to determine the optimum threshold values from synthetic/real-life gray-scale images.
- The complexity increases with the increase of level of thresholding.
- The multi-level thresholding computationally takes more time when the level of thresholding increases.
- As a sequel to the proposed techniques, the experiments with five others techniques has also been conducted to prove the effectiveness of the proposed techniques.

Table 5.13 Consensus results of multi-level thresholding for Jetplane, Jungle, Pinecone, Car, Woman, and House.

Image	K	(1*)	(2*)	(3*)	(4*)	(5*)	(6*)
Jetplane	2	-	-	-	-	-	151
							{g}
	3	-	-	-	-	-	111,172
							{g},{g}
	4	143,186,188,190	89,141	140,187	88	-	-
		{f},{e},{c},{f}	{c,e},{a,c}	{b,d,e},{a,b,d}	{a,b,d,f}	-	-
	5	80,81,83,86,122,124,130,132,203	82,128,171,172,174,202	201	-	-	-
		{e},{f},{b},{c},{e},{f},{d},{c},{f}	{a,d},{a,b},{a,e},{b,d},{c,f},{a,d}	{b,c,e}	-	-	-
Pinecone	2	-	-	-	-	-	135
							{g}
	3	-	-	-	-	-	117,160
							{g},{g}
	4	111,140,141,142,178,179	107,177,180	108,139	-	-	-
		{e},{d},{f},{e},{b},{d}	{b,c},{a,c},{e,f}	{a,d,e},{a,b,c}	-	-	-
	5	99,102,122,123,149,150,151,152,186,188,189,191	125,126,153,187	-	101	-	-
		{a},{b},{d},{a},{a},{d},{f},{b},{a},{b},{f},{e}	{c,e},{b,f},{c,e},{c,d}	-	{c,d,e,f}	-	-
Car	2	-	-	-	-	-	152
							{g}
	3	-	-	-	-	-	107,186
							{g},{g}
	4	80,81,83,141,142,144,216	214	82,143,213	-	-	-
		{f},{b},{e},{f},{d},{e},{e}	{b,d}	{a,c,d},{a,b,d},{a,c,f}	-	-	-
	5	79,120,122,170,171,172,173,174,176,225,229	77,223,224	78	121	-	-
		{a},{d},{e},{a},{f},{c},{b},{d},{e},{b},{f}	{c,f},{c,e},{a,d}	{b,d,e}	{a,b,c,f}	-	-
Woman	2	-	-	-	-	-	124
							{g}
	3	-	-	-	-	-	108,157
							{g},{g}
	4	94,97,132,135,165	95,96,133,134,168	167	-	-	-
		{f},{c},{f},{c},{f}	{b,d},{a,e},{d,e},{d,b},{c,e}	{a,b,d}	-	-	-
	5	49,54,57,59,100,101,102,104,139,141,169,170	56,93,138,140	-	168	-	-
		{f},{b},{c},{e},{f},{d},{b},{e},{f},{e},{a},{c}	{a,d},{a,c},{a,f},{c,d}	-	{b,d,e,f}	-	-
House	2	-	-	-	-	-	146
							{g}
	3	-	-	-	-	-	95,154
							{g},{g}
	4	77,80,82,88,108,155,156	81,111	112	157	-	-
		{e},{b},{d},{a},{e},{e},{f}	{b,c},{a,f}	{b,c,d}	{a,b,c,d}	-	-
	5	81,84,113,155,201,202,205,207,209,213	78,80,110,111,156	154	-	-	-
		{e},{c},{c},{e},{b},{c},{b},{a},{f},{e}	{d,f},{a,b},{a,b},{d,f},{e,f}	{a,c,d}	-	-	-

a:–Signifies that only QIGA reports   b:–Signifies that only QIPSO reports   c:–Signifies that only QIDE reports   d:–Signifies that only QIACO reports
e:–Signifies that only QISA reports   f:–Signifies that only QITS reports   g:–Signifies that all algorithms report

- The effectiveness of the proposed techniques has been established in a variety of ways, such as finding stability and accuracy, performing a statistical test, called the Friedman test, etc.

Exercise Questions

Multiple Choice Questions

1 Which of the following options is true for the statement "the complexity increases with the increase of level of thresholding"?
 (a) the above statement is always true
 (b) the above statement is occasionally true
 (c) the above statement is false
 (d) none of the above

2 The pixel intensity values in the gray-scale images vary in the range of
 (a) [0,256]
 (b) [1,255]

(c) [0,255]
(d) none of these

3 The Friedman test is basically
(a) a parametric statistical test
(b) a rank-based procedure
(c) both (a) and (b)
(d) none of these

4 New synthetic images can be designed by adding different noise and contrast levels to the different synthetic image
(a) by adding contrast level
(b) by adding different noise
(c) both (a) and (b)
(d) none of these

5 In the proposed techniques, the quantum rotation gate was basically used
(a) before the encoding process
(b) after the encoding process
(c) both (a) and (b)
(d) none of these

6 Quantum orthogonality in the proposed techniques ensures
(a) $|\alpha_{ij}|^2 - |\beta_{ij}|^2 = 1$
(b) $|\alpha_{ij}|^2 \geq 1$
(c) $|\alpha_{ij}|^2 \leq 1$
(d) $|\alpha_{ij}|^2 + |\beta_{ij}|^2 = 1$

7 In QIGA, a population matrix, POP^* is created by applying the condition given by
(a) $|\alpha_{ij}|^2 \geq |\beta_{ij}|^2, i = 1, 2, \cdots, L$
(b) $rand(0, 1) > |\beta_i|^2, i = 1, 2, \cdots, L$
(c) $rand(0, 1) > |\alpha_i|^2, i = 1, 2, \cdots, L$
(d) none of the above

8 Quantum mutation, followed by quantum crossover, can be found in
(a) QIPSO
(b) QIDE
(c) QIGA
(d) QITS

9 The concept of the pheromone matrix is used in
(a) QIPSO
(b) QISA
(c) QIGA
(d) QIACO

10 A memory is invoked in
(a) QITS
(b) QISA
(c) QIDE
(d) QIACO

Short Answer Questions

1 What do you mean by multi-level image thresholding? Illustrate with examples.

2 Why does the complexity of image thresholding increase with the increase of level of thresholding?

3 Why are the accuracy and stability important in judging the performance of an algorithm?

4 How is QIPSO measured?

5 Discuss the importance of the pheromone trial in QIACO.

6 What do you mean by annealing? How is it important in designing QISA?

7 How is the result of the Friedman's test interpreted? Explain with examples.

Long Answer Questions

1 State the algorithmic steps of QIACO with proper illustrations. Find its time complexity.

2 Discuss in detail, the algorithmic steps of QIGA. Derive its time complexity.

3 Compare the time complexities of QIGA, QIPSO, QIACO, QIDE, QISA, and QITS.

4 Discuss the steps of quantum mutation and quantum crossover which occur in QIDE.

Table 5.14 Best results of QIPSO, TSMO, MTT, MBF, PSO, and GA for multi-level thresholding for Image1, ImageN30C30, ImageN30C80, ImageN80C30, and ImageN80C80.

Image1

K	QIPSO			TSMO			MTT		
	θ	v_{best}	t	θ	v_{best}	t	θ	v_{best}	t
2	149	4970.47	0.02	149	4970.47	1.25	149	4970.47	2.20
3	95,183	5624.56	2.13	95,182	5624.47	2.38	96,183	5624.36	3.42
4	64,128,197	5850.62	2.60	62,122,192	5847.87	3.01	61,131,202	5846.98	6.19
5	51,105,162,213	5962.95	4.44	46,98,157,211	5959.67	5.04	62,115,172,222	5958.40	8.34

K	MBF			PSO			GA		
	θ	v_{best}	t	θ	v_{best}	t	θ	v_{best}	t
2	149	4970.47	2.38	149	4970.47	1.53	149	4970.47	2.17
3	94,183	5624.49	3.25	94,181	5624.22	8.15	95,181	5624.06	12.24
4	64,124,194	5849.28	7.15	73,131,201	5845.54	26.20	57,130,199	5844.00	32.18
5	53,112,163,214	5960.64	9.10	58,111,168,211	5957.98	34.02	48,97,147,211	5955.64	50.27

ImageN30C30

K	QIPSO			TSMO			MTT		
	θ	v_{best}	t	θ	v_{best}	t	θ	v_{best}	t
2	138	5710.32	0.03	138	5710.32	1.13	138	5710.32	2.14
3	107,199	6677.16	2.31	108,198	6677.05	2.40	107,197	6677.02	3.16
4	57,123,199	7027.44	3.12	56,126,194	7026.77	3.15	64,112,195	7026.16	7.52
5	60,123,190,237	7131.38	4.39	55,115,181,246	7129.09	4.42	70,118,183,246	7129.19	8.43

K	MBF			PSO			GA		
	θ	v_{best}	t	θ	v_{best}	t	θ	v_{best}	t
2	138	5710.32	2.05	138	5710.32	1.59	138	5710.32	2.40
3	109,200	6677.10	3.28	112,199	6676.98	9.45	109,201	6676.95	11.52
4	65,119,200	7027.12	8.10	53,109,195	7025.62	25.15	55,124,189	7025.67	34.14
5	60,133,183,231	7129.79	9.18	73,137,196,242	7128.16	37.19	79,117,195,233	7128.22	51.28

ImageN30C80

K	QIPSO			TSMO			MTT		
	θ	v_{best}	t	θ	v_{best}	t	θ	v_{best}	t
2	133	7340.33	0.02	133	7340.33	1.15	133	7340.33	2.10
3	124,240	7466.66	2.10	124,242	7466.27	2.25	113,239	7466.25	3.13
4	108,216,243	7510.42	3.12	111,221,247	7509.12	3.41	113,210,242	7508.98	7.52
5	27,123,219,244	7541.84	4.09	34,111,213,248	7538.89	1.14	29,118,210,249	7538.25	8.40

K	MBF			PSO			GA		
	θ	v_{best}	t	θ	v_{best}	t	θ	v_{best}	t
2	133	7340.33	2.11	133	7340.33	1.45	133	7340.33	2.36
3	127,240	7466.37	3.42	129,238	7466.09	9.18	129,238	7466.09	11.30
4	94,216,241	7508.10	6.54	112,214,237	7507.56	23.18	112,222,251	7506.26	33.22
5	26,112,219,249	7539.87	9.18	19,135,223,247	7536.02	35.16	33,135,210,251	7534.30	52.12

ImageN80C30

K	QIPSO			TSMO			MTT		
	θ	v_{best}	t	θ	v_{best}	t	θ	v_{best}	t
2	138	5528.15	0.02	138	5528.15	1.27	138	5528.15	4.14
3	110,201	6488.54	2.12	106,202	6488.26	2.10	115,203	6488.19	3.13
4	62,127,202	6822.51	3.24	59,132,206	6820.36	3.40	60,123,206	6820.46	7.55
5	61,125,187,237	6936.49	4.20	57,133,183,232	6935.15	4.57	55,113,195,242	6934.54	8.30

K	MBF			PSO			GA		
	θ	v_{best}	t	θ	v_{best}	t	θ	v_{best}	t
2	138	5528.15	2.12	138	5528.15	1.42	138	5528.15	1.59
3	107,202	6488.37	3.14	116,200	6488.01	8.18	110,199	6487.93	10.53
4	58,136,203	6821.44	7.39	61,109,204	6819.73	24.22	67,142,206	6818.33	32.26
5	57,123,182,233	6935.85	9.28	55,113,195,242	6934.54	34.56	75,137,184,242	6933.59	49.28

ImageN80C80

K	QIPSO			TSMO			MTT		
	θ	v_{best}	t	θ	v_{best}	t	θ	v_{best}	t
2	142	5678.56	0.02	142	5678.56	1.42	142	5678.56	4.14
3	114,218	6075.08	2.12	115,217	6075.01	2.23	119,219	6074.84	3.10
4	75,159,227	6186.48	3.15	74,168,225	6184.20	3.10	66,160,225	6183.27	8.16
5	58,124,182,228	6244.68	4.25	43,125,184,232	6242.93	4.52	50,136,180,231	6241.99	9.43

K	MBF			PSO			GA		
	θ	v_{best}	t	θ	v_{best}	t	θ	v_{best}	t
2	142	5678.56	5.11	142	5678.56	2.05	142	5678.56	2.16
3	116,218	6075.05	3.49	117,216	6074.68	10.18	119,220	6074.56	15.03
4	71,156,227	6185.43	7.11	50,130,218	6182.32	25.11	71,135,221	6180.01	33.12
5	54,113,180,227	6243.43	10.10	66,138,180,223	6240.21	35.17	56,145,212,245	6237.27	52.14

Table 5.15 Best results of QIPSO, TSMO, MTT, MBF, PSO, and GA for multi-level thresholding for Lena, B2, Barbara, Boat, and Cameraman.

Lena

K	QIPSO			TSMO			MTT		
	θ	V_{best}	t	θ	V_{best}	t	θ	V_{best}	t
2	118	1424.30	0.02	118	1424.30	1.13	118	1424.30	2.15
3	99,157	1763.29	2.13	99,158	1763.21	2.18	98,156	1763.19	3.10
4	83,125,168	1913.74	3.32	87,130,168	1911.28	3.40	89,132,173	1910.58	7.49
5	79,115,146,180	1970.47	4.46	73,115,147,183	1967.26	4.21	75,115,141,178	1967.38	9.01

K	MBF			PSO			GA		
	θ	V_{best}	t	θ	V_{best}	t	θ	V_{best}	t
2	118	1424.30	2.18	118	1424.30	1.42	118	1424.30	2.19
3	99,158	1763.21	3.07	100,157	1763.04	9.55	98,158	1762.94	11.11
4	87,132,171	1911.00	8.01	85,134,177	1907.88	23.13	77,118,168	1906.10	32.15
5	74,115,145,181	1968.58	9.22	81,112,149,185	1964.86	19.11	86,116,150,178	1963.08	51.07

B2

K	QIPSO			TSMO			MTT		
	θ	V_{best}	t	θ	V_{best}	t	θ	V_{best}	t
2	128	3108.56	0.03	128	3108.56	1.33	128	3108.56	2.15
3	93,175	3612.92	2.37	92,175	3612.78	2.44	91,174	3612.51	3.22
4	75,138,185	3756.72	3.00	78,143,189	3754.20	3.18	72,133,187	3753.73	7.54
5	75,135,179,215	3839.29	4.34	73,135,179,209	3835.94	5.03	68,138,185,216	3835.40	10.12

K	MBF			PSO			GA		
	θ	V_{best}	t	θ	V_{best}	t	θ	V_{best}	t
2	128	3108.56	2.01	128	3108.56	1.51	128	3108.56	2.55
3	95,175	3612.81	3.12	93,177	3612.14	11.14	96,177	3611.94	11.30
4	76,136,188	3754.86	7.30	72,133,187	3753.73	24.14	82,144,186	3752.88	31.15
5	74,137,182,210	3836.46	10.01	75,134,185,212	3834.43	33.17	78,141,183,225	3832.85	50.10

Barbara

K	QIPSO			TSMO			MTT		
	θ	V_{best}	t	θ	V_{best}	t	θ	V_{best}	t
2	115	1550.91	0.02	115	1550.91	1.18	115	1550.91	2.31
3	79,135	1912.75	2.10	79,137	1912.35	2.14	77,134	1912.29	3.49
4	73,121,164	2046.10	3.01	71,122,168	2044.69	3.18	73,125,167	2044.15	8.11
5	61,97,131,170	2112.66	4.01	67,101,136,173	2110.80	4.42	59,97,127,165	2109.32	9.70

K	MBF			PSO			GA		
	θ	V_{best}	t	θ	V_{best}	t	θ	V_{best}	t
2	115	1550.91	2.16	115	1550.91	2.15	115	1550.91	3.08
3	80,135	1912.48	3.17	79,133	1911.69	9.53	77,132	1911.03	12.10
4	74,120,164	2045.74	7.43	75,127,166	2041.97	25.16	69,116,164	2042.56	31.20
5	65,102,136,175	2110.69	10.14	62,90,129,168	2108.22	25.18	59,98,131,174	2107.33	51.20

Boat

K	QIPSO			TSMO			MTT		
	θ	V_{best}	t	θ	V_{best}	t	θ	V_{best}	t
2	91	2270.87	0.06	91	2270.87	1.36	91	2270.87	2.11
3	73,135	2538.32	2.40	73,134	2538.17	2.50	74,135	2538.13	2.58
4	57,109,147	2631.60	3.27	53,101,146	2629.09	3.38	61,113,153	2628.07	8.10
5	48,94,132,160	2683.90	4.16	53,102,131,161	2680.33	4.21	55,95,133,165	2680.05	9.44

K	MBF			PSO			GA		
	θ	V_{best}	t	θ	V_{best}	t	θ	V_{best}	t
2	91	2270.87	2.40	91	2270.87	1.55	91	2270.87	2.41
3	72,134	2538.26	3.49	71,133	2537.88	10.18	73,133	2537.70	14.48
4	48,102,144	2629.42	8.09	58,100,142	2627.26	23.11	64,115,145	2626.01	34.52
5	54,102,139,165	2681.63	10.12	49,84,128,157	2678.10	37.10	42,88,129,151	2677.34	51.02

Cameraman

K	QIPSO			TSMO			MTT		
	θ	V_{best}	t	θ	V_{best}	t	θ	V_{best}	t
2	88	3147.03	0.08	88	3147.03	1.17	88	3147.03	2.05
3	70,143	3515.77	2.32	69,141	3515.31	2.51	68,144	3515.25	3.05
4	45,100,147	3585.41	3.46	39,93,142	3583.16	3.13	56,118,151	3583.18	7.50
5	39,90,137,167	3642.45	4.30	44,88,138,170	3639.76	4.32	29,89,138,167	3639.30	9.52

K	MBF			PSO			GA		
	θ	V_{best}	t	θ	V_{best}	t	θ	V_{best}	t
2	88	3147.03	2.29	88	3147.03	1.34	88	3147.03	2.14
3	70,144	3515.55	3.47	68,140	3514.59	10.20	74,143	3514.44	12.19
4	43,92,145	3584.12	7.41	33,88,144	3582.59	22.26	60,126,157	3581.87	33.20
5	41,98,141,172	3640.03	10.09	42,91,143,167	3638.74	34.31	34,90,142,165	3637.76	52.16

Table 5.16 Best results of QIPSO, TSMO, MTT, MBF, PSO, and GA for multi-level thresholding for Jetplane, Pinecone, Car, Woman, and House.

Jetplane

K	QIPSO			TSMO			MTT		
	θ	U_{best}	t	θ	U_{best}	t	θ	U_{best}	t
2	151	1764.79	0.05	151	1764.79	1.24	151	1764.79	2.30
3	111,172	1929.18	3.21	110,171	1929.16	3.20	113,172	1929.05	3.14
4	88,140,187	2007.81	3.27	94,138,184	2005.24	3.42	86,141,191	2005.99	7.52
5	83,128,172,201	2054.58	4.30	87,128,169,205	2050.55	4.22	83,121,161,198	2051.18	8.29

K	MBF			PSO			GA		
	θ	U_{best}	t	θ	U_{best}	t	θ	U_{best}	t
2	151	1764.79	2.08	151	1764.79	1.57	151	1764.79	2.28
3	111,172	1929.18	3.41	110,173	1928.72	9.25	106,168	1928.21	15.14
4	80,135,185	2005.50	8.43	79,137,184	2004.86	24.40	94,137,184	2004.92	36.10
5	81,124,174,203	2052.32	9.30	80,125,160,196	2049.96	25.05	85,121,166,203	2050.29	50.18

Pinecone

K	QIPSO			TSMO			MTT		
	θ	U_{best}	t	θ	U_{best}	t	θ	U_{best}	t
2	135	752.67	0.04	135	752.67	1.114	135	752.67	2.18
3	117,160	945.81	2.01	119,163	945.21	2.15	118,159	945.28	3.24
4	107,139,178	1027.76	2.34	104,134,172	1025.90	3.11	109,141,187	1024.12	8.12
5	102,126,152,188	1068.08	4.42	100,123,155,194	1065.36	4.49	98,119,150,181	1064.52	9.11

K	MBF			PSO			GA		
	θ	U_{best}	t	θ	U_{best}	t	θ	U_{best}	t
2	135	752.67	3.01	135	752.67	1.55	135	752.67	2.45
3	118,160	945.64	3.26	114,158	944.28	8.40	121,162	943.56	12.04
4	110,140,181	1026.64	8.18	103,134,179	1023.85	22.30	108,146,182	1022.80	35.16
5	97,126,155,185	1065.22	10.10	99,131,158,191	1063.58	34.18	96,120,154,195	1062.17	52.11

Car

K	QIPSO			TSMO			MTT		
	θ	U_{best}	t	θ	U_{best}	t	θ	U_{best}	t
2	152	4260.51	0.06	152	4260.51	1.34	152	4260.51	2.28
3	107,186	4666.19	2.12	108,184	4665.46	2.29	105,185	4665.92	3.50
4	81,143,214	4901.31	3.06	80,138,213	4899.04	3.10	85,147,212	4898.14	7.32
5	78,121,173,225	4996.49	4.41	73,124,176,224	4993.24	4.48	76,124,169,222	4994.18	9.12

K	MBF			PSO			GA		
	θ	U_{best}	t	θ	U_{best}	t	θ	U_{best}	t
2	152	4260.51	2.10	152	4260.51	1.50	152	4260.51	2.35
3	104,183	4665.55	3.17	111,188	4664.93	10.10	108,190	4664.89	16.24
4	81,146,214	4900.49	8.12	81,148,210	4895.98	25.22	88,148,214	4894.19	37.14
5	80,119,174,229	4993.50	9.32	77,122,168,229	4992.25	38.18	75,117,178,223	4991.58	50.19

Woman

K	QIPSO			TSMO			MTT		
	θ	U_{best}	t	θ	U_{best}	t	θ	U_{best}	t
2	124	1298.23	0.04	124	1298.23	1.34	124	1298.23	3.06
3	108,157	1529.55	2.18	109,158	1529.33	2.29	107,156	1529.37	3.23
4	95,134,167	1606.17	2.32	92,131,163	1604.12	3.12	90,130,166	1604.13	8.44
5	54,102,139,168	1666.05	4.27	58,99,140,171	1663.74	4.58	58,99,140,171	1663.74	9.47

K	MBF			PSO			GA		
	θ	U_{best}	t	θ	U_{best}	t	θ	U_{best}	t
2	124	1298.23	2.07	124	1298.23	1.42	124	1298.23	2.14
3	107,157	1529.48	3.47	105,156	1528.68	9.50	105,154	1527.91	14.38
4	98,138,168	1605.38	7.33	91,135,165	1603.43	34.05	98,135,172	1602.39	36.10
5	58,103,143,173	1664.12	10.14	52,103,143,167	1662.86	34.50	57,103,145,175	1661.79	51.09

House

K	QIPSO			TSMO			MTT		
	θ	U_{best}	t	θ	U_{best}	t	θ	U_{best}	t
2	146	1653.59	0.03	146	1653.59	1.13	146	1653.59	3.08
3	95,154	1962.94	2.13	97,157	1962.02	2.27	92,154	1961.61	3.22
4	81,112,157	2005.43	3.50	88,117,160	2003.39	3.18	86,112,154	2003.28	9.01
5	80,110,155,202	2042.11	4.42	86,113,158,206	2040.49	5.02	75,107,156,201	2040.65	8.42

K	MBF			PSO			GA		
	θ	U_{best}	t	θ	U_{best}	t	θ	U_{best}	t
2	146	1653.59	5.40	146	1653.59	2.01	146	1653.59	2.40
3	95,154	1962.94	3.14	95,157	1962.07	9.13	92,154	1961.61	16.55
4	82,108,155	2004.03	8.49	85,119,159	2002.21	25.34	85,110,152	2001.62	33.06
5	82,111,155,210	2041.57	8.51	86,116,159,202	2039.85	37.38	82,110,162,204	2038.83	52.11

Table 5.17 Average values of fitness ($\mathcal{U}_{avg}^{*}$) and standard deviation (σ) of QIPSO, TSMO, MTT, MBF, PSO, and GA for multi-level thresholding for Image1, ImageN30C30, ImageN30C80, ImageN80C30, and ImageN80C80.

Image1											
K	QIPSO		TSMO		MTT		MBF		PSO		GA	
	$\mathcal{U}_{avg}^{*}$	σ	$\mathcal{U}_{avg}^{*}$	σ	$\mathcal{U}_{avg}^{*}$	σ	$\mathcal{U}_{avg}^{*}$	σ	$\mathcal{U}_{avg}^{*}$	σ	$\mathcal{U}_{avg}^{*}$	σ
2	4970.47	0	4970.47	0	4970.47	0	4970.47	0	4970.47	0	4970.47	0
3	5624.56	0	5624.43	0.082	5624.31	0.91	5624.44	0.086	5624.15	0.160	5623.97	0.175
4	5850.58	0.047	5847.45	0.466	5846.70	0.250	5848.31	0.621	5844.02	1.95	5841.57	1.52
5	5962.89	0.150	5957.78	1.605	5956.24	1.726	5958.39	1.715	5952.80	2.063	5952.76	2.935

ImageN30C30											
K	QIPSO		TSMO		MTT		MBF		PSO		GA	
	$\mathcal{U}_{avg}^{*}$	σ	$\mathcal{U}_{avg}^{*}$	σ	$\mathcal{U}_{avg}^{*}$	σ	$\mathcal{U}_{avg}^{*}$	σ	$\mathcal{U}_{avg}^{*}$	σ	$\mathcal{U}_{avg}^{*}$	σ
2	5710.32	0	5710.32	0	5710.32	0	5710.32	0	5710.32	0	5710.32	0
3	6677.16	0	6677.04	0.014	6676.98	0.034	6677.03	0.050	6676.97	0.014	6676.96	0.015
4	7027.27	0.081	7025.19	0.630	7025.74	0.468	7026.71	0.595	7024.54	0.713	7025.27	0.692
5	7131.24	0.116	7128.18	1.421	7128.00	1.567	7128.26	1.648	7126.03	2.185	7126.00	2.081

ImageN30C80											
K	QIPSO		TSMO		MTT		MBF		PSO		GA	
	$\mathcal{U}_{avg}^{*}$	σ	$\mathcal{U}_{avg}^{*}$	σ	$\mathcal{U}_{avg}^{*}$	σ	$\mathcal{U}_{avg}^{*}$	σ	$\mathcal{U}_{avg}^{*}$	σ	$\mathcal{U}_{avg}^{*}$	σ
2	7340.33	0	7340.33	0	7340.33	0	7340.33	0	7340.33	0	7340.33	0
3	7466.66	0	7466.21	0.083	7466.20	0.080	7466.21	0.094	7466.06	0.032	7466.04	0.031
4	7510.27	0.092	7508.75	0.557	7508.53	0.491	7507.56	0.534	7507.13	0.589	7505.00	0.690
5	7541.73	0.152	7536.75	1.467	7536.47	1.407	7538.32	1.663	7533.37	2.114	7532.40	2.074

ImageN80C30											
K	QIPSO		TSMO		MTT		MBF		PSO		GA	
	$\mathcal{U}_{avg}^{*}$	σ	$\mathcal{U}_{avg}^{*}$	σ	$\mathcal{U}_{avg}^{*}$	σ	$\mathcal{U}_{avg}^{*}$	σ	$\mathcal{U}_{avg}^{*}$	σ	$\mathcal{U}_{avg}^{*}$	σ
2	5528.15	0	2270.87	0	2270.87	0	2270.87	0	2270.87	0	2270.87	0
3	6488.54	0	6488.23	0.050	6488.17	0.033	6488.27	0.083	6487.98	0.038	6487.88	0.100
4	6822.43	0.094	6819.75	0.671	6819.77	0.692	6820.48	0.767	6818.92	0.837	6817.80	0.907
5	6936.25	0.165	6933.60	1.304	6933.59	1.207	6933.92	1.562	6931.91	2.184	6931.07	2.341

ImageN80C80											
K	QIPSO		TSMO		MTT		MBF		PSO		GA	
	$\mathcal{U}_{avg}^{*}$	σ	$\mathcal{U}_{avg}^{*}$	σ	$\mathcal{U}_{avg}^{*}$	σ	$\mathcal{U}_{avg}^{*}$	σ	$\mathcal{U}_{avg}^{*}$	σ	$\mathcal{U}_{avg}^{*}$	σ
2	5678.56	0	5678.56	0	5678.56	0	5678.56	0	5678.56	0	5678.56	0
3	6075.08	0	6074.90	0.086	6074.93	0.092	6074.79	0.096	6074.53	0.087	6074.50	0.097
4	6186.37	0.083	6183.14	0.682	6182.77	0.414	6184.51	0.672	6180.94	1.056	6179.26	1.188
5	6244.37	0.167	6241.78	1.194	6240.73	1.271	6241.98	1.354	6238.06	1.961	6234.60	2.304

Table 5.18 Average values of fitness ($\mathcal{V}_{avg}$) and standard deviation (σ) of QIPSO, TSMO, MTT, MBF, PSO, and GA for multi-level thresholding for Lena, B2, Barbara, Boat, and Cameraman.

Lena											
K	QIPSO		TSMO		MTT		MBF		PSO		GA	
	$\mathcal{V}_{avg}$	σ	$\mathcal{V}_{avg}$	σ	$\mathcal{V}_{avg}$	σ	$\mathcal{V}_{avg}$	σ	$\mathcal{V}_{avg}$	σ	$\mathcal{V}_{avg}$	σ
2	1424.30	0	1424.30	0	1424.30	0	1424.30	0	1424.30	0	1424.30	0
3	1763.29	0	1763.18	0.057	1763.16	0.051	1763.17	0.058	1762.94	0.092	1762.87	0.109
4	1913.50	0.039	1910.66	0.493	1909.98	0.545	1910.74	0.335	1906.98	0.729	1905.18	1.159
5	1969.23	0.112	1965.91	1.107	1966.00	1.133	1966.08	1.368	1962.83	2.064	1960.39	2.156

B2											
K	QIPSO		TSMO		MTT		MBF		PSO		GA	
	$\mathcal{V}_{avg}$	σ	$\mathcal{V}_{avg}$	σ	$\mathcal{V}_{avg}$	σ	$\mathcal{V}_{avg}$	σ	$\mathcal{V}_{avg}$	σ	$\mathcal{V}_{avg}$	σ
2	3108.56	0	3108.56	0	3108.56	0	3108.56	0	3108.56	0	3108.56	0
3	3612.92	0	3612.31	0.071	3612.24	0.076	3612.33	0.091	3611.48	0.119	3610.62	0.116
4	3756.50	0.049	3754.65	0.372	3752.95	0.569	3754.70	0.364	3752.62	0.967	3751.89	1.212
5	3838.56	0.116	3834.37	1.111	3833.95	1.506	3834.62	1.376	3833.04	2.145	3829.04	2.114

Barbara											
K	QIPSO		TSMO		MTT		MBF		PSO		GA	
	$\mathcal{V}_{avg}$	σ	$\mathcal{V}_{avg}$	σ	$\mathcal{V}_{avg}$	σ	$\mathcal{V}_{avg}$	σ	$\mathcal{V}_{avg}$	σ	$\mathcal{V}_{avg}$	σ
2	1550.91	0	1550.91	0	1550.91	0	1550.91	0	1550.91	0	1550.91	0
3	1912.75	0	1912.62	0	1912.58	0	1912.71	0.016	1912.75	0.028	1912.75	0.025
4	2045.92	0.055	2045.01	0.507	2044.88	0.625	2045.19	0.570	2040.56	1.181	2040.96	1.113
5	2112.00	0.165	2108.605	1.259	2108.65	1.046	2109.08	1.073	2105.27	2.042	2104.54	2.078

Boat											
K	QIPSO		TSMO		MTT		MBF		PSO		GA	
	$\mathcal{V}_{avg}$	σ	$\mathcal{V}_{avg}$	σ	$\mathcal{V}_{avg}$	σ	$\mathcal{V}_{avg}$	σ	$\mathcal{V}_{avg}$	σ	$\mathcal{V}_{avg}$	σ
2	2270.87	0	2270.87	0	2270.87	0	2270.87	0	2270.87	0	2270.87	0
3	2538.32	0	2538.14	0.050	2538.09	0.044	2538.13	0.078	2537.78	0.110	2537.61	0.128
4	2631.54	0.037	2628.41	0.425	2627.42	0.699	2628.40	0.506	2625.81	1.327	2624.67	1.293
5	2683.11	0.118	2679.19	1.206	2678.65	1.283	2678.80	1.515	2675.67	2.067	2675.06	2.031

Cameraman											
K	QIPSO		TSMO		MTT		MBF		PSO		GA	
	$\mathcal{V}_{avg}$	σ	$\mathcal{V}_{avg}$	σ	$\mathcal{V}_{avg}$	σ	$\mathcal{V}_{avg}$	σ	$\mathcal{V}_{avg}$	σ	$\mathcal{V}_{avg}$	σ
2	3147.03	0	3147.03	0	3147.03	0	3147.03	0	3147.03	0	3147.03	0
3	3515.77	0	3515.26	0.057	3515.22	0.032	3515.53	0.063	3514.54	0.072	3514.18	0.093
4	3585.34	0.029	3582.92	0.560	3582.78	0.422	3583.11	0.551	3581.23	1.237	3580.78	1.265
5	3642.23	0.123	3638.22	1.373	3638.22	1.243	3637.74	1.865	3634.65	2.349	3635.31	2.263

Table 5.19 Average values of fitness ($\mathcal{U}_{avg}$) and standard deviation (σ) of QIPSO, TSMO, MTT, MBF, PSO, and GA for multi-level thresholding for Jetplane, Pinecone, Car, Woman, and House.

	Jetplane										
K	QIPSO		TSMO		MTT		MBF		PSO		GA	
	$\mathcal{U}_{avg}$	σ	$\mathcal{U}_{avg}$	σ	$\mathcal{U}_{avg}$	σ	$\mathcal{U}_{avg}$	σ	$\mathcal{U}_{avg}$	σ	$\mathcal{U}_{avg}$	σ
2	1764.79	0	1764.79	0	1764.79	0	1764.79	0	1764.79	0	1764.79	0
3	1929.18	0	1929.14	0.037	1929.03	0.019	1929.15	0.050	1928.66	0.097	1928.12	0.090
4	2007.80	0.032	2004.77	0.486	2004.85	0.608	2004.80	0.519	2003.84	1.275	2004.17	1.069
5	2054.26	0.128	2048.86	1.851	2049.72	1.497	2049.25	1.802	2046.90	2.565	2047.67	2.384

	Pinecone										
K	QIPSO		TSMO		MTT		MBF		PSO		GA	
	$\mathcal{U}_{avg}$	σ	$\mathcal{U}_{avg}$	σ	$\mathcal{U}_{avg}$	σ	$\mathcal{U}_{avg}$	σ	$\mathcal{U}_{avg}$	σ	$\mathcal{U}_{avg}$	σ
2	752.67	0	752.67	0	752.67	0	752.67	0	752.67	0	752.67	0
3	945.81	0	945.18	0.054	945.19	0.068	945.57	0.091	944.20	0.099	940.50	0.114
4	1027.71	0.042	1025.09	0.595	1023.37	0.491	1025.95	0.502	1022.67	1.229	1020.64	1.253
5	1067.59	0.113	1063.63	1.067	1063.25	1.124	1063.16	1.250	1061.39	2.036	1059.58	2.195

	Car										
K	QIPSO		TSMO		MTT		MBF		PSO		GA	
	$\mathcal{U}_{avg}$	σ	$\mathcal{U}_{avg}$	σ	$\mathcal{U}_{avg}$	σ	$\mathcal{U}_{avg}$	σ	$\mathcal{U}_{avg}$	σ	$\mathcal{U}_{avg}$	σ
2	4260.51	0	4260.51	0	4260.51	0	4260.51	0	4260.51	0	4260.51	0
3	4666.19	0	4665.44	0.056	4665.88	0.064	4665.51	0.086	4664.86	0.120	4664.86	0.117
4	4900.70	0.043	4898.27	0.576	4898.43	0.516	4899.43	0.628	4894.30	2.190	4892.93	2.382
5	4995.87	0.127	4991.67	1.323	4992.17	1.534	4991.49	1.416	4989.75	2.387	4988.89	2.255

	Woman										
K	QIPSO		TSMO		MTT		MBF		PSO		GA	
	$\mathcal{U}_{avg}$	σ	$\mathcal{U}_{avg}$	σ	$\mathcal{U}_{avg}$	σ	$\mathcal{U}_{avg}$	σ	$\mathcal{U}_{avg}$	σ	$\mathcal{U}_{avg}$	σ
2	1298.23	0	1298.23	0	1298.23	0	1298.23	0	1298.23	0	1298.23	0
3	1529.55	0	1529.31	0.053	1529.34	0.065	1529.45	0.063	1528.52	0.110	1527.83	0.166
4	1605.40	0.046	1603.21	0.637	1603.20	0.661	1604.12	0.599	1602.19	1.184	1601.51	1.298
5	1664.89	0.142	1662.24	1.300	1661.93	1.624	1662.39	1.509	1660.11	1.970	1659.41	2.284

	House										
K	QIPSO		TSMO		MTT		MBF		PSO		GA	
	$\mathcal{U}_{avg}$	σ	$\mathcal{U}_{avg}$	σ	$\mathcal{U}_{avg}$	σ	$\mathcal{U}_{avg}$	σ	$\mathcal{U}_{avg}$	σ	$\mathcal{U}_{avg}$	σ
2	1653.59	0	1653.59	0	1653.59	0	1653.59	0	1653.59	0	1653.59	0
3	1962.94	0	1961.98	0.046	1961.54	0.086	1962.87	0.060	1962.00	0.130	1961.50	0.129
4	2005.36	0.043	2002.92	0.677	2002.96	0.686	2003.02	0.757	2000.66	1.061	1999.91	1.162
5	2041.39	0.125	2039.08	1.167	2038.99	1.254	2038.95	1.425	2037.20	2.506	2036.52	2.536

Table 5.20 Data sets for QIPSO, TSMO, MTT, MBF, PSO, and GA used in the Friedman test for K = 4 and 5, respectively. The rank for each method is shown in parentheses in this statistical test.

SN	Image	QIPSO	TSMO	MTT	MBF	PSO	GA
				For K = 4			
1	Image1	5850.622(1)	5847.876(3)	5846.982(4)	5849.289(2)	5845.543(5)	5844.000(6)
2	ImageN30C30	7027.449(1)	7026.770(3)	7026.163(4)	7027.123(2)	7025.629(6)	7025.679(5)
3	ImageN30C80	7510.425(1)	7509.129(2)	7508.101(4)	7508.986(3)	7507.566(5)	7506.266(6)
4	ImageN80C30	6822.516(1)	6820.365(4)	6820.461(3)	6821.443(2)	6819.732(5)	6818.339(6)
5	ImageN80C80	6186.481(1)	6184.204(3)	6183.273(4)	6185.431(2)	6182.320(5)	6180.015(6)
6	Lena	1913.793(1)	1911.289(2)	1910.581(4)	1911.009(3)	1907.880(5)	1906.109(6)
7	B2	3756.729(1)	3754.204(3)	3753.736(4)	3754.866(2)	3753.736(5)	3752.882(6)
8	Barbara	2046.106(1)	2044.691(3)	2044.156(4)	2045.745(2)	2041.976(6)	2042.560(5)
9	Boat	2631.609(1)	2629.086(3)	2628.075(4)	2629.425(2)	2627.264(5)	2626.013(6)
10	Cameraman	3585.419(1)	3583.167(4)	3583.185(3)	3584.124(2)	3582.596(5)	3581.875(6)
11	Jetplane	2007.816(1)	2005.240(3)	2005.993(4)	2005.506(2)	2004.863(6)	2004.924(5)
12	Pinecone	1027.767(1)	1025.907(3)	1024.120(4)	1026.640(2)	1023.858(5)	1022.806(6)
13	Car	4901.312(1)	4899.044(4)	4898.142(3)	4900.499(2)	4895.981(5)	4894.197(6)
14	Woman	1606.179(1)	1604.177(3)	1604.132(4)	1605.381(2)	1603.435(5)	1602.394(6)
15	House	2005.433(1)	2003.397(4)	2003.285(3)	2004.030(2)	2002.218(5)	2001.624(6)
	Average rank	1.00	3.13	3.73	2.13	5.20	5.80
				For K = 5			
SN	Image	QIPSO	TSMO	MTT	MBF	PSO	GA
1	Image1	5962.951(1)	5959.670(3)	5958.404(4)	5960.642(2)	5957.982(5)	5955.644(6)
2	ImageN30C30	7131.386(1)	7129.097(4)	7129.196(3)	7129.791(2)	7128.161(6)	7128.224(5)
3	ImageN30C80	7541.848(1)	7538.890(3)	7538.253(4)	7539.879(2)	7536.023(5)	7534.305(6)
4	ImageN80C30	6936.493(1)	6935.152(3)	6934.546(4.5)	6935.852(2)	6934.546(4.5)	6933.595(6)
5	ImageN80C80	6244.686(1)	6242.936(3)	6241.998(4)	6243.436(2)	6240.214(5)	6237.274(6)
6	Lena	1970.475(1)	1967.264(4)	1967.388(3)	1968.580(2)	1964.868(5)	1963.080(6)
7	B2	3839.299(1)	3835.943(3)	3835.405(4)	3836.464(2)	3834.437(5)	3832.857(6)
8	Barbara	2112.660(1)	2110.801(2)	2109.320(4)	2110.694(3)	2108.221(5)	2107.333(6)
9	Boat	2683.900(1)	2680.339(3)	2680.059(4)	2681.635(2)	2678.109(5)	2677.347(6)
10	Cameraman	3642.459(1)	3639.763(3)	3639.304(4)	3640.038(2)	3638.745(5)	3637.765(6)
11	Jetplane	2054.581(1)	2050.559(4)	2051.189(3)	2052.320(2)	2049.960(6)	2050.298(5)
12	Pinecone	1068.083(1)	1065.369(2)	1064.528(4)	1065.226(3)	1063.588(5)	1062.174(6)
13	Car	4996.493(1)	4993.242(4)	4994.185(2)	4993.500(3)	4992.253(5)	4991.586(6)
14	Woman	1666.051(1)	1663.744(3.5)	1663.744(3.5)	1664.123(2)	1662.866(5)	1661.793(6)
15	House	2042.110(1)	2040.490(4)	2040.652(3)	2041.571(2)	2039.855(5)	2038.839(6)
	Average rank	1.00	3.23	3.60	2.20	4.90	5.86

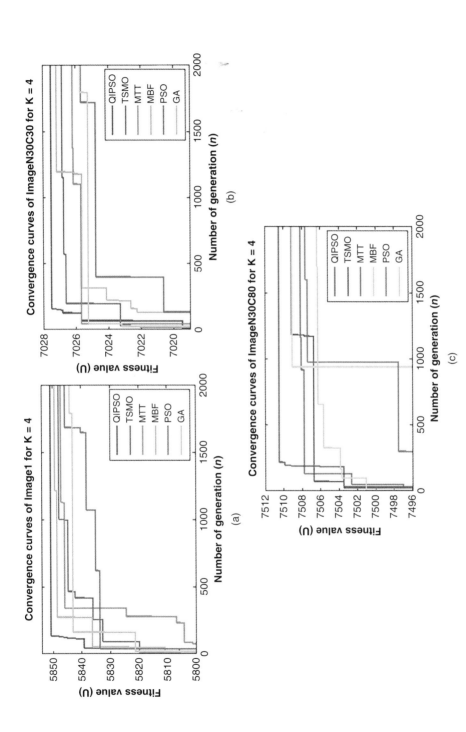

Figure 5.15 For $K = 4$, convergence curves of the comparable algorithms (a) for Image1, (b) for ImageN30C30, (c) for ImageN30C80, (d) for ImageN80C30, and (e) for ImageN80C80.

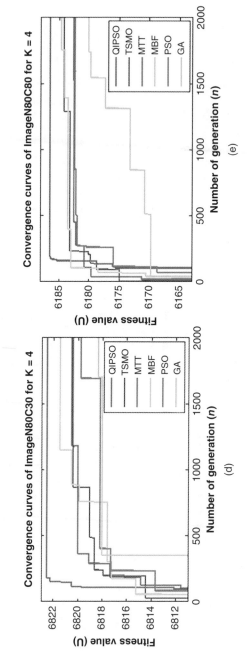

Figure 5.15 (Continued)

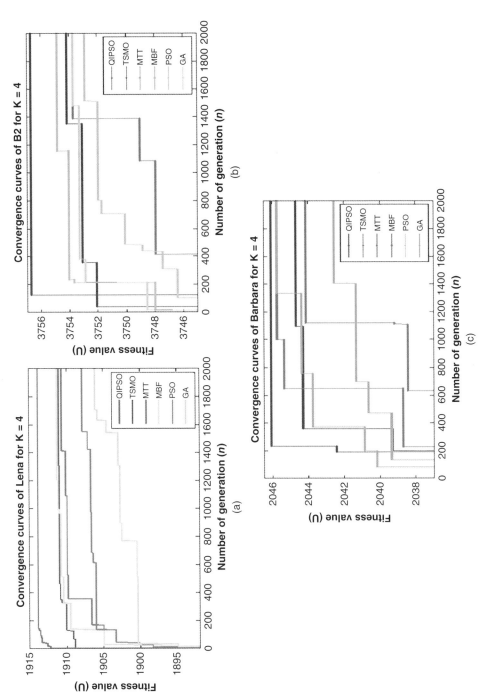

Figure 5.16 For $K = 4$, convergence curves of the comparable algorithms (a) for Lena, (b) for B2, (c) for Barbara, (d) for Boat, and (e) for Cameraman.

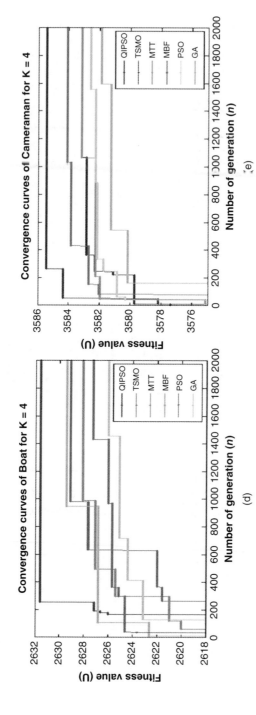

Figure 5.16 (*Continued*)

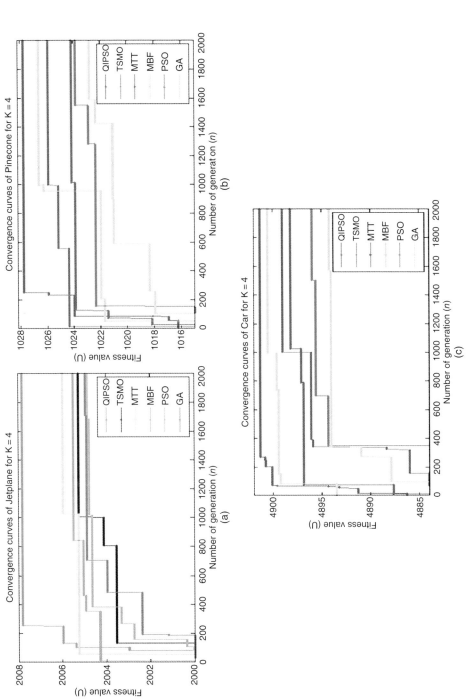

Figure 5.17 For $K = 4$, convergence curves of the comparable algorithms (a) for Jetplane, (b) for Pinecone, (c) for Car, (d) for Woman, and (e) for House.

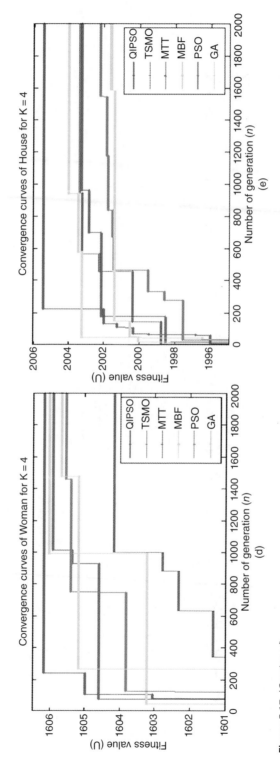

Figure 5.17 (*Continued*)

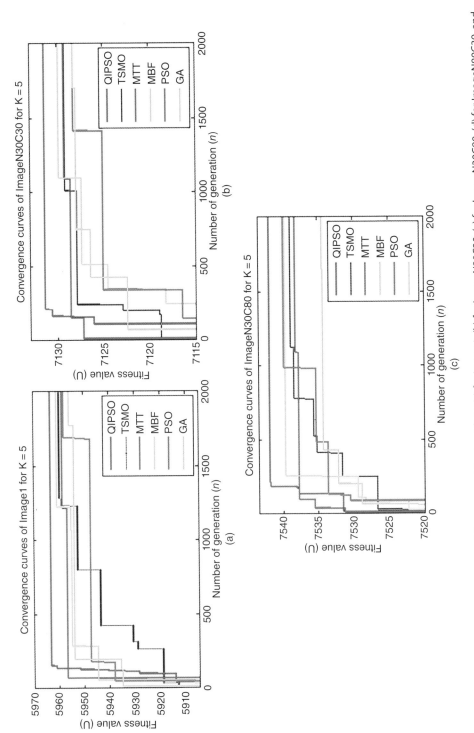

Figure 5.18 For $K = 5$, convergence curves of the comparable algorithms (a) for Image1, (b) for ImageN80C30, (c) for ImageN30C30, (d) for ImageN30C80, and (e) for ImageN80C80.

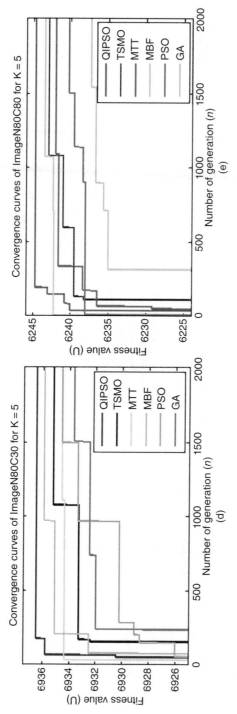

Figure 5.18 *(Continued)*

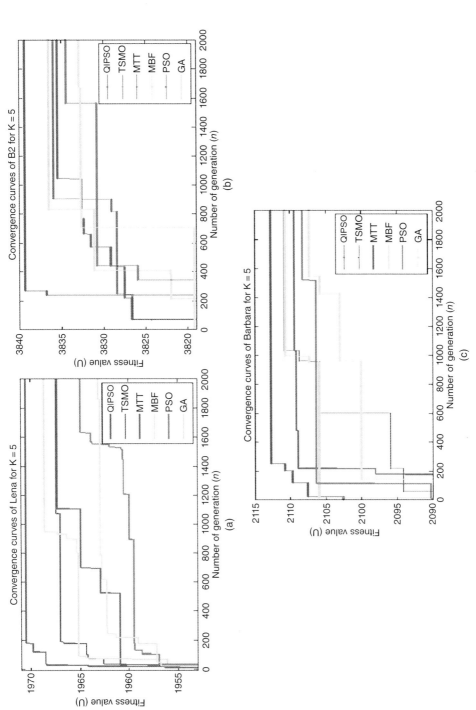

Figure 5.19 For $K = 5$, convergence curves of the comparable algorithms (a) for Lena, (b) for B2, (c) for Barbara, (d) for Boat, and (e) for Cameraman.

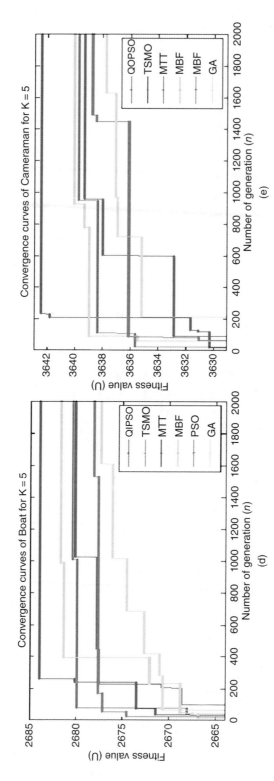

Figure 5.19 (*Continued*)

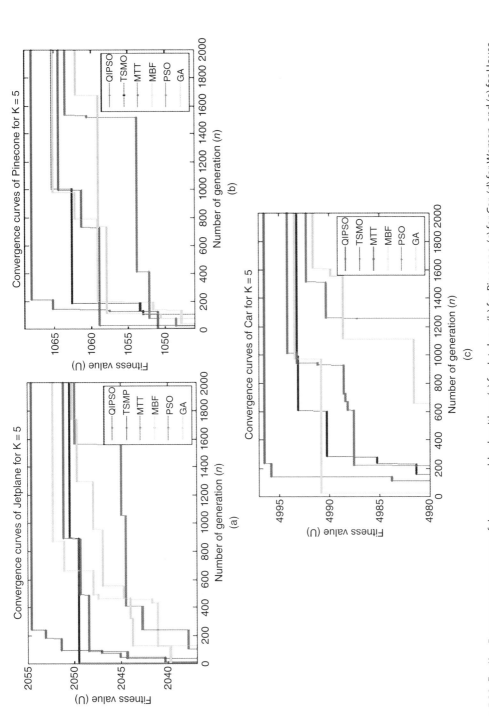

Figure 5.20 For $K = 5$, convergence curves of the comparable algorithms (a) for Jetplane, (b) for Pinecone, (c) for Car, (d) for Woman, and (e) for House.

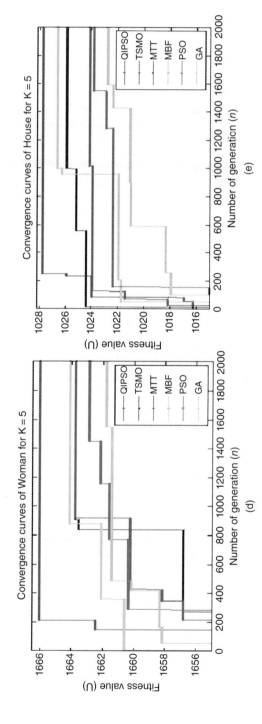

Figure 5.20 (Continued)

Table 5.21 Best results of QIACO, ACO, QISA, and SA for multi-level thresholding.

B2						
K	QIACO			ACO		
	θ	$\mathcal{F}_{best}$	t	θ	$\mathcal{F}_{best}$	t
2	128	3108.569	17.17	128	3108.569	45.10
3	93,175	3612.926	17.19	93,175	3612.926	52.13
4	75,139,185	3756.698	18.13	78,134,186	3754.372	54.51
5	73,138,181,218	3839.032	18.73	70,142,180,215	3836.817	59.75
K	QISA			SA		
	θ	$\mathcal{F}_{best}$	t	θ	$\mathcal{F}_{best}$	t
2	128	3108.569	19.91	128	3108.569	49.07
3	93,175	3612.926	32.85	93,175	3612.926	58.46
4	75,138,186	3756.749	44.64	76,136,188	3754.866	70.48
5	76,135,179,217	3838.779	55.68	67,137,179,216	3836.837	82.05
Barbara						
K	QIACO			ACO		
	θ	$\mathcal{F}_{best}$	t	θ	$\mathcal{F}_{best}$	t
2	115	1550.915	19.28	115	1550.915	44.10
3	79,135	1912.750	19.49	79,135	1912.750	46.13
4	73,121,165	2046.140	20.41	73,123,166	2045.669	50.51
5	60,96,132,172	2111.902	21.88	66,92,132,171	2109.688	52.75
K	QISA			SA		
	θ	$\mathcal{F}_{best}$	t	θ	$\mathcal{F}_{best}$	t
2	115	1550.915	19.85	115	1550.915	48.07
3	79,135	1912.750	33.18	79,135	1912.750	60.18
4	73,121,164	2046.106	43.22	74,120,164	2045.745	87.48
5	65,102,133,173	2111.287	56.96	68,105,139,174	2109.061	92.05

Table 5.22 Average fitness ($\mathcal{F}_{avg}$) and standard deviation (σ) values of QIACO, ACO, QISA, and SA for multi-level thresholding.

B2								
K	QIACO		ACO		QISA		SA	
	$\mathcal{F}_{avg}$	σ	$\mathcal{F}_{avg}$	σ	$\mathcal{F}_{avg}$	σ	$\mathcal{F}_{avg}$	σ
2	3108.56	0	3108.56	0	3108.56	0	3108.56	0
3	3612.91	0.035	3612.89	0.043	3612.92	0.032	3612.87	0.043
4	3756.65	0.039	3756.62	0.082	3756.63	0.079	3756.59	0.093
5	3839.89	0.144	3834.45	0.158	3839.15	0.139	3834.503	0.182
Barbara								
K	QIACO		ACO		QISA		SA	
	$\mathcal{F}_{avg}$	σ	$\mathcal{F}_{avg}$	σ	$\mathcal{F}_{avg}$	σ	$\mathcal{F}_{avg}$	σ
2	1550.91	0	1550.91	0	1550.91	0	1550.91	0
3	1912.71	0.046	1912.65	0.082	1912.69	0.056	1912.65	0.089
4	2046.04	0.069	2045.93	0.151	2045.95	0.096	2045.53	0.152
5	2111.36	0.149	2110.22	0.181	2111.38	0.111	2110.26	0.195

Table 5.23 Results of the two-tailed t-test.

B2						
K	1 & 2	1 & 3	1 & 4	2 & 4	3 & 4	3 & 2
2	NaN	NaN	NaN	NaN	NaN	NaN
3	0.0451	0.0423	0.0465	0.0476	0.0429	0.0411
4	0.0376	0.0370	0.0387	0.0439	0.0446	0.0474
5	0.0272	0.0391	0.0170	0.0343	0.0148	0.0140
Barbara						
K	1 & 2	1 & 3	1 & 4	2 & 4	3 & 4	3 & 2
2	NaN	NaN	NaN	NaN	NaN	NaN
3	0.0342	0.0398	0.0314	0.0452	0.0340	0.0324
4	0.0311	0.0343	0.0317	0.0414	0.0363	0.0321
5	0.0331	0.0326	0.0341	0.0412	0.0320	0.0334
1:→ QIACO		2:→ ACO		3:→ QISA		4:→ SA

Coding Examples

In this chapter, the functionality of the proposed techniques has been extended to the multi-level domain. Six novel quantum inspired meta-heuristics have been introduced for this purpose. Of these six techniques, the code for the proposed QIACO, QIDE, and QITS has been presented in Figures 5.21–5.25. Note that, the code for *insert* and *generate* functions of QIACO has already been presented in Figure 4.57 in Chapter 4. The code for common functions has been presented for one technique only.

```
Func¹

function [tr,fv]=QIACO(n,K)
%n-Number of elements,K-level of thresholding (e.g., for bi-level
                              image thresholding, K=2)
tic ; %Computational time calculation
t = 0 ;
ntrials = 0 ;
L=256;
His=insert(L);%Histogram of the gray scale image is calculated
M=Generate(L,n);%Function used for generating initial population
tao=GeneratePHERO(L,n);%Function used for generating pheromone
                              matrix
s=M;
maxf=0;
T=zeros(1,K);
rho=rand;

for k=1:50
   q0=0.5;
   taoc=max(tao');
   for i=1:n
   for j=1:L
      if(rand>q0)
         s(i,j)=taoc(i);
      else
         s(i,j)=rand;
      end
   end
   end
   [tr,fv,pos]=Fitness(L,s,His,K,n); %Function used for computing
                              fitness of individuals

   if(maxf<fv(1))
      maxf=fv(1);
      T=tr;
      t = toc;
   end
   [tao]=Update(tao,s,L,n,rho,pos); %Function used for updating
                              pheromone matrix

   ntrials = ntrials + 1 ;
   t = toc ;
end
end
```

```
Func²

function tao=GeneratePHERO(L,n)
for i=1:n
for j=1:L
   tao1(i,j)=rand;
end
end
tao=tao1;
end
```

```
Func³

function [tr,fv,pos]=Fitness(L,s,His,K,n)
[tr,fv,pos]=thcal(L,s,His,n,K);
end
```

```
Func⁴

function [tr,fv,pos]=thcal(L,s,His,n,K)
mx=0; j=1; p=1;
T=zeros(1,K);
P=partition(K,s,L); %Function used for partitioning the pixels
for i=1:n
trl,fvl]=thck(L,His,K,P); %Function used for threshold values
                              using Otsu's method

f(j)=fvl(1);
if(mx<f(j))
   mx=fvl(1);
   T=trl;
   p=j;
end
j=j+1;
end
tr=T;
fv=mx;
pos=p;
end
```

Figure 5.21 Sample code of Func¹, Func², Func³, and Func⁴ for QIACO.

```
Func⁵

function P=partition(K,s,L)
f=0;
st=zeros(1,K-1);
j=1;
while(1)
   p=randint(1,1,[1,L-1]);
   pm=rand;%probability factor
   if(pm>s(p)*s(p) && p-=f)
      st(j)=p;
      f=p;
      j=j+1;
   end
   if(j==K)
      break;
   end
end
P=st;
end
```

```
Func⁶

function [tao]=Update(tao,s,L,n,rho,pos)
f=zeros(n,L);
f(pos,:)=s(pos,:);
tao1=rho*tao+(1-rho)*f;
tao=tao1;
end
```

```
Func⁷

function [trl,fvl]=thck(L,His,K,P)
im=His;
w=zeros(1,K);
muk=zeros(1,K);
mu=0;
P=sort(P);
sigb=0;
   for j=1:K
      if(j==1)
         for i=1:P(j)
            w(j)=w(j)+im(i);
            muk(j)=muk(j)+i*im(i);
         end
      end
      if(j==K)
         for i=(P(j-1)+1):L
            w(j)=w(j)+im(i);
            muk(j)=muk(j)+i*im(i);
         end
      end
      if(j>1 && j<K)
         for i=(P(j-1)+1):P(j)
            w(j)=w(j)+im(i);
            muk(j)=muk(j)+i*im(i);
         end
      end
   end
```

```
for i=1:K
   if(w(i)~=0)
      muk(i)=muk(i)/w(i);
   else
      muk(i)=0;
   end
end

for i=1:K
mu=mu+w(i)*muk(i);
end

for i=1:K
   if(w(i)~=0)
      sigb=sigb+w(i)*(muk(i)-mu)*(muk(i)-mu);
   else
      sigb=0;
   end
end
fvl=sigb;
trl=P;
end
```

Figure 5.22 Sample code of Func⁵, Func⁶ and Func⁷ for QIACO.

```
Func¹
function [tr,fv]=QIDE(n,F,cr,K)
%n-Number of chromosomes, F-Mutation probability, Cr-Crossover
probability, K-level of thresholding
tic ; %Computational time calculation
t = 0 ;
ntrials = 0 ;
L=256;
His=insert(L); %Histogram of the gray scale image is calculated
M=Generate(L,n); %Function used for generating initial population
T=zeros(1,K-1);
s=M;
[tr,fv,fb]=Fitness(L,s,His,K,n); %Function used for computing fitness
of individuals
maxf=fb;
fv2=fv;
for i=1:50
BK=M;
MU=mute(M,F,L,n); %Function used for mutation operation
C=cross(MU,BK,cr,L,n); %Function used for crossover operation
s=C;
[tr,fv,fb]=Fitness(L,s,His,K,n);
S=selection(C,BK,n,L,fv,fv2); %Function used for selecting best
                             individuals
if(maxf<fb)
    maxf=fb;
    T=tr;
    t = toc;
end
fv2=fv;
M=S;
end
ntrials = ntrials + 1 ;
t = toc ;
t = t / ntrials
pfile(T,maxf,t,K);
end
```

```
Func²
function [tr,fv,fb]=thcal(L,s,His,n,K)
mx=0;
j=1;
T=zeros(1,K);
P=partition(K,s,L); %Function used for partitioning the pixels
for i=1:n
[tr1,fv1]=thck(L,His,K,P); %Function used for threshold values
                          using Otsu's method
f(j)=fv1(1);
if(mx<f(j))
    mx=fv1(1);
    T=tr1;
end
j=j+1;
end
tr=T;
fv=f;
fb=mx;
end
```

Figure 5.23 Sample code of Func¹ and Func² for QIDE.

```
Func³
function C=cross(MU,BK,cr,L,n)
for i=1:n
for j=1:L
    ct=randi(L);
    rd=rand;
    if(j==ct || rd<=cr)
        MU(i,j)=MU(i,j);
    elseif(j~=ct || rd>cr)
        MU(i,j)=BK(i,j);
    end
end
end
C=MU;
end


Func⁴
function MU=mute(M,F,L,n)
bk=M;
for i=1:n
while 1
 r1=randi(n);
 r2=randi(n);
 r3=randi(n);
 if i~=r1 && i~=r2 && i~=r3 && r1~=r2 && r1~=r3 && r2~=r3
 break
 end
end
for j=1:L
M(i,j)=bk(r1,j)+F*(bk(r2,j)-bk(r3,j));
```

```
if(M(i,j)>1)
    M(i,j)=rand;
end
end
end
MU=M;
end


Func⁵
function S=selection(C,BK,n,L,fv,fv2)
s1=zeros(n,L);
for i=1:n
if(fv(i)>fv2(i))
    s1(i,:)=C(i,:);
else
    s1(i,:)=BK(i,:);
end
end
S=s1;
end
```

Figure 5.24 Sample code of Func³, Func⁴, and Func⁵ for QIDE.

```
Func¹

function [tr,fv]=QITS(K) %K-level of thresholding
tic ; %Computational time calculation
t = 0 ;
ntrials = 0 ;
L=256;
pop=[];
His=insert(L);%Histogram of the gray scale image is calculated
M=Generate(L);%Function used for generating initial population
d=40;
maxf=0;
vbest1=zeros(1,K);
pop=zeros(d,K);
s=M;
[tr,fv]=Fitness(L,s,His,K); %Function used for computing
                               fitness of individual

vb=tr;
fv2=fv(1);
c=1;
for i=1:50
    for j=1:d
        b=Perturbation(s,L); %Function used for modifying
                                characteristics of  individual
        s=b;
        [tr,fv]=Fitness(L,s,His,K);
        if(j==1)
            maxf=fv(1);
            vbest=tr;
        end
        for k=1:d
            for u=1:(K-1)
                if(pop(k,u)~=tr(1,u))
                    p=0;
                    break;
                else
                    p=1;
                end
            end
        end
```

```
        if((p==0) && (maxf<fv(1)))
            maxf=fv(1);
            vbest1=tr;
            for u=1:(K-1)
                pop(c,u)=tr(1,u);
            end
            c=c+1;
        end
    end
if(maxf<fv2(1))
    maxf=fv2(1);
    vbest=vb;
else
    vbest=vbest1;
end
    fv2=maxf;
    vb=vbest;
end
    T=vbest
    ntrials = ntrials + 1 ;
    t = toc ;
end

Func²

function M=Generate (L)
for i=1:L
    a(i)=rand;
end
b=sqrt(1-a.^2);
M=b;
end

Func³

function [tr,fv]=Fitness(L,s,His,K)
P=partition(K,s,L);
```

```
[tr1,fv1]=thck(L,His,K,P);
tr=tr1;
fv=fv1;
end

Func⁴

function b=Perturbation(s,L)
r1=randi(L);
r2=randi(L);
if(r1>r2)
    for i=r2:r1
        s(i)=rand;
        s(i)=sqrt(1-s(i).^2);
    end
end
if(r2>r1)
    for i=r1:r2
        s(i)=rand;
        s(i)=sqrt(1-s(i).^2);
    end
end
b=s;
end
```

Figure 5.25 Sample code of Func¹, Func², Func³, and Func⁴ for QITS.

6

Quantum Behaved Meta-Heuristics for True Color Multi-Level Image Thresholding

6.1   Introduction

In Chapter 4, two novel quantum inspired techniques with regard to binary image thresholding, were presented [28, 61, 64, 68]. Chapter 5 concentrated on functional modification of the quantum inspired meta-heuristic techniques as an attempt to extend them to the multi-level and gray-scale domains. Thereby, six novel quantum inspired meta-heuristic techniques are introduced in this chapter [63, 67].

Parallel extensions to these quantum inspired classical algorithms are introduced in this chapter. The introduction of this approach uses the thresholding of color image information. Application of these versions of quantum inspired techniques is exhibited through the thresholding of multi-level and color images.

In line with the objectives of Chapters 4 and 5, six new quantum inspired meta-heuristic algorithms for multi-level color image thresholding, are introduced in this chapter [62, 65, 70, 72]. The proposed algorithms are capable of handling pure color image information by adopting an approach of processing the component level information of the test image. Apposite modifications in the framework of the proposed approaches, as mentioned in Chapter 5, are, however, required to enable the frameworks of the proposed systems to act in response to true color image information. The proposed algorithms are called the Quantum Inspired Particle Swarm Optimization, the Quantum Inspired Ant Colony Optimization, the Quantum Inspired Differential Evolution, the Quantum Inspired Genetic Algorithm, the Quantum Inspired Simulated Annealing, and the Quantum Inspired Tabu Search.

As a sequel to the proposed algorithms, separate experiments have also been carried out with the Backtracking search optimization algorithm (BSA) [45], the Composite DE (CoDE) method [270], the classical genetic algorithm, the classical simulated annealing, the classical differential evolution, and finally the classical particle swarm optimization. Using multiple approaches, the workability and effectiveness of the proposed algorithms are validated on 24 true color images with a multi-level flavor visually and quantitatively. Thereafter, the accuracy and stability of the proposed algorithms are confirmed by computing the mean and standard deviation of the fitness values of each algorithm. Moreover, the quality of thresholding for each algorithm is quantified with respect to the values of peak signal-to-noise ratio (PSNR) at different levels. Afterwards, the statistical superiority of the proposed algorithms is exhibited by conducting the statistical test, referred to as the Friedman test among them. The proposed Quantum Inspired

Quantum Inspired Meta-heuristics for Image Analysis, First Edition.
Sandip Dey, Siddhartha Bhattacharyya, and Ujjwal Maulik.

Ant Colony Optimization is found to outperform the others. Implementation results are presented with an adequate attempt at parameter tuning of the participating algorithms.

The chapter is presented in the following order. Section 6.2 presents a brief summary of the color images. In Sections 6.3–6.8, six quantum inspired meta-heuristic techniques for multi-level colour image thresholding are introduced and discussed. Experimental results comparing the corresponding classical algorithms and two other comparable algorithms for thresholding of several real-life true color images, are reported in Section 6.9. A brief concluding discussion is given in Section 6.10. At the end of this chapter (Section 6.11), the chapter summary and a set of exercise questions of different types are included. Finally, a few sample codings of various techniques are given.

6.2 Background

Color images are usually the natural expansions of bi-level and multi-level gray-scale images. A pure color image demonstrates information in three prime color components (red, green and blue) and their possible admixtures. These primary color components as well as their admixtures appear either in two intensity values, namely, 0 and 255. Thus, from this perspective, a pure color image is able to be viewed as an extended version of a binary image. With reference to true color images, the color spectrum is formed using the intensity information of all the potential 256 levels (from 0 to 255) of these primary color components and all their possible combinations of admixtures. Thus, a true color image is basically characterized by the extension of a multi-level image exhibiting intensity levels from 0 to 255. Typical applications of color image segmentation include the marker-based watershed algorithm-based remote sensing image segmentation approach [245], multimedia-based segmentation approach [131], multispectral data management systems [269], GIS [163], and the contrast-based approach [43], to name but a few.

6.3 Quantum Inspired Ant Colony Optimization

In this section, the proposed Quantum Inspired Ant Colony Optimization (QIA-COMLTCI) is presented in detail. This technique uses Li's method [158] as an objective function to determine the optimal threshold values. The outline of the proposed QIACOMLTCI is illustrated in Algorithm 8.

Initially, we randomly select pixel intensity values from the test image to form starting population (P) of S strings. As the proposed algorithm deals with color images, we choose the length of each particle in P as $\mathcal{L} = 3 \times \lfloor \sqrt{L} \rfloor$, $L = 255$. The concept of qubit is used to assign a real random value between (0,1) for pixel encoding in P, which forms P'. Thereafter, a rotation gate is used to update P' for fast convergence. In the next step, P' goes through another property of QC, called quantum orthogonality. This produces P''. Then, we use a probability criterion as $(\beta_i)^2 < random(0, 1)$ to segregate the pixel intensity values of each primary color component into a number of classes where i denotes the position of qubit in P''. After fitness computation, the best string, its fitness value, and the threshold values for each color component are recorded. Initially, a pheromone matrix (τ_{PH}) of the identical dimension of P is constructed. This matrix is used at each successive generation to update the population. At each generation, τ_{PH} is updated by

$$\tau_{ij} = \rho\tau_{ij} + (1 - v)bs \tag{6.1}$$

where bs represents the best string in population.

The proposed algorithm is executed for $\mathcal{G}$ number of generations. The optimum threshold values of the primary color components are given in $\theta_R = T_R$, $\theta_G = T_G$ and $\theta_B = T_B$, respectively.

6.3.1 Complexity Analysis

The worst case time complexity of the proposed algorithm is described through the following steps.

1. The time complexity to generate the initial population (P) becomes $O(S \times \mathcal{L})$, where S represents the size of each string.
2. To perform the pixel encoding phase, the time complexity turns into $O(S \times \mathcal{L})$.
3. First, qubits are updated using the quantum rotation gate and thereafter, quantum orthogonality is performed for each member in the population. For both cases, the time complexity becomes $O(S \times \mathcal{L})$.
4. The desired number of threshold values is found from P^*. The time complexity to produce P^* turns out to be $O(S \times \mathcal{L})$.
5. The time complexity for fitness computation turns into $O(S \times C)$. Note that C represents the number of classes to be computed.
6. Like P, the time complexity to construct the pheromone matrix (τ_{PH}), becomes $O(S \times \mathcal{L})$.
7. P'' is updated at each generation using τ_{PH}. To update P'', the time complexity becomes $O(S \times \mathcal{L})$.
8. The time complexity analysis of steps 23 and 24 have already been described in the above steps.
9. Again, τ_{PH} is updated at each generation. For that, the time complexity turns out to be $O(S \times \mathcal{L})$.
10. The algorithm is executed for $\mathcal{G}$ number of generations. Hence, the time complexity to execute all these steps turns into $O(S \times \mathcal{L} \times \mathcal{G})$.

Therefore, keeping in mind the above discussion, the overall worst case time complexity of QIACOMLTCI happens to be $O(S \times \mathcal{L} \times \mathcal{G})$.

6.4 Quantum Inspired Differential Evolution

In this section, a variant version of quantum inspired meta-heuristic technique (QIDEMLTCI), called Quantum Inspired Differential Evolution is introduced. Like QIACOMLTCI, the proposed approach also exploits the features of QC in the framework of a popular meta-heuristic algorithm, called Differential Evolution. We present the proposed QIDEMLTCI in Algorithm 9. The first eight steps of the proposed approach have already been illustrated in Section 6.3. The remaining phases of the proposed approach are discussed in this subsection.

Algorithm 8: Steps of QIACOMLTCIor multi-level color image thresholding

---

Input: Number of generation: $\mathcal{G}$
Population size: S
Persistence of trials: ν
Priori defined number: ι
Number of thresholds: K
Output: The optimal threshold values of three primary components: θ_R, θ_G and θ_B

1: Initialize the population (P) of S strings for primary components of the color image (red, green and blue) with a randomly selected gray-level pixel intensity value from the given image. The length of each string in P is taken as $\mathcal{L} = 3 \times \lfloor \sqrt{L} \rfloor$. Note that the length of each primary color component is taken as L, $L = 255$ (the maximum possible pixel intensity value of any individual primary color component).

2: Using the qubit concept in QC, a real random number between (0,1) is assigned to each pixel encoded string in P to generate P'.

3: For fast convergence, the quantum rotation gate is employed to update P' using Eq. (1.10).

4: The property of QC, called *quantum orthogonality*, is performed in P', which produces P''.

5: For each primary color component, the required number of classes (C) is formed from the corresponding position of each particle in P. Note that the selection of the threshold values in (C) is performed on the basis of $(\beta_i)^2 < random(0,1)$ in P'' where i represents the position of qubit in P''. It produces P^*.

6: Use P^* to compute the fitness of each string using Eq. (2.49) as the objective function.

7: Record the best string in $bs \in P''$. For all color components, the threshold values and their corresponding fitness values are recorded in T_R, T_G and $T_B \in P''$ and in X_R, X_G and X_B, respectively.

8: Set $X = X_R + X_G + X_B$.

9: A pheromone matrix, τ_{PH} having identical dimension of P is produced.

10: $i = 1$.

11: **while** $i \leq \mathcal{G}$ **do**

12: $\quad$ $i = i + 1$.

13: $\quad$ **for** all $k \in P''$ **do**

14: $\quad\quad$ **for** all j^{th} position in k **do**

15: $\quad\quad\quad$ **if** $(random(0,1) > \iota)$ **then**

16: $\quad\quad\quad\quad$ Set $P''_{ij} = \arg\max \tau_{ij}$.

17: $\quad\quad\quad$ **else**

18: $\quad\quad\quad\quad$ Set $P''_{ij} = random(0,1)$.

19: $\quad\quad\quad$ **end if**

20: $\quad\quad$ **end for**

21: $\quad$ **end for**

22: **end while**

23: Repeat steps 3–5 to update P' and P^*.

24: Repeat step 6 to calculate the fitness of each string using P^*.

25: Repeat step 7 to update the best string along with the threshold values and fitness for each basic color component.

26: Repeat step 8.

27: **for** all $k \in P''$ **do**

28: $\quad$ **for** all j^{th} position in k **do**

29: $\quad\quad$ Update pheromone matrix, $\tau_{PH}(k, j)$.

30: $\quad$ **end for**

31: **end for**

32: Report the optimal threshold values, $\theta_R = T_R$, $\theta_G = T_G$ and $\theta_B = T_R$, respectively.

Algorithm 9:   Steps of QIDEMLTCIor multi-level color image thresholding

---

Input: Number of generation: $\mathcal{G}$
Population size: S
Scaling factor: ζ
Crossover probability: φ
Number of thresholds: K

Output: The optimal threshold values of three primary components: θ_R, θ_G and θ_B

1: The first eight steps of QIDEMLTCIare identical to the steps (steps 1–8) in Algorithm 8.
2: $B = P''$.
3: $i = 1$.
4: **while** $i \leq \mathcal{G}$ **do**
5:      $i = i + 1$.
6:      **for** all $k \in P''$ **do**
7:          **for** all j^{th} position in k **do**
8:              Select three positive integers (say, v_1, v_2 and v_3) from $[1, S]$ where
9:              $v_1 \neq v_2 \neq v_3 \neq k$.
10:             Set $P''(k) = B(v_1) + \zeta * (B(v_2) - B(v_3))$.
11:         **end for**
12:     **end for**
13:     **for** all $k \in P''$ **do**
14:         **for** all j^{th} position in k **do**
15:             Select a positive integer (say, ε) from $[1, S]$.
16:             **if** $(j \neq \varepsilon$ and $random(0, 1) > \varphi)$ **then**
17:                 $P''_{kj} = B_{kj}$.
18:             **end if**
19:         **end for**
20:     **end for**
21:     Repeat steps 3–5 of Algorithm 8 to update P' and P^*.
22:     Repeat step 6 of Algorithm 8 to compute the fitness value of each vector in P^*.
23:     Repeat step 7 of Algorithm 8 to find the best vector, its corresponding thresholds and fitness values for each primary component.
24:     Repeat step 8 of Algorithm 8.
25: **end while**
26: Report the optimal threshold values, $\theta_R = T_R$, $\theta_G = T_G$ and $\theta_B = T_B$, respectively.

Afterward, mutation and crossover operators are successively applied to get the population diversity. Three vectors, $B(v_1)$, $B(v_2)$, and $B(v_3)$, are randomly selected from P'' to accomplish quantum mutation. A new mutant vector is formed by computing simple arithmetic. At first, ζ (scaling factor) is multiplied with the weighted difference between the second and third vectors and then the resultant vector is added to the first vector. Here, ζ is selected from $[0, 1]$. Similarly, a positive integer $\varepsilon \in [1, S]$ is randomly selected for each vector location in P'' to perform quantum crossover. This particular location (say, j) undergoes crossover based on the criteria given by $j \neq \varepsilon$ and $random(0, 1) > \varphi$. Here, φ is called the crossover probability, whose value is selected from $[0.5, 1]$. Thereafter, the algorithm follows some certain steps to create the population for the next generation. The proposed method is run for $\mathcal{G}$ number of generations. The optimum threshold values for three primary color components are reported in $\theta_R = T_R$, $\theta_G = T_G$, and $\theta_B = T_B$, respectively.

6.4.1 Complexity Analysis

The worst case time complexity analysis of QIDEMLTCI is described in this subsection. The time complexity for the first eight steps of QIACOMLTCI (described in Section 6.3.1) is identical to QIDEMLTCI. The remaining parts of the time complexity of QIDEMLTCI are discussed below.

1. For each quantum mutation, quantum crossover and selection mechanism, the time complexity becomes $O(S \times \mathcal{L})$.
2. The algorithm runs for $\mathcal{G}$ number of generations. To run these steps, the time complexity turns into $O(S \times \mathcal{L} \times \mathcal{G})$.

Therefore, summarizing the above discussion, the overall time complexity (worst case) of QIDEMLTCI happens to be $O(S \times \mathcal{L} \times \mathcal{G})$.

6.5 Quantum Inspired Particle Swarm Optimization

In this section, the features of QC and PSO have been used together to construct a novel quantum inspired meta-heuristic technique (QIPSOMLTCI). We call the proposed algorithm the Quantum Inspired Particle Swarm Optimization. The proposed algorithm is presented in Algorithm 10. The first eight phases of QIACOMLTCI and QIPSOMLTCI are identical to each other, which have already been described in Section 6.3. The remaining phases of the proposed algorithm are briefly discussed below.

The population of particles is updated at each generation. The velocity and location of each particle are updated using the formulae as given in Eq. (4.11). QIPSOMLTCI has been executed for $\mathcal{G}$ number of generations. The primary color component-wise optimum threshold values of the test image is reported in $\theta_R = T_R$, $\theta_G = T_G$, and $\theta_B = T_B$, respectively.

6.5.1 Complexity Analysis

The worst case time complexity of the proposed algorithm is described in the following steps. Like QIDEMLTCI, the time complexity for the preceding eight steps of QIACOMLTCI (described in Section 6.3.1) and QIPSOMLTCI is identical. The time complexity analysis for the remaining steps of QIPSOMLTCI is presented in this section.

1. The population is updated at each generation. Each particle changes its position through steps 6 and 7. To perform each step, the time complexity turns into $O(S \times \mathcal{L})$.
2. The time complexity of steps 8 and 9 has already been described in the relevant part of QIACOMLTCI (described in Section 6.3.1).
3. This proposed algorithm is executed for $\mathcal{G}$ number of generations. So, the time complexity to run all the steps in QIPSOMLTCI becomes $O(S \times \mathcal{L} \times \mathcal{G})$.

Therefore, considering the above discussion, the overall worst case time complexity of QIPSOMLTCI happens to be $O(S \times \mathcal{L} \times \mathcal{G})$.

Algorithm 10: Steps of QIPSOMLTCI for multi-level color image thresholding

---

Input: Number of generation: $\mathcal{G}$
      Population size: S
      Acceleration coefficients: ξ_a and ξ_b
      Inertia weight: ω
      Number of thresholds: K
Output: The optimal threshold values of three primary components: θ_R, θ_G and θ_B

1: Steps 1–8 of QIPSOMLTCI are identical to the steps (steps 1–8) in Algorithm 8.
2: $i = 1$.
3: **while** $i \leq \mathcal{G}$ **do**
4:    $i = i + 1$.
5:   **for** all $k \in P''$ **do**
6:      Compute particle velocity using Eq. (4.11).
7:      Update particle position using Eq. (4.11).
8:      Repeat steps 3–5 of Algorithm 8 to update P' and P^*.
9:      Repeat step 6 of Algorithm 8 to compute the fitness value of each particle in P^*.
10:     Repeat step 8.
11:     Repeat step 7 to find the best particle. For all color components, the threshold values and their corresponding fitness values are also recorded.
12:   **end for**
13: **end while**
14: Report the optimal threshold values, $\theta_R = T_R$, $\theta_G = T_G$ and $\theta_B = T_B$, respectively.

6.6 Quantum Inspired Genetic Algorithm

A novel Quantum Inspired Genetic Algorithm (QIGAMLTCI) is presented in this section. The basic steps of the proposed approach is illustrated in Algorithm 11. The working steps of the proposed QIGAMLTCI are described below.

Initially, we create a population, *POP* comprising the randomly chosen pixel intensity values from the input image. *POP* contains three distinct sets of intensity values, each of size $\mathcal{B}$, for three different primary components of the color image. The length of each chromosome in each set in *POP* is considered as $\mathcal{L} = \sqrt{L}$, where L denotes the maximum pixel intensity value among all primary components in the color image. Afterwards, the pixel intensity values in *POP* are encoded with a random real number between (0,1) using the theory of qubits to create *POP′*. After pixel encoding with a real number, the quantum rotation gate is employed to update *POP′*. To find the basic feature in QC, the elements in *POP′* undergo quantum orthogonality, which produces *POP″*. A probability constraint is imposed to find a predefined number of thresholds as the pixel intensity value to produce *POP**. Subsequently, *POP** is used to evaluate the fitness of each participating chromosome using Eqs. (2.50) and (2.56) for Kapur's method [140] and Huang's method [130], respectively. QIGAMLTCI iteratively and successively employs selection, crossover, and mutation, three basic genetic operators for population diversity and better fitness. The selected values of (P_c) and (P_μ) are 0.90 and 0.01, respectively.

Algorithm 11: Steps of QIGAMLTCIor multi-level color image thresholding

---

Input: Number of generations: G

Population size for each primary color component: B

Crossover probability: P_c

Mutation probability: P_μ

Number of thresholds: K

Output: The optimal threshold values of three primary color components: θ_R, θ_G and θ_B

1: Select the pixel values randomly to generate a population of initial chromosomes for each primary component (red, green and blue) of the color image, POP, where POP includes three different sets of B number of chromosomes for three basic components. It should be noted that the length of each chromosome is $\mathcal{L} = \sqrt{L}$, where L denotes the maximum pixel intensity value among all color components in the color image.

2: Use the theory of qubits to assign a positive real random number < 1 to each pixel encoded in POP. Let it create POP'.

3: Use the quantum rotation gate to update POP' using Eq. (1.10).

4: POP' undergoes a *quantum orthogonality* to create POP''.

5: Find K number of thresholds as pixel intensity values from each of the different sets of chromosomes, for each individual color component in POP, satisfying the corresponding value in $POP'' > random(0, 1)$. Let it produce POP^*.

6: Evaluate the fitness of each chromosome in POP^* using the objective function.

7: Store the best chromosome for the red component in $b_R \in POP''$, for the green component in $b_G \in POP''$ and for the blue component in $b_B \in POP''$ along with their corresponding threshold values in T_R, T_G and $T_B \in POP^*$. Finally, the fitness values of the best chromosome of each different color component are saved in F_R, F_G and F_B, respectively.

8: Record the summation of F_R, F_G and F_B in F.

9: The tournament selection strategy is applied to refill the chromosome pool.

10: **For** $i = 1$ to G **do**

11:   **for** each individual set of color component **do**

12:     **repeat**

13:       Randomly select two different chromosomes, a and b from the respective chromosome pool.

14:       **if** $random(0, 1) < P_c$ **then**

15:         Randomly select one crossover point, $p \in [2, \mathcal{L} - 1]$.

16:       **end if**

17:       a and b are crossed over at p.

18:     **until** B number of chromosomes are picked up.

19:   **end for**

20:   **for** all $c \in POP''$ **do**

21:     **for** each position in c **do**

22:       **if** $random(0, 1) < P_\mu$ **then**

23:         Alter the number in the particular location with another randomly generated number between $(0, 1)$.

24:       **end if**

25:     **end for**

26:   **end for**

27:   Repeat steps 3 to 6.

28:   Record the best chromosome and its threshold value for the red component in $b_{R1} \in POP''$, $T_{R1} \in POP^*$, for the green component in $b_{G1} \in POP''$, $T_{G1} \in POP^*$ and finally, for the blue component in $b_{B1} \in POP''$, $T_{B1} \in POP^*$, respectively. The fitness values of the above best chromosomes are also recorded in F_{R1}, F_{G1} and F_{B1}, respectively.

29:   Record the addition of F_{R1}, F_{G1} and F_{B1} in F_L.

30:   Compare fitness values of F and F_L and record the best of them in F. The best chromosomes of the red, green and blue components and their threshold values are stored in $b_R, b_G, b_B \in POP''$ and T_R, T_G and $T_B \in POP^*$ respectively.

31: **end for**

32: The optimum threshold values for three different color components are reported in $\theta_R = T_R, \theta_G = T_G$ and $\theta_B = T_B$, respectively.

6.6.1 Complexity Analysis

To analyze the worst case time complexity of the proposed QIGAMLTCI algorithm, the following steps are described:

1. In QIGAMLTCI, the population comprises three distinct sets of $\mathcal{B}$ number of chromosomes for each primary color component. Therefore, the total population size in *POP* becomes $3 \times \mathcal{B} = \mathcal{U}$ (say). To produce the initial population, the time complexity turns into $O(\mathcal{U} \times \mathcal{L})$. Note that each participating chromosome is of length $\mathcal{L} = \sqrt{L}$, where L is the maximum pixel intensity among each of primary component.

2. In the second step, a real number is assigned to each pixel encoded as the population of chromosomes. To evaluate this encoding process, the time complexity turns into $O(\mathcal{U} \times \mathcal{L})$.

3. To update *POP′* using the quantum rotation gate, the complexity becomes $O(\mathcal{U} \times \mathcal{L})$.

4. To perform quantum orthogonality, the time complexity becomes $O(\mathcal{U} \times \mathcal{L})$.

5. The time complexity to produce *POP** turns into $O(\mathcal{U} \times \mathcal{L})$.

6. For fitness computation in QIGAMLTCI, time complexity turns out to be $O(\mathcal{U} \times K)$.

7. For selection step, time complexity becomes $O(\mathcal{U})$.

8. To perform crossover and mutation in QIGAMLTCI, the time complexity is $O(\mathcal{U} \times \mathcal{L})$.

9. Hence, to run QIGAMLTCI for a predefined number of generations, the time complexity becomes $O(\mathcal{U} \times \mathcal{L} \times \mathcal{G})$, where $\mathcal{G}$ represents number of generations.

Therefore, summarizing the above facts, the overall worst case time complexity for the proposed QIGAMLTCI happens to be $O(\mathcal{U} \times \mathcal{L} \times \mathcal{G})$.

6.7 Quantum Inspired Simulated Annealing

A new Quantum Inspired Simulated Annealing for multi-level thresholding for color image (QISAMLTCI) is presented in this section. The outline of the proposed Quantum Inspired Simulated Annealing is shown in Algorithm 12. The details of QISAMLTCI are illustrated below.

Initially, a population, *P*, of three configurations (one for each primary component of the input image), is generated by randomly selected pixel intensity values from the image. The length of each configuration is considered as $\mathcal{L} = \sqrt{L}$, where L is the maximum pixel intensity value measured from the primary components in the input image. For encoding purposes, a real value between (0,1) is randomly picked and assigned to each pixel encoded in *P* using the principle of qubits, which forms *P′*. Afterwards, two quantum features are applied in *P′* successively. At first, *P′* is updated using the quantum rotation gate, then it passes through quantum orthogonality, which produces *P″*. Applying a defined probability criteria, QISAMLTCI determines a predefined number of thresholds as the pixel intensity value. QISAMLTCI starts executing with a predefined high temperature, $\mathcal{T}_{start}$, and becomes cooled iteratively using the reduction factor, ϱ. At each temperature, QISAMLTCI is run for $\imath$ number of iterations. QISAMLTCI stops executing when the temperature crosses the predefined value, $\mathcal{T}_{final}$. At each iteration, the participating configuration, $\mathcal{U}$, is randomly perturbed at different points, which produces W. Later, steps 2–5 are repeated, which results in W^*. W is indubitably accepted as the new configuration for the next iteration if $V(W^*) - V(X) > 0$, where X is created by using steps 2–5. W may also be accepted for the next iteration with a probability (usually it uses Boltzmann's distribution) given by $\exp(-(V(X) - V(W^*)))/\mathcal{T}$. The selected values of the initial temperature ($\mathcal{T}_{start}$) and the final temperature ($\mathcal{T}_{final}$) are 1000 and 0.5, respectively. The reduction factor (ϱ) was chosen as 0.95. K represents the number of thresholds to be determined. The proposed QISAMLTCI reports the optimized threshold values for three primary color components of the color image in θ_R, θ_G and θ_B, respectively.

6.7.1 Complexity Analysis

The worst case time complexity of the proposed QISAMLTCI algorithm is analyzed through the following steps. Unlike the proposed QIGAMLTCI, this technique deals with a single configuration for each primary color component. Therefore, the effective population size in P is 3. Hence, to generate an initial population, the time complexity becomes $O(\mathcal{L})$, where L has the same specification as QIGAMLTCI. To execute steps 2–5, QISAMLTCI follows the same procedure as QIGAMLTCI. The time complexity for performing these steps turns out to be $O(\mathcal{L})$.

In addition to the steps described above, the following steps should also be addressed to evaluate the worst case time complexity of the proposed QISAMLTCI.

1. The time complexity for fitness computation turns into $O(K)$.
2. Suppose the inner loop is run for ι number of times. To determine the time complexity at each such run, the description is basically confined to the following steps.
   - To disturb the configuration, the time complexity leads to $O(\mathcal{L})$.
   - To execute steps 2–5, the time complexity becomes $O(\mathcal{L})$ as described earlier.
   - Again, for the fitness computation of the new configuration, the time complexity leads to $O(K)$.
3. Let us presume that the outer loop is run for M_i times. Therefore, the time complexity to run steps 10–29 in the proposed algorithm becomes $\iota \times M_i$.

Therefore, analyzing the steps discussed above, the overall worst case time complexity of QISAMLTCI turns out to be $O(\mathcal{L} \times \iota \times M_i)$.

6.8 Quantum Inspired Tabu Search

In this section, a new Quantum Inspired Tabu Search Algorithm for Multi-Level Thresholding for True Color Images (QISAMLTCI) is presented. The features of QC are used in collaboration with the conventional meta-heuristic algorithm, Tabu Search to form the aforesaid algorithm. The details of the proposed Quantum Inspired Tabu Search are outlined in Algorithm 13. Unlike QISAMLTCI, we choose the string length for this proposed approach as $\mathcal{L} = \sqrt{L_l \times L_w}$, where L_l and L_w represent the length and height of the input image. The working procedure of the first seven steps of the proposed QITSMLTCI have already been discussed in the previous subsection. We discuss the remaining parts of the proposed algorithm in this subsection. A tabu memory denoted by *mem*, is introduced and is set a null value before starting execution. At each generation, it explores the neighbors of the best string for each primary color component. The best string of

---

Algorithm 12: Steps of QISAMLTCIor multi-level color image thresholding

---

Input: Initial temperature: T_{start}
Final temperature: T_{final}
Reduction factor: ϱ
Number of iterations: ι
Number of thresholds: K
Output: The optimal threshold values of three primary color components: θ_R, θ_G and θ_B

1: To generate an initial configuration, P, the pixel intensity values are chosen from the input image for each color component separately at random. Note that the length of each configuration is $\mathcal{L} = \sqrt{L}$, where L signifies the maximum pixel intensity value among the color components in the color image.

2: A positive real random number between $(0,1)$ is assigned to each pixel encoded in P using the concept of qubits. Let it produce P'.

3: P' is updated using the quantum rotation as given in Eq. (1.10).

4: P' endures a *quantum orthogonality* to produce P''.

5: For each color component in the color image, find K number of thresholds as pixel intensity values from the different set of configurations. It should satisfy the corresponding value in $P'' > random(0, 1)$. Let it generate P^*.

6: Evaluate the fitness of the configuration in P^* using the objective function. Let it be symbolized by $\mathcal{V}(P^*)$.

7: Save the best configuration for the red component in $b_R \in P''$, for the green component in $b_G \in P''$ and for the blue component in $b_B \in P''$ and their corresponding threshold values in T_R, T_G, and $T_B \in P^*$. Lastly, the fitness values of the best configuration of each color component are recorded in F_R, F_G, and F_B, respectively.

8: Add F_R, F_G, and F_B and save in F.

9: $T = T_{start}$.

10: **Repeat**

11:    **for** each configuration in P **do**

12:       **if** configuration is for the red component, **then**

13:          $U = P_{red}$ and $X = P^*_{red}$.

14:       **else if** configuration is for the green component, **then**

15:          $U = P_{green}$ and $X = P^*_{green}$.

16:       **else**

17:          $U = P_{blue}$ and $X = P^*_{blue}$.

18:       **end if**

19:       **for** $k = 1$ to ι **do**

20:          Perturb U. Let it produce W.

21:          Repeat steps 2–5 to create W^*.

22:          Evaluate fitness of the configuration W^* using the objective function.

23:          **if** $(\mathcal{V}(W^*) - \mathcal{V}(X) > 0)$ **then**

24:             Set $U = W, X = W^*$ and $\mathcal{V}(X) = \mathcal{V}(W^*)$.

25:          **else**

26:             Set $U = W, X = W^*$ and $\mathcal{V}(X) = \mathcal{V}(W^*)$ with probability $\exp(-(\mathcal{V}(X) - \mathcal{V}(W^*)))/T$.

27:          **end if**

28:       **end for**

29:    **end for**

30:    $T = T \times \varrho$.

31: **until** $T >= T_{final}$

32: Report the optimal threshold values for the primary color components in $\theta_R = X_{red}, \theta_G = X_{green}$ and $\theta_B = X_{blue}$, respectively.

---

the strings in the neighbors is determined for each component level execution. If this is not found in *mem*, then the tabu memory is updated with the best string for each component. This algorithm is run for $\mathcal{G}$ number of generations. Note that we apply FIFO strategy to eliminate a string from the list.

Algorithm 13: Steps of QITSMLTCI for multi-level color image thresholding

---

Input: Number of generation: $\mathcal{G}$
Number of thresholds: K
Output: The optimal threshold values of three primary color components: θ_R, θ_G and θ_B

1: The pixel intensity values are selected from the color test image at random. These pixels are used to form the initial population, P, which contains one string for each single color component. The length of each string is taken as $\mathcal{L} = \sqrt{L_l \times L_w}$, where L_l and L_w are the length and height of the input image.

2: Using the theoretical concept of the basic unit of QC, a random positive floating number between $(0,1)$ is assigned to each element in P. Let it create P'.

3: Each element in P' endures *quantum orthogonality* to ensure a fundamental property of QC. Let it create P''.

4: Three different sets of a predefined number of threshold values are found for three individual color components in P''. It should satisfy the relation $P'' > random(0, 1)$ to acquire the desired threshold values. Let it create P^*.

5: Evaluate the fitness values of the strings in P^* for each different color component using Eq. (2.42).

6: Save the string as the best string for each individual color component in B_r, B_g and B_b, their respective threshold values in T_r, T_g, and T_b and the fitness values in F_r, F_g, and F_b, respectively.

7: Record the sum of the best fitness value of each color component in F.

8: Initialize tabu memory, $mem = \phi$.

9: **for** $j = 1$ to $\mathcal{G}$ **do**

10:    Set $F_r = B_r$, $F_g = B_g$, and $F_b = B_b$.

11:    A set of three different subsets of string is formed by exploring the neighbors of the best strings for each individual color component. Let it create V_N.

12:    For each string, $v \in V_N$, if the fitness value is better than the corresponding fitness value in V_r, V_g, or V_b and $v \notin mem$, update V_r, V_g, or V_b by v.

13:    The tabu memory is updated with the best individual string for each color component as $mem = mem \cup V_r \cup V_g \cup V_b$.

14:    Repeat steps 3–5 to compute the fitness values of strings in V_r, V_g and V_b. Record the fitness values in F_{r1}, F_{g1}, and F_{b1} with the threshold values in T_{r1}, T_{g1}, and T_{b1}, respectively.

15:    Repeat step 7 to record the sum of F_{r1}, F_{g1}, and F_{b1} in F_l.

16:    The value of F is compared with the value of F_l and record the respective threshold values in T_{rs}, T_{gs}, and T_{bs}, respectively.

17: **end for**

18: Report the optimal threshold values, $\theta = T_{rs}$, T_{gs}, and T_{bs}.

6.8.1 Complexity Analysis

The worst case time complexity analysis of QITSMLTCI is presented in this subsection.

1. Since the tabu search works on a single string, the population, P, is initially populated with three different strings for three distinct color components of the test image. Therefore, time complexity to create the population, P turns out to be $O(3 \times \mathcal{L}) = O(\mathcal{L})$.

2. The time complexity to complete the pixel encoding scheme as described in step 2 is $O(\mathcal{L})$.
3. To perform quantum orthogonality as explained in step 3, the time complexity becomes $O(\mathcal{L})$.
4. To execute step 4, the time complexity to produce P^* turns into $O(\mathcal{L})$.
5. To compute the fitness in QITSMLTCI, the time complexity grows to be $O(\mathcal{L})$.
6. After finding the best string at every generation, the neighbors of the string are explored. To find the set of neighbors, the time complexity becomes $O(\mathcal{U})$, where $\mathcal{U}$ represents the number of neighbors.
7. To update the best string as explained in step 12, the time complexity turns into $O(\mathcal{U} \times \mathcal{L})$.
8. Hence, to run QITSMLTCI for a preset number of generations, the time complexity becomes $O(\mathcal{U} \times \mathcal{L} \times \mathcal{G})$. Here, $\mathcal{G}$ denotes the number of generations.

Therefore, the overall worst case time complexity (with reference to all the steps stated above) happens to be $O(\mathcal{U} \times \mathcal{L} \times \mathcal{G})$.

6.9 Implementation Results

We have applied the proposed algorithms to determine the optimum threshold values (θ) from several real-life true color images. The effectiveness of the proposed algorithms for real-life true color image thresholding is presented at various levels with two sets of color images of different dimensions. The first set of test images, namely, Airplane, Fruits, House, Sailboat, London, Barbara, Lostlake, Anhinga, Barn, Tahoe, Tulips and Barge, each of dimensions 256×256, are presented in Figures 6.1(a)–(l). Thereafter, another set of test images, namely, Baboon, Bird, Elephant, Seaport, Montreal, Monolake, Mona Lisa, Manhattan, Lighthouse, Daisies, Lena, and Peppers, each of dimensions 512×512, are shown in Figures 6.2(a)–(l). The experiments have been conducted for the proposed QIPSOMLTCI, QIACOMLTCI, QIDEMLTCI, QIGAMLTCI, and QISAMLTCI using several thresholding methods as objective functions. Implementation results are presented with an amalgamation of the parameter adjustments of the proposed algorithms. The results are reported in two different phases in this chapter.

At the outset, results of QIPSOMLTCI, QIACOMLTCI and QIDEMLTCI with objective function given by Li's method [158], are presented for the test images of Figures 6.1(a)–(j)). Later, in the subsequent phase, results of QIGAMLTCI, QISAMLTCI, QIPSOMLTCI, and QIDEMLTCI with objective functions offered by Kapur's method [140] and Huang's method [130], are reported separately for the test images of Figures 6.2(a)–(j)).

Experiments have been conducted for their corresponding classical algorithms, and also for two recently proposed row-genetics based optimization algorithms, called Composite DE (CoDE) method [270] and the Backtracking search optimization algorithm (BSA) [45]. CoDE is a very popular method which is designed by using the basic strategies of DE. As the performance of DE is greatly influenced by the proper tuning of control parameters and appropriate selection of trial vector generation strategies, CoDE applies three different settings of control parameter and three different trial vector generation strategies together. This method basically analyses different aspects of

Figure 6.1 Original test images (a) Airplane, (b) Fruits, (c) House, (d) Sailboat, (e) London, (f) Barbara, (g) Lostlake, (h) Anhinga, (i) Barn, (j) Tahoe, (k) Tulips, and (l) Barge.

improving the effectiveness of DE. For this reason, it incorporates the aforementioned facets to introduce a more powerful method than DE. This method syndicates them for generating trial vectors at random.

BSA is a popular and efficient evolutionary algorithm. It was basically designed to solve numerical optimization problems. BSA employs a single control parameter, but it does not show excessive sensitivity to its initial selected value to perform. BSA generates a trial population and at each generation, the population is guided by two new genetic operators, namely, crossover and mutation. BSA incorporates and controls two important features, called search-direction matrix and search-space boundaries, which help it to strengthen the exploitation and exploration ability. BSA uses experiences from preceding generations, and it retains a memory to store the best individual.

Due to the similarities between the proposed quantum inspired meta-heuristics and the row-genetics-based optimization algorithms, namely, CoDE [270] and BSA [45], it can be stated that all the methods are purely population-based and explore the search space depending on a set of parameter settings. While CoDE [270] and BSA [45] rely on different heuristic strategies to determine the best fit solutions, the proposed quantum inspired methods attain the optimal solutions by running across generations.

Figure 6.2 Original test images (a) Baboon, (b) Bird, (c) Elephant, (d) Seaport, (e) Montreal, (f) Monolake, (g) Mona Lisa, (h) Manhattan, (i) Lighthouse, (j) Daisies, (k) Lena, and (l) Peppers.

As the part of the comparative research, results for the abovementioned classical algorithms, CoDE [270] and BSA [45], as regards to color image thresholding, are also presented in this chapter.

6.9.1 Experimental Results (Phase I)

In this phase, we demonstrate the application of QIPSOMLTCI, QIACOMLTCI, and QIDEMLTCI with multi-level thresholding and color images. Here, we present the implementation results taken as a whole, on the basis of the following aspects.

1. The experimental results of multi-level thresholding of the proposed algorithms including the classical algorithms and other two comparable algorithms.
2. The stability and accuracy of all participating algorithms.
3. Evaluation of performance for each comparable algorithm.

As the different meta-heuristic methods are the backbone of the proposed algorithms, an ideal set of parameter selection accelerates their performance. After tuning the requisite parameters several times, the final list of parameter settings for these proposed approaches and their respective counterparts are shown in Table 6.1. Each algorithm has been executed for 40 different runs at different levels of thresholds.

Table 6.1 Parameters specification for QIPSOMLTCI, PSO, QIACOMLTCI, ACO, QIDEMLTCI, and DE.

QIPSOMLTCI	PSO
Population size: $S = 20$	Population size: $S = 20$
Acceleration coefficients: $\xi = 0.5$	Acceleration coefficients: $\iota = 0.45$
Inertia weight: $\omega = 0.4$	Inertia weight: $\omega = 0.45$
No. of classes: $K = 1, 3, 5, 7$	No. of classes: $K = 1, 3, 5, 7$
QIACOMLTCI	ACO
Population size: $S = 20$	Population size: $S = 20$
Priori defined number: $\iota = 0.4$	Priori defined number: $\iota = 0.4$
Persistence of trials: $v = 0.25$	Persistence of trials: $v = 0.25$
No. of classes: $K = 1, 3, 5, 7$	No. of classes: $K = 1, 3, 5, 7$
QIDEMLTCI	DE
Population size: $S = 20$	Population size: $S = 20$
Scaling factor: $\zeta = 0.2$	Scaling factor: $\zeta = 0.2$
Crossover probability: $\omega = 0.9$	Crossover probability: $\omega = 0.9$
No. of classes: $K = 1, 3, 5, 7$	No. of classes: $K = 1, 3, 5, 7$

The best results of each algorithm are reported in Tables 6.2–6.9, which include the number of thresholds (K), the optimum threshold value of each primary color component (θ_R), (θ_G) and (θ_B), and the fitness value ($\mathcal{U}_{best}$) for $K = 1, 3, 5$, and 7. The computational time (t) (in seconds) and PSNR are then reported in Tables 6.10–6.11 for each level. Thereafter, the Friedman test [94, 95] has been conducted to establish the statistical superiority among them. As an outcome, it finds the mean rank of each single algorithm, which confirms the better performing algorithm among them. So, we arrange 10 different data sets (best results among different runs) for $K = 3, 5$, and 7 to perform this test. The corresponding results of the Friedman test [94, 95] are then presented in Table 6.14.

The convergence curves for the participating algorithms are presented in Figures 6.5–6.8 for $K = 5$ and 7. This curve for each algorithm is illustrated using different color combinations. It can easily be noted from each curve that the proposed approaches take the least number of generations to converge. Hence, the effectiveness of the proposed algorithms is visually established in this manner.

As QIACOMLTCI outperforms the other algorithms and each of the proposed algorithm, reports the identical or very close threshold values to each other, only the images after thresholding for QIACOMLTCI are reported as the visual representation of the test results, which are shown in Figures 6.3–6.4.

6.9.1.1 The Stability of the Comparable Algorithms

The average fitness value ($\mathcal{U}_{avg}$) and standard deviation (σ) of fitness over 40 runs are reported in Tables 6.12 and 6.13 for each level of threshold. It can be noted that the proposed and other comparable algorithms find almost the identical threshold value among the different runs for the lowest level of thresholding. This results in a

Table 6.2 Best results obtained from QIPSOMLTCI and PSO for multi-level thresholding of Airplane, Fruits, House, Sailboat, and London.

Airplane

K	QIPSOMLTCI				PSO			
	θ_R	θ_G	θ_B	U'_{best}	θ_R	θ_G	θ_B	U'_{best}
1	138	136	165	4.697	138	136	165	4.697
3	111,162,183	45,74,147	62,109,145	1.773	65,126,169	55,101,165	103,142,174	1.829
5	36,52,105,143,154	12,61,82,117,186	83,117,129,144,170	1.305	60,99,132,148,172	34,56,84,122,172	105,108,144,172,190	1.389
7	32,56,59,77,95,126,160	10,44,64,85,90,127,143	66,92,111,147,154,175,218	1.172	31,63,86,107,130,158,181	23,41,68,75,107,118,169	91,108,124,134,138,174,224	1.340

Fruits

K	QIPSOMLTCI				PSO			
	θ_R	θ_G	θ_B	U'_{best}	θ_R	θ_G	θ_B	U'_{best}
1	213	99	104	15.402	213	99	104	15.402
3	123,199,253	60,133,227	70,124,203	5.640	101,150,220	79,99,217	65,129,229	5.901
5	66,143,197,245,249	32,68,122,185,220	35,61,110,143,226	3.292	23,107,136,163,233	90,109,162,217,245	27,69,110,140,241	3.665
7	8,33,102,131,195,236,245	22,64,136,153,181,227,241	55,89,119,142,151,188,251	2.705	18,53,117,137,160,213,253	9,34,74,88,158,192,249	38,57,64,98,126,167,224	3.083

House

K	QIPSOMLTCI				PSO			
	θ_R	θ_G	θ_B	U'_{best}	θ_R	θ_G	θ_B	U'_{best}
1	137	139	145	5.100	137	139	145	5.100
3	94,135,175	76,131,153	65,116,175	1.962	76,132,189	38,90,170	85,151,193	2.015
5	82,125,140,184,196	37,73,130,179,207	62,77,118,153,193	1.364	69,87,121,144,179	44,90,162,205,231	78,81,95,123,156	1.488
7	45,94,105,112,149,195,229	30,39,79,119,161,181,228	51,59,75,85,141,195,216	1.261	27,30,73,99,102,150,189	83,89,117,132,174,208,219	90,113,137,177,187,214,230	1.324

Sailboat

K	QIPSOMLTCI				PSO			
	θ_R	θ_G	θ_B	U'_{best}	θ_R	θ_G	θ_B	U'_{best}
1	127	99	104	10.424	127	99	104	10.424
3	117,134,190	51,90,157	43,77,144	3.696	144,146,148	8,67,78	44,84,156	3.838
5	85,105,135,156,207	25,61,100,137,184	42,60,98,131,156	2.501	50,59,85,113,175	29,50,102,167,205	94,101,117,126,167	2.670
7	55,68,73,77,146,174,198	36,73,98,117,165,197,240	49,77,127,148,157,184,234	2.089	43,70,97,117,138,155,159	39,64,74,137,148,185,219	37,70,86,128,133,204,206	2.203

London

K	QIPSOMLTCI				PSO			
	θ_R	θ_G	θ_B	U'_{best}	θ_R	θ_G	θ_B	U'_{best}
1	93	81	95	8.198	93	81	95	8.198
3	81,133,186	42,101,204	38,99,199	3.487	73,95,136	28,93,160	31,56,84	3.559
5	51,97,167,206,221	16,38,88,165,235	36,65,98,142,237	2.026	29,51,65,80,205	45,72,115,136,226	26,54,93,183,214	2.196
7	22,72,94,128,130,197,244	34,46,91,126,141,224,247	35,63,71,77,98,169,228	1.598	44,83,87,105,142,168,249	44,70,95,109,135,182,209	21,44,76,109,112,154,251	1.853

Table 6.3 Best results obtained from QIPSOMLTCI and PSO for multi-level thresholding of Barbara, Lostlake, Anhinga, Barn, and Tahoe.

Barbara

K	QIPSOMLTCI				PSO			
	θ_R	θ_G	θ_B	V_{best}	θ_R	θ_G	θ_B	V_{best}
1	116	127	91	12.731	116	127	91	12.731
3	53,150,166	16,110,178	62,73,146	4.406	60,112,156	39,79,126	39,79,127	4.602
5	47,58,103,119,186	52,105,141,159,197	60,111,159,168,208	2.801	8,19,31,104,146	36,53,83,111,203	34,43,88,101,182	3.036
7	24,57,80,103,119,149,225	24,35,63,113,119,157,205	47,77,128,138,158,174,188	2.303	13,52,83,125,131,137,157	27,49,67,95,114,158,169	2,13,34,57,71,109,142	2.562

Lostlake

K	QIPSOMLTCI				PSO			
	θ_R	θ_G	θ_B	V_{best}	θ_R	θ_G	θ_B	V_{best}
1	123	63	92	26.042	123	63	92	26.042
3	14,77,137	24,58,134	38,106,221	5.544	29,65,136	31,88,240	66,191,236	5.707
5	35,67,88,164,185	25,51,87,148,248	34,42,100,186,210	3.455	43,45,163,211,222	17,33,102,166,214	25,40,90,129,190	3.688
7	46,57,73,87,101,143,205	47,86,102,133,198,232,246	25,38,70,104,145,190,245	2.744	20,22,74,139,160,184,202	26,58,72,79,119,151,244	29,58,68,93,206,213,241	2.872

Anhinga

K	QIPSOMLTCI				PSO			
	θ_R	θ_G	θ_B	V_{best}	θ_R	θ_G	θ_B	V_{best}
1	106	109	72	14.883	106	109	72	14.883
3	45,94,177	65,146,230	59,212,218	4.785	69,113,188	91,121,210	50,72,223	4.839
5	49,85,117,165,233	36,77,121,201,243	22,76,103,147,250	2.764	81,96,123,191,243	58,92,119,152,248	33,68,86,138,249	3.060
7	28,74,94,108,135,199,247	31,71,102,130,203,247,250	21,33,44,79,130,139,220	2.146	48,79,155,201,215,227,237	27,45,67,101,128,146,240	44,71,103,133,154,167,215	2.304

Barn

K	QIPSOMLTCI				PSO			
	θ_R	θ_G	θ_B	V_{best}	θ_R	θ_G	θ_B	V_{best}
1	63	116	41	6.355	63	116	41	6.355
3	65,89,183	62,86,173	44,64,145	3.210	60,103,108	60,88,135	30,123,170	3.304
5	46,69,83,104,175	68,88,136,142,169	36,61,83,119,168	2.343	22,49,78,105,185	44,63,71,101,132	12,31,55,76,128	2.611
7	48,59,90,97,111,221,241	42,51,91,99,120,158,167	23,36,50,61,81,113,118	2.054	17,39,67,85,109,118,190	42,66,73,104,114,140,185	24,46,53,64,158,194,211	2.373

Tahoe

K	QIPSOMLTCI				PSO			
	θ_R	θ_G	θ_B	V_{best}	θ_R	θ_G	θ_B	V_{best}
1	67	82	122	18.886	67	82	122	18.886
3	46,63,165	38,88,135	54,95,166	7.689	35,57,103	83,125,227	61,135,195	7.785
5	13,34,67,106,171	12,44,93,147,158	34,55,62,123,201	4.777	59,134,191,209,220	49,94,145,173,239	9,47,71,155,159	4.924
7	12,22,61,88,94,116,197	14,42,49,86,95,179,225	25,29,77,80,90,141,211	3.797	20,56,74,110,123,143,216	34,52,89,152,156,176,204	7,63,153,153,163,182,222	4.044

Table 6.4 Best results obtained from QIACOMLTCI and ACO for multi-level thresholding of Airplane, Fruits, House, Sailboat, and London.

Airplane

	QIACOMLTCI				ACO			
K	θ_R	θ_G	θ_B	U_{best}	θ_R	θ_G	θ_B	U_{best}
1	138	136	165	4.697	138	136	165	4.697
3	95,138,179	65,114,174	113,170,191	1.766	57,134,166	110,178,233	78,182,197	1.830
5	65,102,123,127,170	58,107,126,141,185	106,123,148,172,191	1.293	40,44,101,200,225	55,65,88,156,239	118,175,183,206,224	1.387
7	52,83,123,153,181,185,229	39,72,101,118,155,176,184	67,80,96,100,136,149,182	1.139	29,64,70,87,131,163,186	43,86,116,161,165,182,190	110,126,133,158,180,194,223	1.348

Fruits

	QIACOMLTCI				ACO			
K	θ_R	θ_G	θ_B	U_{best}	θ_R	θ_G	θ_B	U_{best}
1	213	99	104	15.402	213	99	104	15.402
3	109,196,245	61,145,198	74,131,204	5.591	117,119,154	95,179,239	77,205,252	5.848
5	47,120,205,215,253	40,79,137,176,225	43,71,104,154,232	3.255	11,127,203,209,249	37,64,111,122,203	72,110,147,160,213	3.686
7	24,62,112,121,207,225,253	22,55,97,162,212,233,249	53,87,117,121,163,197,220	2.645	16,110,169,171,241,246,251	30,55,128,147,172,204,251	33,59,84,90,100,140,214	2.876

House

	QIACOMLTCI				ACO			
K	θ_R	θ_G	θ_B	U_{best}	θ_R	θ_G	θ_B	U_{best}
1	137	139	145	5.100	137	139	145	5.100
3	109,135,182	73,148,218	75,137,198	1.923	28,77,134	79,160,210	89,124,164	2.001
5	44,84,105,147,178	52,85,138,185,212	76,125,156,162,205	1.341	62,69,105,138,227	78,135,145,155,176	79,109,115,127,138	1.473
7	88,113,122,122,141,198,203	64,75,84,128,146,146,155	53,76,95,110,127,188,208	1.208	50,94,141,143,168,191,211	37,67,94,110,137,150,225	104,116,138,179,201,212,219	1.300

Sailboat

	QIACOMLTCI				ACO			
K	θ_R	θ_G	θ_B	U_{best}	θ_R	θ_G	θ_B	U_{best}
1	127	99	104	10.424	127	99	104	10.424
3	105,137,180	44,76,138	28,56,145	3.692	110,124,169	49,116,203	47,113,222	3.781
5	65,97,123,132,167	32,64,96,155,218	8,39,72,123,168	2.471	63,145,157,163,181	48,63,106,142,165	28,59,105,177,188	2.668
7	66,86,89,129,162,198,219	25,43,59,101,135,156,203	31,45,59,97,144,185,233	2.018	88,92,101,128,166,196,207	44,68,87,96,105,129,148	9,20,29,60,94,130,170	2.188

London

	QIACOMLTCI				ACO			
K	θ_R	θ_G	θ_B	U_{best}	θ_R	θ_G	θ_B	U_{best}
1	93	81	95	8.198	93	81	95	8.198
3	38,71,104	110,119,248	51,94,238	3.470	73,204,233	94,139,199	40,94,216	3.531
5	52,81,112,139,228	88,105,142,231,241	47,76,154,163,202	1.982	77,214,214,215,239	182,213,220,223,237	35,69,101,140,238	2.149
7	29,64,102,133,179,184,206	52,56,70,106,175,201,241	38,47,53,55,81,177,212	1.619	63,116,191,223,224,237,247	30,95,140,182,190,210,241	23,54,88,127,156,187,234	1.889

Table 6.5 Best results obtained from QIACOMLTCI and ACO for multi-level thresholding of Barbara, Lostlake, Anhinga, Barn, and Tahoe.

Barbara

K	QIACOMLTCI				ACO			
	θ_R	θ_G	θ_B	V_{best}	θ_R	θ_G	θ_B	V_{best}
1	116	127	91	12.731	116	127	91	12.731
3	49,115,177	47,81,125	57,118,154	4.412	80,98,140	108,124,204	66,167,208	4.541
5	44,87,126,185,212	47,69,104,129,164	3,44,151,155,190	2.871	19,58,93,129,138	82,91,133,143,179	20,95,103,140,152	3.045
7	29,40,58,93,134,148,185	46,69,85,112,131,173,220	11,42,66,74,92,118,164	2.315	31,45,55,98,123,157,184	15,20,31,52,84,103,200	19,85,98,116,151,153,200	2.473

Lostlake

K	QIACOMLTCI				ACO			
	θ_R	θ_G	θ_B	V_{best}	θ_R	θ_G	θ_B	V_{best}
1	123	63	92	26.042	123	63	92	26.042
3	33,51,141	48,91,162	36,81,195	5.554	4,64,141	30,65,215	37,71,173	5.711
5	38,84,91,141,228	28,76,136,140,214	31,46,87,149,252	3.457	63,129,156,189,197	19,63,82,127,250	37,74,108,214,251	3.691
7	37,86,106,139,200,202,208	41,86,95,161,202,215,252	41,64,88,139,177,224,248	2.710	18,40,64,68,86,94,124	13,33,49,75,156,197,224	24,31,59,79,179,200,245	2.851

Anhinga

K	QIACOMLTCI				ACO			
	θ_R	θ_G	θ_B	V_{best}	θ_R	θ_G	θ_B	V_{best}
1	106	109	72	14.883	106	109	72	14.883
3	83,137,231	44,106,147	48,139,240	4.772	46,113,239	64,122,241	59,166,253	4.855
5	46,49,83,106,187	53,149,166,218,247	45,67,135,214,225	2.751	79,120,153,168,253	49,57,98,158,241	43,85,179,179,215	2.888
7	42,79,114,130,139,185,238	36,69,87,114,127,149,172	33,50,64,101,141,159,231	2.069	60,73,104,165,175,190,226	72,93,132,178,181,248,251	46,52,59,82,123,216,242	2.256

Barn

K	QIACOMLTCI				ACO			
	θ_R	θ_G	θ_B	V_{best}	θ_R	θ_G	θ_B	V_{best}
1	63	116	41	6.355	63	116	41	6.355
3	68,114,213	31,80,157	41,59,145	3.193	40,70,108	55,101,207	45,56,75	3.300
5	43,56,95,110,203	50,75,130,153,216	12,19,41,61,107	2.384	29,70,98,138,150	52,79,123,159,214	23,43,48,68,165	2.512
7	25,69,88,99,114,114,168	37,61,92,92,147,191,205	23,59,93,105,138,151,210	2.007	34,37,59,82,107,167,248	53,63,69,94,103,132,195	21,48,63,84,98,104,125	2.345

Tahoe

K	QIACOMLTCI				ACO			
	θ_R	θ_G	θ_B	V_{best}	θ_R	θ_G	θ_B	V_{best}
1	67	82	122	18.886	67	82	122	18.886
3	44,73,148	47,86,156	57,112,180	7.695	51,109,207	62,111,230	65,85,160	7.795
5	26,52,140,217,237	46,53,99,160,198	59,72,108,167,192	4.753	3,57,64,135,234	25,74,139,170,181	15,54,115,161,216	4.931
7	15,33,71,128,165,193,223	44,60,81,155,208,212,219	9,60,90,177,179,189,209	3.738	33,110,130,177,199,216,233	2,19,65,73,100,153,243	50,60,98,132,183,186,215	3.981

Table 6.6 Best results obtained from QIDEMLTCI and DE for multi-level thresholding of Airplane, Fruits, House, Sailboat, and London.

Airplane

K	QIDEMLTCI				DE			
	θ_R	θ_G	θ_B	U_{best}	θ_R	θ_G	θ_B	U_{best}
1	138	136	165	4.697	138	136	165	4.697
3	85,122,174	54,118,179	117,125,178	1.775	155,187,249	30,150,254	66,171,175	1.847
5	50,89,108,154,183	47,63,98,130,167	63,110,126,155,176	1.301	48,72,117,145,170	54,75,139,146,181	106,110,134,202,213	1.379
7	39,93,100,104,125,134,179	27,38,68,127,168,171,245	73,82,100,111,187,215,223	1.188	61,94,109,141,155,185,199	6,13,53,80,87,131,168	71,74,92,117,137,162,182	1.419

Fruits

K	QIDEMLTCI				DE			
	θ_R	θ_G	θ_B	U_{best}	θ_R	θ_G	θ_B	U_{best}
1	213	99	104	15.402	213	99	104	15.402
3	113,195,248	63,137,225	64,133,213	5.603	34,95,237	64,148,215	33,102,215	5.956
5	70,141,204,234,251	10,50,138,186,253	55,101,153,173,215	3.265	37,118,160,215,236	16,24,70,140,243	84,107,184,192,251	3.719
7	14,47,113,118,140,184,231	25,39,99,112,153,180,244	43,78,130,147,178,189,245	2.709	23,82,113,142,216,234,252	30,47,98,130,184,202,243	58,64,67,129,140,144,220	2.886

House

K	QIDEMLTCI				DE			
	θ_R	θ_G	θ_B	U_{best}	θ_R	θ_G	θ_B	U_{best}
1	137	139	145	5.100	137	139	145	5.100
3	96,130,199	67,144,221	73,121,180	1.932	100,137,199	75,185,223	80,147,170	2.010
5	91,109,143,190,193	50,79,146,157,216	83,88,133,158,200	1.358	92,151,197,200,226	64,114,174,194,203	58,65,98,147,177	1.486
7	45,55,63,128,134,136,174	18,61,68,116,162,206,235	62,72,96,122,149,172,205	1.242	44,60,63,67,115,165,173	47,76,83,145,152,158,215	46,83,129,140,146,198,202	1.318

Sailboat

K	QIDEMLTCI				DE			
	θ_R	θ_G	θ_B	U_{best}	θ_R	θ_G	θ_B	U_{best}
1	127	99	104	10.424	127	99	104	10.424
3	102,133,179	36,68,142	40,85,150	3.697	169,195,254	82,84,238	24,188,217	3.972
5	82,120,131,151,176	29,50,61,106,164	36,53,95,130,175	2.477	90,123,140,192,200	30,57,97,117,171	26,48,99,124,167	2.757
7	79,97,120,131,171,195,200	24,39,69,114,148,204,237	23,48,56,94,114,138,138	2.035	46,48,65,74,85,96,233	42,61,80,103,135,175,222	47,53,66,107,109,116,163	2.207

London

K	QIDEMLTCI				DE			
	θ_R	θ_G	θ_B	U_{best}	θ_R	θ_G	θ_B	U_{best}
1	93	81	95	8.198	93	81	95	8.198
3	39,99,172	72,111,145	43,75,167	3.466	126,186,229	81,175,181	118,170,203	3.543
5	15,73,98,159,173	90,163,174,184,236	34,91,135,148,256	2.037	44,95,152,177,196	42,76,91,116,214	49,88,95,160,203	2.170
7	24,52,92,121,162,164,235	18,29,75,81,107,191,241	62,86,152,178,191,197,210	1.630	47,71,84,112,129,203,240	32,55,68,84,112,147,216	29,59,83,92,148,183,194	1.797

Table 6.7 Best results obtained from QIDEMLTCI and DE for multi-level thresholding of Barbara, Lostlake, Anhinga, Barn, and Tahoe.

Barbara

K	QIDEMLTCI				DE			
	θ_R	θ_G	θ_B	V'_{best}	θ_R	θ_G	θ_B	V'_{best}
1	116	127	91	12.731	116	127	91	12.731
3	34,94,138	85,132,167	36,54,155	4.413	30,74,151	65,112,122	37,72,130	4.566
5	20,45,105,140,224	20,59,97,109,218	38,80,107,132,213	2.886	181,214,223,239,252	46,50,81,123,159	63,111,167,215,245	3.329
7	27,70,115,129,135,149,216	39,81,115,149,173,192,210	4,52,71,97,123,154,188	2.317	27,68,83,94,136,157,169	17,25,33,77,110,181,187	49,63,66,118,164,188,207	2.479

Lostlake

K	QIDEMLTCI				DE			
	θ_R	θ_G	θ_B	V'_{best}	θ_R	θ_G	θ_B	V'_{best}
1	123	63	92	26.042	123	63	92	26.042
3	44,114,159	37,105,164	40,103,216	5.602	48,58,130	26,85,163	34,98,231	5.725
5	42,9*,107,113,223	43,70,142,229,248	30,113,146,204,234	3.490	63,129,156,189,197	19,63,82,127,250	37,74,108,214,251	3.691
7	33,78,106,143,180,209,235	21,51,132,136,169,215,237	45,51,64,108,191,203,253	2.716	45,78,83,83,119,198,228	80,136,146,149,164,237,242	17,39,56,87,147,197,247	2.826

Anhinga

K	QIDEMLTCI				DE			
	θ_R	θ_G	θ_B	V'_{best}	θ_R	θ_G	θ_B	V'_{best}
1	106	109	72	14.883	106	109	72	14.883
3	82,139,233	48,115,173	74,100,229	4.797	82,143,237	74,76,166	28,42,109	4.851
5	92,111,161,169,214	65,76,95,138,239	40,86,143,174,226	2.795	98,160,207,213,239	77,105,195,205,231	78,91,149,169,236	2.952
7	52,68,7e,110,131,173,252	57,92,124,134,182,225,252	26,65,74,100,148,201,249	2.085	85,90,158,169,219,253,255	20,122,126,175,191,213,254	35,55,124,125,134,235,243	2.320

Barn

K	QIDEMLTCI				DE			
	θ_R	θ_G	θ_B	V'_{best}	θ_R	θ_G	θ_B	V'_{best}
1	63	116	41	6.355	63	116	41	6.355
3	50,97,138	46,91,106	18,88,125	3.203	59,104,250	66,91,202	9,55,121	3.318
5	67,83,120,149,204	19,78,132,180,220	55,115,140,163,212	2.378	98,173,212,215,222	44,148,219,223,248	30,53,72,101,124	2.749
7	14,51,65,77,100,177,199	40,63,71,76,94,139,210	18,26,42,59,91,171,185	2.019	149,159,213,214,238,245,251	114,189,208,209,249,254,255	11,105,123,211,219,230,255	2.452

Tahoe

K	QIDEMLTCI				DE			
	θ_R	θ_G	θ_B	V'_{best}	θ_R	θ_G	θ_B	V'_{best}
1	67	82	122	18.886	67	82	122	18.886
3	62,78,191	21,83,151	50,130,166	7.689	153,170,219	58,127,131	206,223,235	7.993
5	25,43,72,122,237	52,81,92,156,220	26,55,76,122,186	4.756	36,54,88,113,227	11,63,97,150,218	24,27,69,146,195	4.936
7	14,20,36,59,98,212,233	30,76,92,115,139,170,230	29,35,81,103,157,161,218	3.831	4,13,16,34,47,102,209	47,50,96,102,167,217,239	12,38,49,130,141,161,220	4.033

Table 6.8 Best results obtained from CoDE and BSA for multi-level thresholding of Airplane, Fruits, House, Sailboat, and London.

Airplane

K	CoDE				BSA			
	θ_R	θ_G	θ_B	U_{best}	θ_R	θ_G	θ_B	U_{best}
1	139	136	174	4.698	139	136	174	4.698
3	25,147,155	64,90,170	82,115,132	1.790	110,127,169	22,91,155	65,88,222	1.805
5	54,101,143,191,201	15, 56, 69,135,137	84, 92,156,194,204	1.329	141,153,160,176,211	20, 42,104,152,170	69, 99,139,152,170	1.335
7	62,81,103,122,145,165,185	31,63,103,113,151,172,181	6,78,110,141,165,178,191	1.205	21,83,88,122,151,155,173	26,34,47,86,120,176,186	62,90,123,149,151,166,211	1.231

Fruits

K	CoDE				BSA			
	θ_R	θ_G	θ_B	U_{best}	θ_R	θ_G	θ_B	U_{best}
1	213	99	104	15.402	213	99	104	15.402
3	39,207,223	86,126,208	56,110,180	5.795	122,196,253	41,77,213	60,120,232	5.819
5	14,72,129,206,243	54,127,180,216,241	8,60,77,137,235	3.606	33,38,136,199,235	2,57,116,195,226	35, 67,117,191,240	3.628
7	76,109,144,190,221,246,250	5,76,91,144,188,201,242	55,58,78,82,89,133,252	2.715	41,76,78,123,139,193,243	20,92,121,133,153,225,250	4,55,77,80,118,216,231	2.762

House

K	CoDE				BSA			
	θ_R	θ_G	θ_B	U_{best}	θ_R	θ_G	θ_B	U_{best}
1	137	139	145	5.100	137	139	145	5.100
3	91,142,215	79,143,255	82,121,179	1.993	88,189,207	11,81,141	65,132,205	1.988
5	46,96,132,140,213	28,41,90,164,211	98,155,172,192,205	1.471	31,34,44,123,148	22,54,84,150,221	67,69,104,158,175	1.429
7	31,81,106,112,137,188,217	48,67,90,100,148,189,235	46,74,81,83,107,115,145	1.274	77,105,143,187,207,238,249	46,77,86,89,141,187,254	71,80,84,111,138,192,237	1.281

Sailboat

K	CoDE				BSA			
	θ_R	θ_G	θ_B	U_{best}	θ_R	θ_G	θ_B	U_{best}
1	127	99	97	10.429	127	99	104	10.424
3	86,105,119	46,93,152	50,72,117	3.737	61,102,149	43,92,177	37,152,209	3.748
5	49,82,94,138,189	51,100,186,214,234	45, 59, 90,192,213	2.646	104,111,146,148,197	17,40,74,116,193	45, 73,128,151,202	2.634
7	87,116,140,163,172,208,234	17,31,63,118,150,156,209	50,83,86,95,118,156,181	2.186	62,64,97,106,117,138,174	10,41,51,101,148,215,231	14,26,40,64,133,209,214	2.194

London

K	CoDE				BSA			
	θ_R	θ_G	θ_B	U_{best}	θ_R	θ_G	θ_B	U_{best}
1	93	81	95	8.198	93	81	95	8.198
3	44,99,234	57,100,220	50,94,212	3.483	50,95,232	52,104,233	49,95,198	3.503
5	23,32,65,98,241	21,24,66,108,159	35,98,130,132,169	2.145	24,53,83,99,241	62,99,101,193,218	30,53,83,111,231	2.165
7	30,39,51,110,174,176,181	39,58,99,119,198,215,241	46,78,99,165,172,216,229	1.706	61,69,75,101,121,181,244	20,53,88,121,174,177,198	48,60,85,95,147,230,238	1.738

Table 6.9 Best results obtained from CoDE and BSA for multi-level thresholding of Barbara, Lostlake, Anhinga, Barn, and Tahoe.

Barbara

K	CoDE				BSA			
	θ_R	θ_G	θ_B	U_{best}	θ_R	θ_G	θ_B	U_{best}
1	106	85	79	12.731	106	85	79	12.731
3	64,92,183	32,72,115	46,113,152	4.489	11,95,124	29,73,128	17,57,76	4.488
5	31,84,111,153,160	32,110,159,181,213	12,49,102,128,136	2.964	24,63,94,118,141	68,77,157,169,200	6,49,80,142,204	2.961
7	25,46,66,128,157,180,215	78,116,123,127,129,166,186	31,80,91,118,130,172,174	2.467	17,60,91,126,171,183,222	58,63,99,113,131,153,155	11,21,45,71,99,169,179	2.461

Lostlake

K	CoDE				BSA			
	θ_R	θ_G	θ_B	U_{best}	θ_R	θ_G	θ_B	U_{best}
1	123	63	92	26.042	123	63	92	26.042
3	23,93,158	43,76,197	44,115,218	5.681	37,41,154	32,100,172	65,185,244	5.673
5	53,64,75,135,158	20,73,113,177,227	35,60,121,200,239	3.548	6,14,45,83,124	19,24,38,64,139	53,65,124,192,249	3.569
7	25,48,95,117,127,141,227	42,46,104,167,177,223,250	32,55,69,71,96,108,198	2.811	45,78,83,83,119,198,228	80,136,146,149,164,237,242	17,39,56,87,147,197,247	2.826

Anhinga

K	CoDE				BSA			
	θ_R	θ_G	θ_B	U_{best}	θ_R	θ_G	θ_B	U_{best}
1	106	109	72	14.883	106	109	72	14.883
3	81,154,249	49,141,232	58,202,245	4.829	95,111,246	114,170,249	49,124,233	4.846
5	54,87,106,144,160	40,72,102,226,242	67,111,142,157,194	2.825	53,66,126,126,234	52,69,97,165,238	65,100,108,161,173	2.844
7	36,52,82,121,165,169,242	22,39,71,81,87,165,169	34,75,105,141,165,173,239	2.146	69,75,138,148,176,183,249	58,66,86,153,228,233,245	48,77,85,135,188,242,251	2.167

Barn

K	CoDE				BSA			
	θ_R	θ_G	θ_B	U_{best}	θ_R	θ_G	θ_B	U_{best}
1	63	116	41	6.355	63	116	41	6.355
3	79,162,195	88,138,222	47,168,213	3.256	68,76,119	104,139,215	27,57,159	3.269
5	41,60,71,102,137	41,60,114,124,138	44,55,123,202,209	2.457	23,76,81,105,214	43,65,101,134,196	54,72,129,160,179	2.476
7	47,50,66,83,87,114,185	53,67,83,102,123,140,176	7,27,39,61,63,143,207	2.278	38,61,84,116,162,247,252	51,65,84,86,90,109,126	24,30,50,56,73,106,211	2.230

Tahoe

K	CoDE				BSA			
	θ_R	θ_G	θ_B	U_{best}	θ_R	θ_G	θ_B	U_{best}
1	67	82	122	18.886	67	82	122	18.886
3	69,114,235	47,78,182	50,140,189	7.701	31,85,237	49,76,130	61,126,168	7.744
5	17,81,84,160,205	21,53,66,144,192	34,48,125,132,212	4.835	19,59,88,156,235	15,69,133,215,240	43,55,55,102,153	4.858
7	36,56,71,93,146,188,233	3,11,67,78,86,92,222	4,29,42,66,113,218,256	3.840	16,33,59,62,76,135,180	7,37,64,69,94,204,232	35,50,70,85,189,204,207	3.883

Table 6.10 Execution time (t) and PSNR of QIPSOMLTCI, PSO, QIACOMLTCI, ACO, QIDEMLTCI, DE, BSA, and CoDE for multi-level thresholding of Airplane, Fruits, House, Sailboat, and London.

Airplane

K	QIPSOMLTCI		PSO		QIACOMLTCI		ACO		QIDEMLTCI		DE		CoDE		BSA	
	PSNR	t	PSNR	t	PSNR	t	PSNR	t	PSNR	t	PSNR	t	PSNR	t	PSNR	t
1	19.962	01.07	19.962	18.02	19.962	01.05	19.962	31.14	19.962	01.09	19.962	33.11	19.666	05.02	19.666	04.56
3	22.682	12.10	22.596	27.06	23.381	09.12	22.389	43.09	23.921	11.12	21.506	46.13	22.432	15.18	18.554	17.17
5	24.633	20.13	22.926	38.15	24.159	22.10	22.349	51.33	24.902	21.05	24.090	53.23	23.440	28.10	24.156	27.02
7	25.279	25.40	25.222	44.21	26.113	32.03	24.361	66.21	25.369	33.17	25.180	63.41	24.391	34.16	24.509	34.22

Fruits

K	QIPSOMLTCI		PSO		QIACOMLTCI		ACO		QIDEMLTCI		DE		CoDE		BSA	
	PSNR	t	PSNR	t	PSNR	t	PSNR	t	PSNR	t	PSNR	t	PSNR	t	PSNR	t
1	15.760	01.32	15.760	18.14	15.760	01.30	15.760	32.20	15.760	01.10	16.272	34.22	15.760	04.11	15.760	04.22
3	22.135	12.11	21.293	28.16	22.588	10.11	18.868	41.18	22.082	12.01	20.311	42.30	18.981	14.21	20.028	16.10
5	24.886	18.30	22.994	37.12	24.533	22.10	23.647	51.33	24.619	21.05	22.141	53.23	23.948	28.29	23.916	30.13
7	26.044	23.19	25.675	49.03	24.929	33.03	24.613	71.34	25.544	34.17	24.736	65.05	23.822	40.02	24.001	38.08

House

K	QIPSOMLTCI		PSO		QIACOMLTCI		ACO		QIDEMLTCI		DE		CoDE		BSA	
	PSNR	t	PSNR	t	PSNR	t	PSNR	t	PSNR	t	PSNR	t	PSNR	t	PSNR	t
1	18.597	01.30	18.597	22.10	18.597	01.25	18.597	30.08	18.597	01.19	18.597	29.21	18.597	04.14	18.597	04.12
3	25.081	09.13	23.118	29.58	23.948	10.18	20.990	42.06	22.506	14.08	22.506	45.14	23.778	16.09	22.535	15.22
5	28.428	19.10	24.739	38.01	26.898	23.07	22.604	58.25	25.480	22.06	25.049	56.31	22.256	29.01	21.002	28.09
7	26.635	22.03	25.670	48.15	26.795	30.15	25.436	69.30	26.176	31.14	25.470	69.40	26.066	37.23	26.034	38.18

Sailboat

K	QIPSOMLTCI		PSO		QIACOMLTCI		ACO		QIDEMLTCI		DE		CoDE		BSA	
	PSNR	t	PSNR	t	PSNR	t	PSNR	t	PSNR	t	PSNR	t	PSNR	t	PSNR	t
1	18.165	01.26	18.165	20.05	18.165	01.15	18.165	28.20	18.165	01.32	18.165	30.17	18.077	05.02	18.165	04.56
3	22.117	10.07	18.414	28.09	22.064	10.18	20.265	42.06	22.467	14.08	15.899	45.14	20.934	13.14	20.754	15.10
5	24.939	18.16	24.332	37.34	24.014	22.11	24.014	54.11	24.628	25.12	24.303	55.58	24.118	28.02	22.747	27.14
7	25.881	24.21	24.971	44.44	27.581	36.10	24.531	66.36	25.771	34.13	23.529	68.41	25.320	40.11	24.855	40.23

London

K	QIPSOMLTCI		PSO		QIACOMLTCI		ACO		QIDEMLTCI		DE		CoDE		BSA	
	PSNR	t	PSNR	t	PSNR	t	PSNR	t	PSNR	t	PSNR	t	PSNR	t	PSNR	t
1	14.829	01.18	14.829	19.22	14.829	01.37	14.829	30.15	14.829	01.07	14.829	32.15	14.829	04.12	14.829	04.01
3	21.293	10.09	19.429	27.29	23.952	10.12	20.078	44.01	23.687	12.13	18.192	45.50	18.159	15.10	18.047	14.44
5	25.702	22.10	22.298	38.10	25.102	23.01	19.094	50.33	24.156	22.01	22.780	52.45	19.897	30.08	17.905	30.12
7	26.143	24.10	19.344	46.17	25.984	33.03	24.225	65.61	24.461	34.05	23.389	64.10	22.848	40.09	24.467	41.56

Table 6.11 Execution time (t) and PSNR of QIPSOMLTCI, PSO, QIACOMLTCI, ACO, QIDEMLTCI, DE, BSA, and CoDE for multi-level thresholding of Barbara, Lostlake, Anhinga, Barn, and Tahoe.

Barbara

K	QIPSOMLTCI		PSO		QIACOMLTCI		ACO		QIDEMLTCI		DE		CoDE		BSA	
	PSNR	t	PSNR	t	PSNR	t	PSNR	t	PSNR	t	PSNR	t	PSNR	t	PSNR	t
1	16.906	01.25	16.906	21.04	16.906	01.25	16.906	31.20	16.906	01.04	16.906	32.28	16.148	04.02	16.148	04.50
3	20.948	12.10	20.832	31.14	21.559	09.58	20.315	42.17	20.916	11.55	20.253	43.01	20.537	14.54	18.118	15.01
5	24.854	19.11	22.206	39.22	23.077	21.44	21.866	54.01	22.673	24.50	19.332	53.10	22.339	30.30	22.569	29.35
7	26.179	23.12	24.151	47.50	26.710	33.03	23.893	65.61	26.645	34.05	24.995	66.01	25.621	00.00	25.652	00.00

Lostlake

K	QIPSOMLTCI		PSO		QIACOMLTCI		ACO		QIDEMLTCI		DE		CoDE		BSA	
	PSNR	t	PSNR	t	PSNR	t	PSNR	t	PSNR	t	PSNR	t	PSNR	t	PSNR	t
1	13.534	01.28	13.534	18.19	13.534	01.40	13.534	32.2	13.534	01.01	13.534	31.40	13.534	04.44	13.534	05.03
3	21.757	12.01	19.364	29.05	22.292	10.01	19.443	44.10	22.404	10.10	21.577	45.14	20.899	14.14	20.873	14.39
5	24.368	17.25	22.202	38.08	24.850	23.02	23.538	52.01	23.538	22.13	23.370	54.18	24.198	31.22	23.640	29.05
7	27.740	22.25	24.302	48.11	26.366	32.12	24.650	65.25	24.821	34.01	24.433	64.31	24.652	40.16	24.433	41.15

Anhinga

K	QIPSOMLTCI		PSO		QIACOMLTCI		ACO		QIDEMLTCI		DE		CoDE		BSA	
	PSNR	t	PSNR	t	PSNR	t	PSNR	t	PSNR	t	PSNR	t	PSNR	t	PSNR	t
1	15.648	01.21	15.648	19.10	15.648	01.28	15.648	33.02	15.648	01.05	15.648	33.24	15.648	04.09	15.648	04.01
3	19.867	11.14	18.590	31.05	21.669	10.14	21.165	42.10	19.352	11.22	18.562	42.28	19.911	12.14	19.916	12.22
5	24.618	19.20	23.538	41.18	23.587	22.10	21.989	54.10	23.558	24.10	22.696	53.13	23.175	30.07	23.698	29.11
7	26.273	21.30	25.834	50.10	26.879	32.05	26.110	70.13	26.727	33.17	22.112	66.14	24.985	39.22	23.867	40.10

Barn

K	QIPSOMLTCI		PSO		QIACOMLTCI		ACO		QIDEMLTCI		DE		CoDE		BSA	
	PSNR	t	PSNR	t	PSNR	t	PSNR	t	PSNR	t	PSNR	t	PSNR	t	PSNR	t
1	13.275	01.15	13.275	21.01	13.275	01.21	13.275	30.15	13.275	01.10	13.275	33.21	13.275	04.33	13.275	04.17
3	21.663	10.11	20.247	29.19	21.524	09.07	20.695	44.11	21.524	10.50	17.830	45.22	19.723	16.27	20.101	15.38
5	26.024	18.12	24.247	38.48	24.994	21.40	23.711	52.30	24.437	22.22	22.106	54.10	23.877	30.12	23.723	31.04
7	25.246	21.40	24.262	46.10	27.476	33.01	20.836	65.11	26.326	32.17	19.626	64.31	25.200	42.07	25.024	41.56

Tahoe

K	QIPSOMLTCI		PSO		QIACOMLTCI		ACO		QIDEMLTCI		DE		CoDE		BSA	
	PSNR	t	PSNR	t	PSNR	t	PSNR	t	PSNR	t	PSNR	t	PSNR	t	PSNR	t
1	15.633	01.30	15.633	22.11	15.633	01.24	15.633	33.01	15.633	01.09	15.633	33.50	15.633	04.11	15.633	04.23
3	22.536	10.47	21.056	30.03	23.452	10.14	21.282	42.10	21.633	11.10	16.501	43.10	21.600	14.25	20.070	14.37
5	25.497	19.03	23.566	41.07	24.139	22.10	23.890	54.10	25.520	24.40	24.051	53.20	22.725	32.28	23.764	31.19
7	25.034	22.10	24.198	48.10	24.217	32.05	24.208	70.21	24.610	35.01	23.360	66.01	21.917	42.05	22.257	41.09

Figure 6.3 For $K = 1, 3, 5,$ and 7, images (a)–(d), for Airplane, (e)–(h), for Fruits, (i)–(l), for House, (m)–(p), for Sailboat and (q)–(t), for London, after using QIACOMLTCI, for multi-level thresholding.

Figure 6.4 For $K = 1, 3, 5,$ and 7, images (a)–(d), for Barbara, (e)–(h), for Lostlake, (i)–(l), for Anhinga, (m)–(p), for Barn and (q)–(t), for Tahoe, after using QIACOMLTCI, for multi-level thresholding.

Table 6.12 Average fitness (V_{avg}) and standard deviation (σ) of QIPSOMLTCI, PSO, QIACOMLTCI, ACO, QIDEMLTCI, DE, BSA, and CoDE for multi-level thresholding of Airplane, Fruits, House, Sailboat, and London.

K	QIPSOMLTCI V_{avg}	σ	PSO V_{avg}	σ	QIACOMLTCI V_{avg}	σ	ACO V_{avg}	σ	QIDEMLTCI V_{avg}	σ	DE V_{avg}	σ	CoDE V_{avg}	σ	BSA V_{avg}	σ
Airplane																
1	4.697	0	4.699	0.001	4.697	0	4.699	0.001	4.697	0	4.704	0.002	4.702	0.002	4.702	0.003
3	1.813	0.011	1.891	0.040	1.803	0.008	1.851	0.046	1.813	0.014	1.885	0.023	1.928	0.044	1.875	0.033
5	1.363	0.022	1.460	0.043	1.332	0.008	1.413	0.113	1.347	0.018	1.450	0.115	1.479	1.445	0.067	0.046
7	1.212	0.016	1.308	0.046	1.207	0.020	1.308	0.128	1.212	0.011	1.283	0.121	1.282	0.055	1.267	0.031
Fruits																
1	15.402	0	15.402	0	15.402	0	15.402	0	15.402	0	15.408	0.004	15.0418	0.010	15.408	0.004
3	5.867	0.020	6.041	0.079	5.864	0.015	5.995	0.059	5.886	0.022	5.981	0.096	6.062	0.081	6.027	0.063
5	3.620	0.022	3.770	0.104	3.562	0.023	3.799	0.132	3.614	0.044	3.826	0.155	3.799	0.084	3.822	0.075
7	2.800	0.019	3.044	0.085	2.757	0.031	2.977	0.111	2.822	0.059	3.055	0.076	2.935	0.097	2.913	0.091
House																
1	5.100	0	5.100	0	5.100	0	5.100	0	5.100	0	5.100	0	5.109	0.003	5.107	0.004
3	1.985	0.010	1.995	0.017	1.935	0.006	2.091	0.121	1.973	0.014	2.118	0.050	2.138	0.081	2.094	0.071
5	1.432	0.010	1.513	0.026	1.410	0.011	1.547	0.104	1.427	0.023	1.519	0.095	1.550	0.050	1.533	0.073
7	1.262	0.029	1.311	0.015	1.268	0.012	1.357	0.109	1.282	0.023	1.318	0.107	1.345	0.036	1.336	0.030
Sailboat																
1	10.424	0	10.431	0.003	10.424	0	10.425	0.002	10.424	0	10.431	0.003	10.446	0.011	10.427	0.002
3	3.756	0.028	3.924	0.076	3.738	0.007	3.862	0.190	3.751	0.017	3.904	0.206	4.078	0.111	3.966	0.117
5	2.578	0.034	2.798	0.074	2.545	0.023	2.692	0.223	2.574	0.034	2.736	0.254	2.970	0.135	2.878	0.126
7	2.155	0.037	2.321	0.075	2.116	0.020	2.352	0.247	2.140	0.037	2.260	0.212	2.447	0.124	2.386	0.110
London																
1	8.198	0	8.198	0.003	8.198	0	8.199	0.001	8.198	0	8.198	0.003	8.210	0.009	8.202	0.005
3	3.494	0.021	3.610	0.054	3.482	0.007	3.551	0.069	3.499	0.012	3.588	0.036	3.639	0.062	3.585	0.059
5	2.139	0.051	2.318	0.080	2.064	0.038	2.190	0.200	2.112	0.043	2.394	0.113	2.458	0.114	2.343	0.113
7	1.679	0.024	1.805	0.065	1.645	0.014	1.844	0.205	1.676	0.024	1.899	0.179	1.904	0.083	1.871	0.095

Table 6.13 Average fitness ($\mathcal{V}_{avg}$) and standard deviation (σ) of QIPSOMLTCI, PSO, QIACOMLTCI, ACO, QIDEMLTCI, DE, BSA, and CoDE for multi-level thresholding of Barbara, Lostlake, Anhinga, Barn, and Tahoe.

K	QIPSOMLTCI		PSO		QIACOMLTCI		ACO		QIDEMLTCI		DE		CoDE		BSA	
	$\mathcal{V}_{avg}$	σ	$\mathcal{V}_{avg}$	σ	$\mathcal{V}_{avg}$	σ	$\mathcal{V}_{avg}$	σ	$\mathcal{V}_{avg}$	σ	$\mathcal{V}_{avg}$	σ	$\mathcal{V}_{avg}$	σ	$\mathcal{V}_{avg}$	σ
Barbara																
1	12.731	0	12.731	0.001	12.731	0	12.731	0.001	12.748	0.011	12.748	0.011	12.765	0.024	12.731	0.001
3	4.480	0.049	4.710	0.077	4.494	0.050	4.823	0.155	4.489	0.035	4.802	0.138	4.788	0.077	4.758	0.144
5	2.947	0.035	3.269	0.120	2.909	0.018	3.089	0.301	2.924	0.042	3.174	0.350	3.322	0.148	3.350	0.189
7	2.632	0.032	2.744	0.150	2.341	0.019	2.695	0.385	2.418	0.044	2.603	0.337	2.765	0.051	2.737	0.174
Lostlake																
1	26.042	0	26.042	0	26.042	0	26.042	0.007	26.042	0	26.042	0.004	26.056	0.016	26.048	0.004
3	5.650	0.044	5.897	0.176	5.599	0.019	5.914	0.471	5.640	0.038	5.809	0.397	6.103	0.190	5.985	0.180
5	3.588	0.037	3.950	0.131	3.542	0.039	3.960	0.349	3.595	0.051	3.897	0.344	4.041	0.188	4.097	0.104
7	2.829	0.042	3.058	0.104	2.745	0.026	3.091	0.362	2.793	0.042	3.153	0.354	3.189	0.114	3.228	0.060
Anhinga																
1	14.883	0	14.883	0	14.883	0	14.886	0.002	14.883	0	14.886	0.002	14.891	0.006	14.889	0.003
3	4.812	0.014	4.971	0.070	4.800	0.011	4.826	0.057	4.827	0.019	4.844	0.065	4.920	0.040	4.944	0.039
5	2.864	0.049	3.086	0.091	2.856	0.047	2.993	0.121	2.859	0.040	3.145	0.086	3.051	0.103	3.026	0.102
7	2.170	0.016	2.409	0.042	2.134	0.037	2.287	0.138	2.182	0.041	2.323	0.131	2.383	0.103	2.372	0.042
Barn																
1	6.355	0	6.365	0.006	6.355	0	6.355	0	6.355	0	6.365	0.006	6.380	0.015	6.364	0.006
3	3.255	0.030	3.438	0.080	3.214	0.024	3.488	0.103	3.253	0.030	3.517	0.110	3.412	0.064	3.555	0.063
5	2.450	0.049	2.672	0.100	2.402	0.025	2.682	0.221	2.449	0.033	2.934	0.080	2.818	0.142	2.713	0.091
7	2.128	0.042	2.343	0.119	2.107	0.037	2.228	0.202	2.117	0.031	2.638	0.102	2.466	0.065	2.346	0.099
Tahoe																
1	18.886	0	18.886	0	18.886	0	18.886	0.002	18.886	0	18.892	0.004	18.903	0.009	18.886	0.003
3	7.741	0.038	7.815	0.030	7.705	0.019	8.111	0.050	7.741	0.033	8.106	0.064	7.934	0.096	7.945	0.096
5	4.367	0.077	4.889	0.064	4.826	0.060	5.590	0.122	4.836	0.059	5.583	0.116	4.968	0.056	4.916	0.095
7	3.900	0.069	3.976	0.143	3.823	0.070	4.100	0.121	3.908	0.067	4.144	0.153	3.951	0.077	3.917	0.080

diminutive deviation for $(\mathcal{U}_{avg})$ and (σ) values for each test image. As we move towards the higher levels, the values of $((\mathcal{U}_{avg})$ and $(\sigma))$ deviate significantly. The stability of an algorithm is inversely proportional to the (σ) value. From Tables 6.12 and 6.13, it can easily be noted that the other comparable algorithms find more (σ) values compared to the proposed algorithms, specially for higher levels of thresholding. Hence, the stability of the proposed algorithms is established. Moreover, no major deviation is found in the fitness values at different runs of the proposed approaches. But in the case of comparable algorithms, the deviation is significant, particularly for the upper levels of thresholding. Hence, the accuracy is also established in favor of the proposed algorithms compared to others.

6.9.2 The Performance Evaluation of the Comparable Algorithms of Phase I

In this subsection, a metric called Peak Signal to Noise Ratio (PSNR) is introduced to evaluate the performance of the participating algorithms from the segmentation viewpoint. Basically, it takes the original image and the image after thresholding for its measurement. The unit of measurement of PSNR is decibels (dB). The higher value of PSNR indicates a better quality of thresholding. Formally, the PSNR is defined by

$$PSNR = 10 \log_{10} \left[\frac{R^2}{RMSE} \right] \tag{6.2}$$

where R denotes the maximum pixel intensity value in the image. RMSE stands for the root mean-squared error, which can also be defined by

$$RMSE = \frac{1}{M_1 \times M_2} \sum_{i=1}^{M_1} \sum_{j=1}^{M_2} (I(i,j) - I_T(i,j))^2 \tag{6.3}$$

where I and I_T represent the original image and its corresponding thresholded version of dimension $M_1 \times M_2$. Since a color image has three primary color components, an average of PSNR value for each such component is considered the overall PSNR value of the color image. The PSNR value for each test image has been reported for $K = 1, 3, 5,$ and 7 in Tables 6.10 and 6.11. It can be noted that the proposed algorithms always possess the highest PSNR value above that of the others for each level of thresholds. As the results prove, the superiority of the proposed approaches is also established from the segmentation point of view.

The computational time of the different algorithms is presented in Tables 6.10 and 6.11. These tables show that the proposed approaches take the least time to compute than the rest. Hence, the efficacy of the proposed algorithms is now also established with regards to computational time at different levels of thresholds.

Moreover, to exhibit the statistical superiority among the different algorithms, the results of the Friedman test [94, 95] are presented in Table 6.14 for $K = 3, 5,$ and 7. For each level of thresholds, the average ranking of QIACOMLTCI is found to be the smallest among other comparable algorithms. Hence, the superiority of QIACOMLTCI is statistically established. Finally, from the convergence curves shown in Figures 6.5–6.8, it can easily be noted that QIACOMLTCI is the fastest converging algorithm among all participating algorithms. Therefore, the efficiency of the proposed approaches is quantitatively and visually established in different aspects.

Table 6.14 Data sets used in QIPSOMLTCI, PSO, QIACOMLTCI, ACO, QIDEMLTCI, DE, CoDE, and BSA for the Friedman test for K = 3, 5, and 7, respectively. The value in parentheses indicates the rank of the relevant algorithm.

SN	Image	(a)	(b)	(c)	(d)	(e)	(f)	(g)	(h)
					For K=3				
1	Airplane	1.773(2)	1.829(6)	1.766(1)	1.830(7)	1.775(3)	1.847(8)	1.790(4)	1.805(5)
2	Fruits	5.640(3)	5.901(7)	5.591(1)	5.848(6)	5.603(2)	5.956(8)	5.795(4)	5.819(5)
3	House	1.962(3)	2.015(8)	1.923(1)	2.001(6)	1.932(2)	2.010(7)	1.993(5)	1.988(4)
4	Sailboat	3.696(2)	3.838(7)	3.692(1)	3.781(6)	3.697(3)	3.972(8)	3.737(4)	3.748(5)
5	London	3.487(4)	3.559(8)	3.470(2)	3.531(6)	3.466(1)	3.543(7)	3.483(3)	3.503(5)
6	Barbara	4.406(1)	4.602(8)	4.412(2)	4.541(6)	4.413(3)	4.566(7)	4.489(5)	4.488(4)
7	Lostlake	5.544(1.5)	5.707(6)	5.554(1.5)	5.711(7)	5.602(3)	5.725(8)	5.681(5)	5.673(4)
8	Anhinga	4.785(2)	4.839(5)	4.772(1)	4.855(8)	4.797(3)	4.851(7)	4.829(4)	4.846(6)
9	Barn	3.210(3)	3.304(7)	3.193(1)	3.300(6)	3.203(2)	3.318(8)	3.256(4)	3.269(5)
10	Tahoe	7.689(1.5)	7.785(6)	7.695(3)	7.795(7)	7.689(1.5)	7.993(8)	7.701(4)	7.744(5)
	Average rank	2.30	6.80	1.45	6.50	2.35	7.60	4.20	4.80
					For K=5				
SN	Image	(a)	(b)	(c)	(d)	(e)	(f)	(g)	(h)
1	Airplane	1.305(3)	1.389(8)	1.293(1)	1.387(7)	1.301(2)	1.379(6)	1.329(4)	1.335(5)
2	Fruits	3.292(3)	3.665(6)	3.255(1)	3.686(7)	3.265(2)	3.719(8)	3.606(4)	3.628(5)
3	House	1.364(3)	1.488(8)	1.341(1)	1.473(6)	1.358(2)	1.486(7)	1.471(5)	1.429(4)
4	Sailboat	2.501(3)	2.670(7)	2.471(1)	2.668(6)	2.477(2)	2.757(8)	2.646(5)	2.634(4)
5	London	2.026(2)	2.196(8)	1.982(1)	2.149(5)	2.037(3)	2.170(7)	2.145(4)	2.165(6)
6	Barbara	2.801(1)	3.036(6)	2.871(2)	3.045(7)	2.886(3)	3.329(8)	2.964(5)	2.961(4)
7	Lostlake	3.455(1)	3.688(6)	3.457(2)	3.691(7.5)	3.490(3)	3.691(7.5)	3.548(4)	3.569(5)
8	Anhinga	2.764(2)	3.060(8)	2.751(1)	2.888(6)	2.795(3)	2.952(7)	2.825(4)	2.844(5)
9	Barn	2.343(1)	2.611(7)	2.384(3)	2.512(6)	2.378(2)	2.749(8)	2.457(4)	2.476(5)
10	Tahoe	4.777(3)	4.924(6)	4.753(1)	4.931(7)	4.756(2)	4.936(8)	4.835(4)	4.858(5)
	Average rank	2.20	7.00	1.40	6.45	2.40	7.45	4.30	4.80
					For K=7				
SN	Image	(a)	(b)	(c)	(d)	(e)	(f)	(g)	(h)
1	Airplane	1.172(2)	1.340(6)	1.139(1)	1.348(7)	1.188(3)	1.419(8)	1.205(4)	1.231(5)
2	Fruits	2.705(2)	3.083(8)	2.645(1)	2.876(6)	2.709(3)	2.886(7)	2.715(4)	2.762(5)
3	House	1.261(3)	1.324(8)	1.208(1)	1.300(6)	1.242(2)	1.318(7)	1.274(4)	1.281(5)
4	Sailboat	2.089(3)	2.203(7)	2.018(1)	2.188(5)	2.035(2)	2.207(8)	2.186(4)	2.194(6)
5	London	1.598(1)	1.853(7)	1.619(2)	1.889(8)	1.630(3)	1.797(6)	1.706(4)	1.738(5)
6	Barbara	2.303(1)	2.562(8)	2.315(2)	2.473(6)	2.317(3)	2.479(7)	2.467(5)	2.461(4)
7	Lostlake	2.744(3)	2.872(8)	2.710(1)	2.851(7)	2.716(2)	2.826(5.5)	2.811(4)	2.826(5.5)
8	Anhinga	2.146(3.5)	2.304(7)	2.069(1)	2.256(6)	2.085(2)	2.320(8)	2.146(3.5)	2.167(5)
9	Barn	2.054(3)	2.373(7)	2.007(1)	2.345(6)	2.019(2)	2.452(8)	2.278(5)	2.230(4)
10	Tahoe	3.797(2)	4.044(8)	3.738(1)	3.981(6)	3.831(3)	4.033(7)	3.840(4)	3.883(5)
	Average rank	2.35	7.40	1.20	6.30	2.50	7.15	4.15	4.95

(a):→ **QIPSOMLTCI**   (b):→ **PSO**   (c):→ **QIACOMLTCI**   (d):→ **ACO**

(e):→ **QIDEMLTCI**   (f):→ **DE**   (g):→ **CoDE**   (h):→ **BSA**

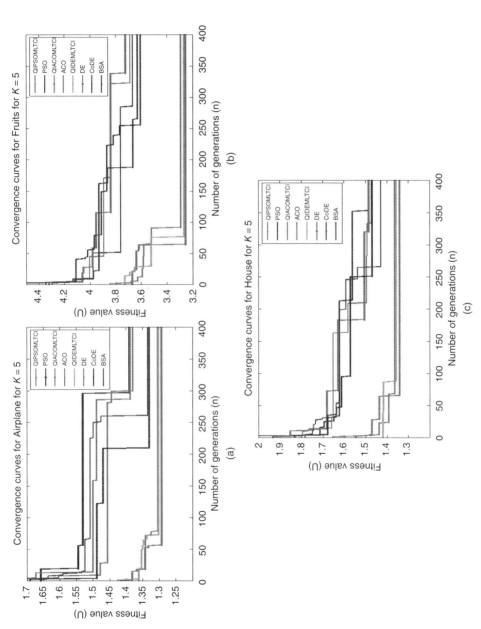

Figure 6.5 Convergence curves for $K = 5$, (a) Airplane, (b) Fruits, (c) House, (d) Sailboat, and (e) London.

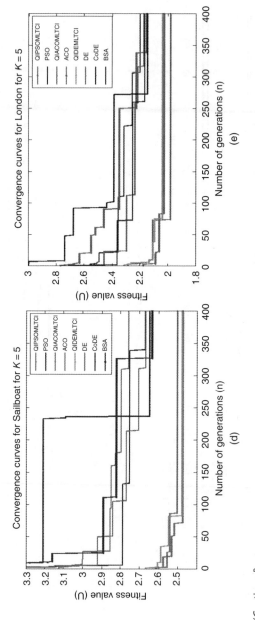

Figure 6.5 (*Continued*)

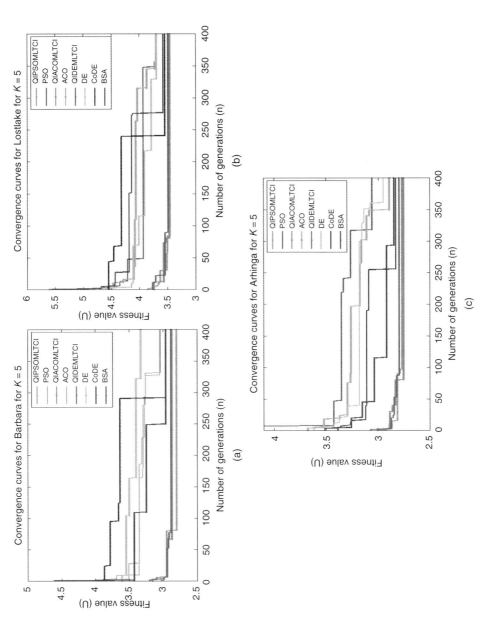

Figure 6.6 Convergence curves for $K = 5$, (a) Barbara, (b) Lostlake, (c) Anhinga, (d) Barn, and (e) Tahoe.

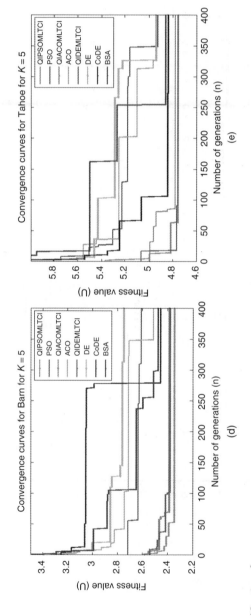

Figure 6.6 (*Continued*)

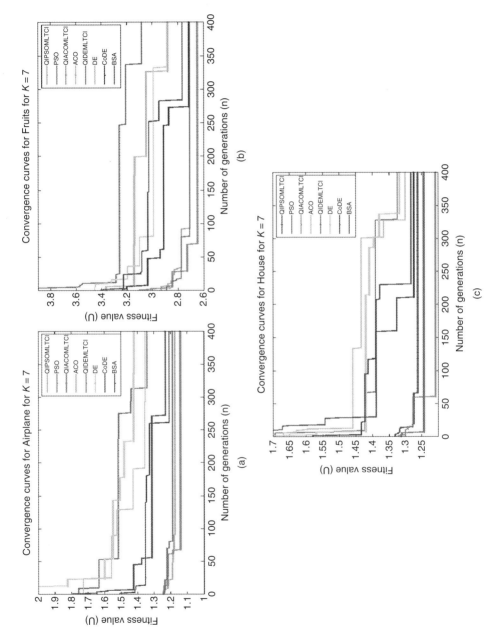

Figure 6.7 Convergence curves for K = 7. (a) Airplane, (b) Fruits, (c) House, (d) Sailboat, and (e) London.

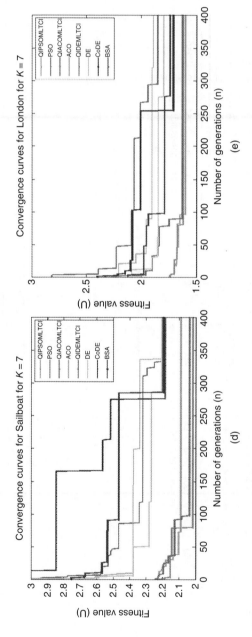

Figure 6.7 *(Continued)*

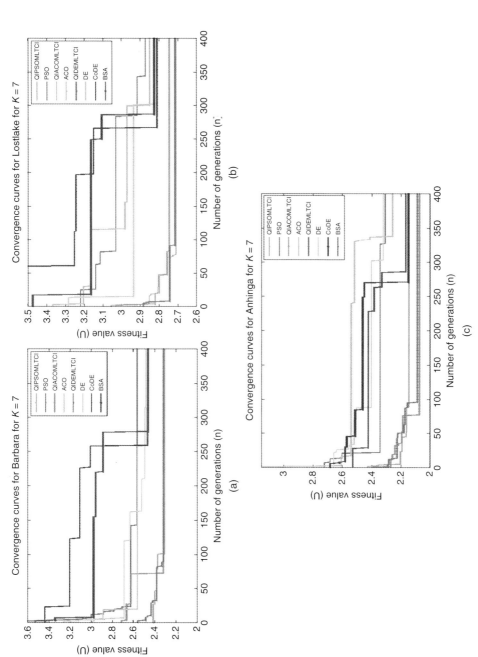

Figure 6.8 Convergence curves for $K = 7$, (a) Barbara, (b) Lostlake, (c) Anhinga, (d) Barn, and (e) Tahoe.

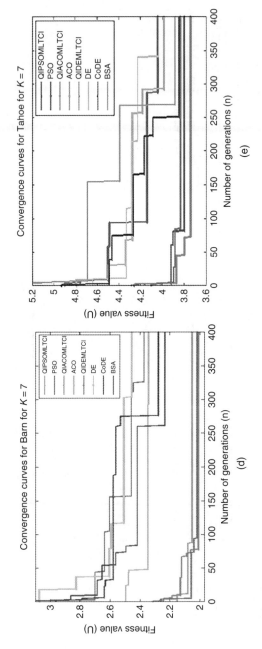

Figure 6.8 (*Continued*)

In another approach, the application of the proposed QIPSOMLTCI and QIDEMLTCI algorithms have been demonstrated on Lena and Peppers using a standard objective function, called Kapur's method [140]. The original test images are presented in Figures 6.2(k) and (l). Each algorithm has been executed for 10 different runs. The best results of each algorithm in terms of number of thresholds (K), component level optimal threshold values ($\theta_R, \theta_G, \theta_B$) of the test image, fitness value ($\mathcal{V}_{best}$), and the execution time (t) (in seconds) are listed in Table 6.15 for $K = 2, 3, 4,$ and 5. As part of the comparative study, the proposed approaches have been compared with their respective classical counterparts. The mean fitness ($\mathcal{V}_{avg}$) and the standard deviation (σ) over all runs are reported in Table 6.16. Moreover, a two-tailed t-test has been carried out at 5% confidence level for each algorithm. The test results are presented in Table 6.17.

For $K = 2$, each algorithm reports identical fitness value. Hence, the σ value is 0 for each case. For $K = 3$, there exists very few deviations for $\mathcal{V}_{avg}$ and σ values. Significant changes of these values are found for $K = 4$ and 5. Since the proposed QIPSOMLTCI possesses the least computational time, this algorithm outperforms others with regard to time efficacy. The results of two-tailed t-test establish the effectiveness of the proposed approaches.

6.9.3   Experimental Results (Phase II)

In this phase, the application of QIGAMLTCI, QISAMLTCI, QIDEMLTCI, and QIPSOMLTCI is exhibited with multi-level thresholding and a set of true color images. The implementation results of this phase are presented in the following two phases.

1. Experimental results of the proposed approaches, their classical counterparts, and other two comparable algorithms.
2. Performance evaluation of each.

The parameter specification for the proposed algorithms and their classical counterparts is listed in Table 6.18. Each algorithm has been executed 30 different times, for $K = 3, 5,$ and 7. These algorithms have been executed using a pair of objective functions, namely, Kapur's method [140] and Huang's method [130], separately. At the outset, using Kapur's method [140], the set of optimal threshold values of the primary color components ($\theta_R, \theta_G, \theta_B$) of the test images and their corresponding fitness values ($\mathcal{V}_{best}$) are reported in Tables 6.19–6.28. Thereafter, the same has been reported in Tables 6.29–6.38 using Huang's method [130]. The computational time (t) (in seconds) for each experiment has been reported in Tables 6.39 and 6.40, respectively. As QIGAMLTCI outperforms other algorithms and the set of identical or very close results are found for each of the proposed approaches for each level of thresholds, only the images after thresholding for QIGAMLTCI are presented to visually represent the test results, which are shown in Figures 6.9 and 6.10 for Kapur's method [140] and in Figures 6.11 and 6.12 for Huang's method [130].

6.9.4   The Performance Evaluation of the Participating Algorithms of Phase II

With reference to the viewpoint of image segmentation, the PSNR values from each algorithm have been reported in Tables 6.41-6.42, respectively. We have conducted two

Table 6.15 Best results obtained for QIPSOMLTCI, PSO, QIDEMLTCI, and DE for multi-level thresholding of color images.

Lena

K	QIPSOMLTCI					PSO				
	θ_R	θ_G	θ_B	v_{best}	t	θ_R	θ_G	θ_B	v_{best}	t
2	167	140	131	26.6464	25.32	167	140	131	26.6464	39.15
3	124,172	76,154	95,139	36.3568	26.21	132,188	76,156	94,150	36.3511	41.10
4	116,163,214	43,112,168	90,126,174	45.0167	27.13	131,169,212	60,132,188	79,125,174	44.8164	42.23
5	114,159,177,210	57,75,115,151	74,109,130,177	52.4867	27.55	84,154,190,221	37,87,174,208	118,137,170,195	51.4777	45.18

Peppers

K	QIPSOMLTCI					PSO				
	θ_R	θ_G	θ_B	v_{best}	t	θ_R	θ_G	θ_B	v_{best}	t
2	99	123	98	26.9378	24.09	99	123	98	26.9378	38.16
3	93,160	67,147	57,107	37.1412	26.19	86,162	77,152	58,107	37.1039	39.10
4	63,108,157	67,127,172	43,93,155	45.8218	27.33	55,106,143	49,111,164	24,80,116	45.5546	42.22
5	54,88,123,143	38,79,129,188	46,87,120,159	53.7731	27.49	76,100,146,194	79,128,175,209	42,85,110,150	53.4157	44.07

Lena

K	QIDEMLTCI					DE				
	θ_R	θ_G	θ_B	v_{best}	t	θ_R	θ_G	θ_B	v_{best}	t
2	167	140	131	26.6464	86.28	167	140	131	26.6464	105.15
3	123,192	73,145	92,142	36.3630	101.12	139,194	71,155	96,135	36.3533	155.10
4	117,159,203	39,91,160	78,124,160	45.0008	116.17	107,136,189	57,89,141	89,148,190	44.6888	187.01
5	109,137,168,213	65,90,120,166	79,100,178,196	52.1019	128.17	79,99,180,227	34,69,99,156	59,83,135,181	51.5233	193.09

Peppers

K	QIDEMLTCI					DE				
	θ_R	θ_G	θ_B	v_{best}	t	θ_R	θ_G	θ_B	v_{best}	t
2	99	123	98	26.9378	90.315	99	123	98	26.9378	110.13
3	97,157	70,141	56,105	37.1117	109.46	98,161	81,149	56,114	37.0668	142.04
4	58,97,162	76,131,187	57,105,142	45.8652	117.20	55,98,160	38,136,182	34,103,153	45.6195	190.25
5	66,108,144,238	51,95,136,181	24,67,121,154	53.9110	130.16	64,113,157,200	38,59,127,175	54,101,135,172	53.8134	197.15

Table 6.16 Mean fitness ($\mathcal{V}_{avg}$) and standard deviation (σ) of QIPSOMLTCI, PSO, QIDEMLTCI, and DE for multi-level thresholding of color images.

	Lena							
K	QIPSOMLTCI		PSO		QIDEMLTCI		DE	
	$\mathcal{V}_{avg}$	σ	$\mathcal{V}_{avg}$	σ	$\mathcal{V}_{avg}$	σ	$\mathcal{V}_{avg}$	σ
2	26.6461	0	26.6461	0	26.9375	0	26.9375	0
3	36.3394	0.018	36.3362	0.020	36.3190	0.054	36.2517	0.079
4	44.8200	0.090	44.7814	0.125	44.7809	0.090	44.7466	0.122
5	51.9988	0.186	51.5488	0.214	51.5745	0.207	51.5926	0.255
	Peppers							
K	QIPSOMLTCI		PSO		QIDEMLTCI		DE	
	$\mathcal{V}_{avg}$	σ	$\mathcal{V}_{avg}$	σ	$\mathcal{V}_{avg}$	σ	$\mathcal{V}_{avg}$	σ
2	26.9377	0	26.9376	0	26.9376	0	26.9375	0
3	37.0793	0.026	37.0686	0.033	37.0725	0.031	37.0406	0.095
4	45.7009	0.070	45.5870	0.205	45.6875	0.059	45.5651	0.215
5	53.5727	0.158	53.3696	0.241	53.5775	0.171	53.4368	0.281

Table 6.17 Results of two-tailed *t*-test for multi-level thresholding of test images.

	Lena					
K	A & B	A & C	A & D	B & D	C & D	C & B
2	NaN	NaN	NaN	NaN	NaN	NaN
3	0.0443	0.0451	0.0449	0.0466	0.0447	0.0436
4	0.0365	0.0368	0.0370	0.0349	0.0341	0.0365
5	0.0286	0.0279	0.0281	0.0272	0.0353	0.0344
	Peppers					
K	A & B	A & C	A & D	B & D	C & D	C & B
2	NaN	NaN	NaN	NaN	NaN	NaN
3	0.0456	0.0448	0.0437	0.0462	0.0437	0.0455
4	0.0294	0.0264	0.0275	0.0323	0.0268	0.0283
5	0.0241	0.0345	0.0255	0.0270	0.0265	0.0251
A:→ QIPSOMLTCI B:→ PSO						
C:→ QIDEMLTCI D:→ DE, NaN:→ No result						

Table 6.18 Parameters specification for QIGAMLTCI, QISAMLTCI, QIPSOMLTCI, QIDEMLTCI, GA, SA, PSO, and DE.

QIGAMLTCI	QISAMLTCI
Number of generations: $G = 100$	Initial temperature: $T_{start} = 100$
Population size: $B = 30$	Final temperature: $T_{final} = 0.9$
Crossover probability: $P_c = 0.9$	Number of iterations: $\iota = 20$
Mutation probability: $P_\mu = 0.01$	Reduction factors: $\varrho = 0.950$
No. of classes: $K = 3, 5, 7$	No. of classes: $K = 3, 5, 7$
QIPSOMLTCI	**QIDEMLTCI**
Number of generations: $G = 100$	Number of Generations: $G = 100$
Population size: $B = 30$	Population size: $B = 30$
Scaling factor: $\eta = 0.2$	Acceleration coefficients: $\varrho_1 = 0.5, \varrho_2 = 0.5$
Crossover probability: $\kappa = 0.950$	Inertia weight: $\varsigma = 0.4$
No. of classes: $K = 3, 5, 7$	No. of classes: $K = 3, 5, 7$
GA	**SA**
Number of generations: $G = 100$	Initial temperature: $T_{start} = 100$
Population size: $B = 30$	Final temperature: $T_{final} = 0.9$
Crossover probability: $P_c = 0.9$	Number of Iterations: $\iota = 20$
Mutation probability: $P_\mu = 0.01$	Reduction factors: $\varrho = 0.950$
No. of classes: $K = 3, 5, 7$	No. of classes: $K = 3, 5, 7$
PSO	**DE**
Number of generations: $G = 100$	Number of generations: $G = 100$
Population size: $B = 30$	Population size: $B = 30$
Scaling factor: $\eta = 0.2$	Acceleration coefficients: $\varrho_1 = 0.5, \varrho_2 = 0.5$
Crossover probability: $\kappa = 0.950$	Inertia weight: $\varsigma = 0.4$
No. of classes: $K = 3, 5, 7$	No. of classes: $K = 3, 5, 7$

different tests separately to establish the superiority of the proposed approaches. These are the Friedman test [94, 95] and a statistical test based on the median-based estimation [98]. The Friedman test [94, 95] has been conducted for both the objective functions using 10 data sets. For $K = 3, 5,$ and 7, this test determines the average ranking for each of the participating algorithms. The data set and computed results have been presented in Tables 6.43 and 6.44, respectively for both of them. We incorporate the null hypothesis (H_0) and the alternative hypothesis (H_1) to affirm a conclusive decision about the best performing algorithm of them all. H_0 confirms that each algorithm performs equally (average ranks equal), on the contrary, H_1 substantiates that there must be a significant change in performance among the participating algorithms (average ranks differ). Since the computed values of average rank for each algorithm are unequal to each other for all level of thresholds, H_1 is accepted. It can be noted that the proposed QIGAMLTCI and QIPSOMLTCI possess the lowest and second lowest average rank for all cases. Hence, QIGAMLTCI can be acknowledged as the best performing algorithm with regard to the value of average ranking.

Table 6.19 Best results obtained for QIGAMLTCI and QISAMLTCI for multi-level thresholding of Baboon, Bird, Elephant, Seaport, and Montreal for Kapur's method [140].

Baboon

K	QIGAMLTCI				QISAMLTCI			
	θ_R	θ_G	θ_B	V_{best}	θ_R	θ_G	θ_B	V_{best}
3	98,173	58,135	94,166	38.1106	96,173	55,131	94,165	38.1053
5	43,93,143,196	48,91,132,171	47,95,143,194	56.4427	25,75,137,182	70,114,144,177	69,115,152,218	55.7638
7	30,65,101,136,171,207	32,64,96,128,155,183	41,80,113,146,179,213	72.5895	21,83,117,156,180,219	33,59,89,122,142,179	36,64,98,127,175,215	71.9987

Bird

K	QIGAMLTCI				QISAMLTCI			
	θ_R	θ_G	θ_B	V_{best}	θ_R	θ_G	θ_B	V_{best}
3	72,148	72,146	71,116	27.7920	70,144	57,140	56,112	27.6512
5	40,101,118,164	40,90,132,165	42,74,103,149	39.7076	39,74,126,154	63,95,118,163	49,80,108,142	39.3170
7	25,52,79,93,173,214	11,66,77,126,153,177	61,102,113,127,156,188	48.1137	31,42,97,137,168,182	27,53,58,74,151,194	3,41,77,111,174,251	48.0243

Elephant

K	QIGAMLTCI				QISAMLTCI			
	θ_R	θ_G	θ_B	V_{best}	θ_R	θ_G	θ_B	V_{best}
3	89,165	85,147	51,118	19.4261	81,150	86,175	87,159	19.4261
5	82,119,149,185	53,90,176,206	46,96,119,180	24.9208	60,122,169,220	51,104,134,174	48,92,141,178	24.2754
7	34,72,102,145,181,225	27,64,98,128,156,200	36,71,92,128,148,175	27.4002	26,53,86,145,182,208	89,101,141,152,152,213	46,51,69,105,118,177	27.3890

Seaport

K	QIGAMLTCI				QISAMLTCI			
	θ_R	θ_G	θ_B	V_{best}	θ_R	θ_G	θ_B	V_{best}
3	81,150	66,137	74,126	19.4944	75,150	66,137	73,120	19.4453
5	24,76,116,153	33,66,101,142	31,64,73,130	26.3884	23,81,103,148	48,101,138,165	28,60,86,128	26.0978
7	24,52,81,112,145,162	38,65,100,138,161,177	16,61,64,72,91,128	30.8059	18,24,81,115,143,162	39,74,102,137,143,205	23,60,64,80,104,132	29.9389

Montreal

K	QIGAMLTCI				QISAMLTCI			
	θ_R	θ_G	θ_B	V_{best}	θ_R	θ_G	θ_B	V_{best}
3	42,136	63,128	46,128	21.3461	54,128	49,128	47,135	20.9971
5	39,88,136,169	36,64,104,136	32,45,94,129	29.2005	50,91,143,168	70,104,130,147	32,46,99,129	28.9368
7	38,91,136,185,198,252	35,59,83,112,162,184	20,100,111,138,164,202	31.57393	30,94,109,141,147,189	44,93,140,148,234,255	40,97,131,156,160,238	31.5541

Table 6.20 Best results obtained for QIGAMLTCI and QISAMLTCI for multi-level thresholding of Monolake, Mona Lisa, Manhattan, Lighthouse, and Daisies for Kapur's method [140].

Monolake

K	QIGAMLTCI				QISAMLTCI			
	θ_R	θ_G	θ_B	V_{best}	θ_R	θ_G	θ_B	V_{best}
3	37,117	81,170	51,142	22.9667	37,99	52,139	54,172	22.5170
5	37,82,117,150	50,89,136,170	54,109,132,192	31.7998	24,78,109,152	51,75,135,187	52,99,140,190	31.1351
7	27,44,82,117,150,191	51,81,112,136,170,198	36,54,108,133,175,210	37.8726	28,57,92,111,133,167	62,74,104,147,178,199	24,46,69,117,132,189	35.6220

Mona Lisa

K	QIGAMLTCI				QISAMLTCI			
	θ_R	θ_G	θ_B	V_{best}	θ_R	θ_G	θ_B	V_{best}
3	101,176	88,163	65,121	35.9646	101,174	89,163	67,119	35.9594
5	32,131,181,219	73,117,159,194	59,96,133,163	53.2803	71,128,172,198	66,118,145,203	47,111,136,163	52.3239
7	64,96,128,161,192,226	48,80,111,140,168,196	38,64,90,116,140,163	68.3283	51,74,107,149,191,234	41,70,108,130,156,190	48,61,81,126,144,164	67.5836

Manhattan

K	QIGAMLTCI				QISAMLTCI			
	θ_R	θ_G	θ_B	V_{best}	θ_R	θ_G	θ_B	V_{best}
3	100,173	63,127	53,140	20.5450	116,214	111,122	141,147	20.4452
5	78,104,141,178	29,64,91,137	28,55,86,132	28.2676	81,104,133,176	24,62,91,143	27,62,82,127	27.5575
7	52,80,102,139,182,212	27,85,118,133,155,236	27,60,119,138,171,214	31.5377	42,58,76,85,133,172	41,76,122,157,190,239	21,34,78,117,130,150	31.4702

Lighthouse

K	QIGAMLTCI				QISAMLTCI			
	θ_R	θ_G	θ_B	V_{best}	θ_R	θ_G	θ_B	V_{best}
3	43,137	63,199	118,173	21.8839	96,156	78,143	97,159	21.7753
5	37,107,137,208	51,66,156,185	108,171,191,208	30.6146	44,71,109,197	42,74,152,204	71,97,150,199	30.5153
7	40,56,76,112,141,164	39,72,79,112,149,216	40,62,96,129,159,196	37.6379	37,66,77,139,148,191	21,67,113,135,159,166	23,43,101,118,172,202	36.0877

Daisies

K	QIGAMLTCI				QISAMLTCI			
	θ_R	θ_G	θ_B	V_{best}	θ_R	θ_G	θ_B	V_{best}
3	67,118	76,146	88,146	24.7691	101,165	72,123	40,165	24.7433
5	48,70,159,186	39,59,123,185	17,27,62,158	34.8079	50,82,130,163	12,78,135,201	60,103,168,215	34.4185
7	49,66,95,135,167,216	40,72,97,119,160,216	21,46,78,116,141,185	42.8365	41,69,110,150,169,234	52,85,107,144,195,224	18,40,87,109,149,195	41.0997

Table 6.21 Best results obtained for QIPSOMLTCI and QIDEMLTCI for multi-level thresholding of Baboon, Bird, Elephant, Seaport, and Montreal for Kapur's method [140].

Baboon

K	QIPSOMLTCI θ_R	θ_G	θ_B	ν_{best}	QIDEMLTCI θ_G	θ_B	ν_{best}	θ_R
3	148,189	53,157	130,205	38.1191	130,134	96,177	38.1170	100,177
5	45,107,172,204	20,90,148,176	28,86,130,165	56.4036	67,133,152,173	40,97,143,193	56.3867	48,125,198,219
7	20,99,124,163,169,212	43,57,124,184,198,217	33,54,86,117,137,165	72.2779	30,50,78,135,198,215	15,30,85,112,140,180	72.1471	41,65,108,151,183,235

Bird

K	QIPSOMLTCI θ_R	θ_G	θ_B	ν_{best}	QIDEMLTCI θ_G	θ_B	ν_{best}	θ_R
3	81,123	149,170	39,75	27.7700	64,131	62,100	27.7692	56,139
5	68,144,167,191	45,80,99,134	28,61,108,167	39.4554	50,70,93,178	57,93,118,211	39.4510	16,28,149,158
7	30,46,86,91,109,245	22,64,96,124,165,239	20,32,44,79,145,191	48.2604	37,92,111,138,159,222	22,52,109,114,129,153	48.1517	37,51,117,136,226,232

Elephant

K	QIPSOMLTCI θ_R	θ_G	θ_B	ν_{best}	QIDEMLTCI θ_G	θ_B	ν_{best}	θ_R
3	96,175	64,142	75,126	19.4261	64,135	57,122	19.4261	74,176
5	42,126,148,206	57,108,133,192	22,78,117,154	12.9244	45,91,135,183	43,98,127,164	24.2858	52,118,176,216
7	33,96,107,153,187,240	54,88,115,154,173,207	19,49,83,100,129,165	27.6376	39,40,81,160,179,206	17,41,82,102,140,202	27.6274	52,116,172,201,219,248

Seaport

K	QIPSOMLTCI θ_R	θ_G	θ_B	ν_{best}	QIDEMLTCI θ_G	θ_B	ν_{best}	θ_R
3	82,150	66,134	73,128	19.4934	61,134	73,128	19.4704	77,150
5	19,82,116,150	38,96,137,148	33,60,80,132	26.2047	47,100,138,163	28,65,74,125	26.2414	23,80,129,152
7	34,62,96,103,135,188	34,50,78,92,130,145	26,47,95,103,148,169	29.3007	38,61,100,134,148,199	23,59,71,84,90,127	30.6428	23,54,78,106,145,161

Montreal

K	QIPSOMLTCI θ_R	θ_G	θ_B	ν_{best}	QIDEMLTCI θ_G	θ_B	ν_{best}	θ_R
3	35,135	119,204	11,115	20.4068	63,112	53,95	20.39435	66,138
5	41,96,136,164	27,71,112,144	31,88,95,133	12.0734	63,104,128,147	22,49,93,132	28.9920	45,96,136,164
7	66,68,85,137,160,210	34,43,99,107,131,179	35,61,112,141,143,167	32.0464	53,74,95,128,136,193	17,36,54,82,151,187	31.7646	60,104,138,159,200,255

Table 6.22 Best results obtained for QIPSOMLTCI and QIDEMLTCI for multi-level thresholding of Monolake, Mona Lisa, Manhattan, Lighthouse, and Daisies for Kapur's method [140].

Monolake

K	QIPSOMLTCI				QIDEMLTCI			
	θ_R	θ_G	θ_B	V_{best}	θ_R	θ_G	θ_B	V_{best}
3	30,100	40,177	98,146	22.9651	33,113	54,168	84,177	22.8912
5	48,111,127,167	75,109,175,209	61,102,134,184	31.2537	70,96,128,174	79,148,175,232	50,109,132,203	31.1196
7	10,21,49,117,123,177	35,62,138,170,191,194	51,75,98,161,201,215	36.4533	36,64,81,103,155,191	30,73,82,130,151,217	36,51,101,127,147,202	36.3196

Mona Lisa

K	QIPSOMLTCI				QIDEMLTCI			
	θ_R	θ_G	θ_B	V_{best}	θ_R	θ_G	θ_B	V_{best}
3	109,179	87,163	74,146	35.9679	135,174	101,160	51,98	35.9708
5	62,82,103,174	48,91,175,207	66,104,123,161	53.1221	69,159,193,225	101,148,165,217	27,49,125,154	53.1210
7	35,116,133,156,188,234	43,79,103,127,159,208	15,46,63,123,145,169	67.7713	17,66,95,125,187,203	69,88,113,141,162,205	36,53,70,81,128,184	67.7204

Manhattan

K	QIPSOMLTCI				QIDEMLTCI			
	θ_R	θ_G	θ_B	V_{best}	θ_R	θ_G	θ_B	V_{best}
3	159,188	87,151	37,130	20.5766	109,170	99,142	46,108	20.5543
5	62,82,107,176	21,47,97,142	28,43,70,121	27.9636	70,156,171,222	49,85,118,154	5,54,100,133	27.8503
7	36,77,126,171,206,255	44,68,101,154,180,224	33,60,79,105,161,175	17.4277	19,66,85,98,154,197	26,36,113,131,144,190	28,31,60,73,136,253	31.8317

Lighthouse

K	QIPSOMLTCI				QIDEMLTCI			
	θ_R	θ_G	θ_B	V_{best}	θ_R	θ_G	θ_B	V_{best}
3	58,168	72,140	61,161	21.9626	108,181	110,147	107,166	21.9579
5	25,128,148,198	83,160,201,221	25,78,123,168	30.5039	41,62,121,166	32,80,126,207	22,78,158,232	30.4884
7	40,56,81,113,145,178	37,72,104,130,171,218	43,58,88,115,148,196	36.8774	9,90,148,191,218,246	37,104,157,178,184,227	20,35,88,150,171,238	36.3339

Daisies

K	QIPSOMLTCI				QIDEMLTCI			
	θ_R	θ_G	θ_B	V_{best}	θ_R	θ_G	θ_B	V_{best}
3	56,114	71,171	97,150	24.7745	61,162	85,178	85,141	24.7852
5	72,135,155,185	51,73,120,209	56,122,150,236	34.5995	49,65,185,217	79,98,149,220	43,74,107,197	34.5201
7	47,73,105,162,174,233	40,105,124,139,169,225	33,90,132,164,206,242	13.8475	42,58,105,123,140,202	25,91,118,158,180,235	25,33,59,90,174,222	41.5488

Table 6.23 Best results obtained for GA and SA for multi-level thresholding of Baboon, Bird, Elephant, Seaport, and Montreal for Kapur's method [140].

K	GA				SA			
	θ_R	θ_G	θ_B	V_{best}	θ_R	θ_G	θ_B	V_{best}
Baboon								
3	115,174	52,126	115,173	37.9784	77,202	114,176	8,244	37.9686
5	35,98,100,216	54,120,168,210	36,46,82,147	55.5483	35,103,154,191	44,95,119,168	88,155,187,230	55.5499
7	93,142,149,207,237,238	12,67,96,119,144,150	6,74,96,177,189,216	71.1919	40,148,171,173,182,190	10,22,94,135,190,228	56,89,113,233,244,247	70.5951
Bird								
3	9,162	7,38	88,196	27.5126	45,119	109,188	22,40	27.4627
5	59,89,128,160	53,91,114,157	30,60,131,157	38.6345	78,80,151,172	1,26,206,211	1,29,65,186	38.0177
7	52,108,192,197,199,242	24,65,92,117,143,178	6,27,165,220,247,250	45.2848	27,47,60,85,110,136	28,77,106,150,162,174	41,69,105,140,182,240	44.2234
Elephant								
3	96,167	64,136	64,122	19.0550	101,179	77,152	59,115	19.0328
5	69,124,167,211	40,87,132,185	48,94,123,152	24.1188	57,99,166,214	56,122,149,197	43,98,144,174	24.1444
7	45,66,108,151,181,222	29,49,97,120,146,199	33,57,98,127,159,175	26.8600	21,29,67,93,125,196	39,69,93,131,172,213	2,59,115,143,158,247	24.4608
Seaport								
3	221,226	67,175	104,117	18.5216	78,142	58,126	37,128	18.8618
5	19,68,103,158	55,83,109,149	22,54,70,134	24.9379	81,104,147,153	35,121,138,204	58,91,155,180	23.4984
7	19,53,95,125,150,183	14,55,73,94,135,155	22,63,72,85,171,187	28.2911	17,40,83,120,134,240	54,90,151,155,200,248	46,76,144,151,156,238	27.3997
Montreal								θ_B´
3	57,117	63,81	33,63	20.3537	70,152	71,90	32,47	20.3604
5	29,107,143,235	67,135,161,204	18,61,72,90	26.7768	50,122,155,206	33,106,133,166	29,65,109,130	26.7635
7	88,123,148,158,184,202	5,60,168,210,223,234	12,54,63,90,134,218	30.1747	46,84,97,138,177,213	21,53,91,109,158,221	13,42,81,105,138,161	30.3882

Table 6.24 Best results obtained for GA and SA for multi-level thresholding of Monolake, Mona Lisa, Manhattan, Lighthouse, and Daisies for Kapur's method [140].

Monolake

K	GA				SA			
	θ_R	θ_G	θ_B	V_{best}	θ_R	θ_G	θ_B	V_{best}
3	50,55	69,130	72,173	22.3014	27,85	86,185	53,165	22.1184
5	36,69,121,166	62,101,132,192	45,89,129,201	30.1623	27,74,98,177	76,104,131,168	45,71,99,154	29.2394
7	14,46,72,82,126,167	46,91,95,177,244,245	136,170,175,202,219,247	34.7246	5,42,149,224,247,248	31,63,126,156,195,210	51,59,60,88,117,200	34.4819

Mona Lisa

K	GA				SA			
	θ_R	θ_G	θ_B	V_{best}	θ_R	θ_G	θ_B	V_{best}
3	126,143	190,207	27,75	35.4958	29,116	148,229	23,75	35.5675
5	3,5,83,155	20,22,226,237	47,61,227,232	52.3380	75,81,133,157	45,145,176,226	26,112,126,213	51.9529
7	56,95,146,179,185,231	35,70,72,82,143,190	43,106,135,163,175,190	65.4327	7,48,57,76,171,238	36,61,80,87,164,201	24,31,67,126,171,212	65.7589

Manhattan

K	GA				SA			
	θ_R	θ_G	θ_B	V_{best}	θ_R	θ_G	θ_B	V_{best}
3	67,143	127,183	89,107	20.1754	125,229	86,126	19,221	20.1180
5	75,103,173,227	84,103,112,115	1,99,150,185	26.1379	76,125,134,218	27,55,95,143	19,31,66,115	26.2211
7	55,97,136,172,189,198	26,63,92,110,140,148	22,44,104,133,145,173	29.9079	70,89,142,148,174,229	22,28,122,213,241,252	5,80,169,178,203,223	29.2814

Lighthouse

K	GA				SA			
	θ_R	θ_G	θ_B	V_{best}	θ_R	θ_G	θ_B	V_{best}
3	53,152	74,99	96,174	21.3411	65,158	212,216	54,231	21.3241
5	14,166,170,196	38,62,147,253	54,100,121,171	29.1642	51,85,115,155	17,122,150,182	19,93,130,155	29.1657
7	81,100,207,229,239,254	38,71,94,150,211,226	61,97,126,143,159,219	34.1485	35,70,75,109,213,248	94,115,181,184,209,248	69,135,137,165,193,240	34.0885

Daisies

K	GA				SA			
	θ_R	θ_G	θ_B	V_{best}	θ_R	θ_G	θ_B	V_{best}
3	55,126	73,127	84,144	24.6400	52,127	69,123	71,141	24.6209
5	54,79,111,153	64,105,127,173	34,68,124,154	34.1841	51,78,107,143	78,119,157,224	42,79,133,149	34.2137
7	52,64,103,132,183,219	59,85,109,157,187,216	29,92,117,143,155,197	40.7998	39,84,147,159,171,221	63,92,127,143,171,223	21,51,62,107,120,140	40.3464

Table 6.25 Best results obtained for PSO and DE for multi-level thresholding of Baboon, Bird, Elephant, Seaport, and Montreal for Kapur's method [140].

Baboon

K	PSO				DE			
	θ_R	θ_G	θ_B	v_{best}	θ_R	θ_G	θ_B	v_{best}
3	101,174	54,111	100,187	37.9606	35,170	65,133	37,90	37.9664
5	31,95,122,217	31,64,127,175	54,125,158,189	55.5318	28,82,119,157	36,77,106,206	15,29,110,216	55.5391
7	16,79,100,157,189,216	19,61,86,105,137,174	54,119,133,155,183,223	70.7130	55,83,110,149,191,235	30,58,78,119,151,191	31,73,121,157,164,223	70.7642

Bird

K	PSO				DE			
	θ_R	θ_G	θ_B	v_{best}	θ_R	θ_G	θ_B	v_{best}
3	79,123	86,123	55,116	27.4770	62,81	101,189	187,243	27.5439
5	6,84,108,166	52,119,154,228	2,15,210,220	38.4960	40,139,186,217	17,28,52,100	53,120,137,183	38.4133
7	57,135,136,139,166,175	34,75,130,137,163,175	30,38,85,174,183,251	44.6108	40,165,183,193,196,226	42,79,111,196,235,236	3,10,195,230,242,246	44.4637

Elephant

K	PSO				DE			
	θ_R	θ_G	θ_B	v_{best}	θ_R	θ_G	θ_B	v_{best}
3	54,73	178,201	2,77	19.3616	32,127	75,89	90,108	19.3727
5	102,137,154,246	55,72,111,216	17,34,112,164	24.1186	71,166,237,250	126,152,165,210	68,128,167,250	24.1182
7	57,77,83,90,201,233	89,143,194,194,242,243	13,109,119,125,151,245	26.3680	101,105,105,135,206,219	31,45,54,119,128,134	9,106,124,218,231,238	26.3799

Seaport

K	PSO				DE			
	θ_R	θ_G	θ_B	v_{best}	θ_R	θ_G	θ_B	v_{best}
3	129,178	70,130	59,238	18.4979	170,233	107,196	46,165	18.5438
5	34,108,202,217	12,100,184,253	48,91,196,249	24.3427	29,55,134,180	187,228,239,245	9,29,69,193	24.7252
7	20,82,140,162,167,248	114,125,198,207,208,254	16,26,61,62,167,232	27.5768	122,187,188,243,251,254	54,81,115,156,180,207	49,50,59,103,143,167	27.4480

Montreal

K	PSO				DE			
	θ_R	θ_G	θ_B	v_{best}	θ_R	θ_G	θ_B	v_{best}
3	53,126	67,118	40,102	20.3471	48,97	61,185	26,80	20.3936
5	49,106,138,160	31,91,133,190	11,67,111,130	26.6425	22,53,75,150	40,97,147,227	39,51,163,254	26.5740
7	5,28,61,220,225,234	11,72,144,176,185,249	49,98,99,233,245,253	30.3928	56,138,171,199,205,225	28,57,100,131,149,218	14,34,38,42,82,233	30.1752

Table 6.26 Best results obtained for PSO and DE for multi-level thresholding of Monolake, Mona Lisa, Manhattan, Lighthouse, and Daisies for Kapur's method [140].

Monolake

K	PSO θ_R	θ_G	θ_B	ν_{best}	DE θ_R	θ_G	θ_B	ν_{best}
3	57,113	199,233	20,91	22.3318	36,104	22,161	45,126	22.4737
5	24,127,170,225	57,153,230,245	55,93,158,204	30.0592	10,113,136,245	6,147,174,235	36,145,178,180	30.0847
7	27,37,60,108,172,190	3,55,72,115,174,216	8,28,40,45,52,173	34.5598	46,64,72,89,125,166	71,117,119,177,186,239	41,75,152,160,169,216	34.4870

Mona Lisa

K	PSO θ_R	θ_G	θ_B	ν_{best}	DE θ_R	θ_G	θ_B	ν_{best}
3	55,147	68,248	60,110	35.6474	97,166	203,214	57,108	35.7945
5	5,36,37,106	50,62,148,164	52,78,96,145	52.1727	112,131,213,232	71,115,149,169	42,83,87,112	52.1185
7	80,107,142,152,158,240	47,83,152,177,180,182	78,88,120,188,205,219	65.4037	5,8,61,112,133,201	52,60,80,102,143,249	2,47,83,124,158,163	65.5240

Manhattan

K	PSO θ_R	θ_G	θ_B	ν_{best}	DE θ_R	θ_G	θ_B	ν_{best}
3	201,216	12,14	46,136	20.1655	79,185	25,221	27,84	20.1957
5	13,39,56,89	153,174,176,238	21,116,179,241	26.0354	44,152,204,216	113,129,151,183	47,114,115,131	26.1850
7	27,108,133,159,192,232	6,47,72,107,126,241	26,37,61,89,129,178	29.2113	60,94,99,143,144,219	58,84,122,147,159,168	30,50,68,85,104,137	29.2357

Lighthouse

K	PSO θ_R	θ_G	θ_B	ν_{best}	DE θ_R	θ_G	θ_B	ν_{best}
3	130,139	7,21	12,168	21.4026	67,246	238,254	18,239	21.4661
5	24,70,110,158	129,174,199,201	57,109,140,188	29.1259	97,129,173,195	46,111,121,237	71,103,110,243	29.1518
7	6,31,43,60,138,228	87,109,113,133,184,229	24,36,99,101,105,144	34.2636	21,38,53,113,146,208	29,70,103,151,173,194	3,50,81,113,209,230	34.0844

Daisies

K	PSO θ_R	θ_G	θ_B	ν_{best}	DE θ_R	θ_G	θ_B	ν_{best}
3	49,97	73,127	83,145	24.6201	88,156	72,119	60,135	24.6329
5	54,108,155,223	68,102,122,166	43,91,131,179	34.1360	51,76,142,192	38,79,127,187	42,69,117,156	34.1273
7	52,75,109,125,147,219	58,73,100,108,154,203	38,70,107,136,150,193	40.3380	32,56,87,102,133,213	27,66,90,126,166,218	20,33,51,86,104,177	40.5699

Table 6.27 Best results obtained for BSA and CoDE for multi-level thresholding of Baboon, Bird, Elephant, Seaport, and Montreal for Kapur's method [140].

Baboon

K	BSA				CoDE			
	θ_R	θ_G	θ_B	$\mathcal{V}_{best}$	θ_R	θ_G	θ_B	$\mathcal{V}_{best}$
3	157,180	28,122	167,232	37.9892	27,30	26,65	152,196	37.9983
5	38,128,162,192	4,101,159,192	3,143,170,199	55.5556	191,192,226,233	63,144,234,250	40,49,105,249	55.4125
7	57,70,77,127,190,210	5,9,138,158,188,254	14,23,92,112,115,233	71.2335	52,120,121,134,137,250	11,36,123,163,194,252	71,105,106,116,127,202	71.1250

Bird

K	BSA				CoDE			
	θ_R	θ_G	θ_B	$\mathcal{V}_{best}$	θ_R	θ_G	θ_B	$\mathcal{V}_{best}$
3	62,154	25,71	56,98	27.5058	68,74	24,117	25,186	27.5091
5	3,40,185,245	40,113,118,122	13,23,137,188	38.5333	67,90,132,161	94,111,153,202	35,200,235,239	38.8943
7	31,59,77,108,135,185	40,74,117,151,168,191	34,71,93,114,143,147	45.4073	28,81,92,126,153,171	58,79,116,122,168,181	59,84,87,116,163,176	45.3422

Elephant

K	BSA				CoDE			
	θ_R	θ_G	θ_B	$\mathcal{V}_{best}$	θ_R	θ_G	θ_B	$\mathcal{V}_{best}$
3	70,149	43,108	88,110	19.3683	71,158	61,132	72,130	19.0709
5	69,76,153,209	54,93,152,220	36,62,194,200	24.4105	71,125,187,221	45,109,134,189	47,87,119,151	24.5744
7	56,94,123,170,187,215	49,77,127,150,189,213	29,76,121,153,167,207	26.4115	53,94,127,157,207,233	26,53,93,123,145,187	63,76,111,121,158,191	26.1802

Seaport

K	BSA				CoDE			
	θ_R	θ_G	θ_B	$\mathcal{V}_{best}$	θ_R	θ_G	θ_B	$\mathcal{V}_{best}$
3	95,139	43,108	54,121	18.9487	68,146	59,112	35,119	18.9448
5	26,59,111,142	49,98,140,179	60,79,91,138	24.1767	22,84,110,119	55,95,134,172	63,91,124,187	24.3885
7	2,17,60,80,149,227	24,41,70,126,144,164	24,36,57,131,228,228	28.8505	75,103,126,144,165,242	50,64,71,100,145,237	38,67,94,118,145,159	28.8068

Montreal

K	BSA				CoDE			
	θ_R	θ_G	θ_B	$\mathcal{V}_{best}$	θ_R	θ_G	θ_B	$\mathcal{V}_{best}$
3	68,104	70,108	15,87	20.3887	92,116	25,119	22,107	20.3728
5	20,90,145,174	35,98,153,184	21,84,144,190	26.9463	75,103,134,162	71,107,128,170	44,69,121,247	26.7383
7	14,163,186,219,234,243	46,49,149,174,214,230	44,54,122,143,161,196	30.1502	76,134,185,193,207,252	22,73,177,188,188,219	41,42,88,91,124,174	30.1179

Table 6.28 Best results obtained for BSA and CoDE for multi-level thresholding of Monolake, Mona Lisa, Manhattan, Lighthouse, and Daisies for Kapur's method [140].

Monolake

K	BSA				CoDE			
	θ_R	θ_G	θ_B	y_{best}	θ_R	θ_G	θ_B	y_{best}
3	42,119	52,78	30,71	22.4351	38,79	16,21	12,194	22.3835
5	183,205,244,255	14,89,140,185	75,90,169,191	30.2282	5,8,23,197	75,140,144,186	40,54,65,243	30.1864
7	3,31 32,115,142,219	15,51,128,173,227,250	25,37,146,152,160,180	35.1292	39,100,101,130,212,247	116,146,196,216,238,249	16,70,82,137,138,250	35.222

Mona Lisa

K	BSA				CoDE			
	θ_R	θ_G	θ_B	y_{best}	θ_R	θ_G	θ_B	y_{best}
3	11,126	42,51	106,230	35.8757	116,184	34,37	34,142	35.9014
5	89,136,185,230	14,79,134,150	30,90,91,167	52.2325	52,69,88,91	95,120,138,226	59,172,173,236	52.1575
7	2,13,84,151,170,177	5,6,17,33,85,163	28,50,85,134,204,210	65.8547	24,29,62,119,173,175	22,33,123,143,168,194	8,26,82,214,221,224	65.6869

Manhattan

K	BSA				CoDE			
	θ_R	θ_G	θ_B	y_{best}	θ_R	θ_G	θ_B	y_{best}
3	47,122	122,214	15,195	20.3019	116,164	48,126	84,111	20.2649
5	105,136,200,230	2,82,84,154	5,35,220,227	26.4056	6,8,135,152	102,103,180,206	14,182,192,248	26.4182
7	44,53,209,211,220,252	9,31,89,115,134,174	65,93,118,157,176,187	30.2522	26,53,90,101,153,212	8,26,39,60,118,141	43,96,116,161,175,188	29.9357

Lighthouse

K	BSA				CoDE			
	θ_R	θ_G	θ_B	y_{best}	θ_R	θ_G	θ_B	y_{best}
3	39,107	29,50	201,211	21.6006	24,105	20,109	29,218	21.6263
5	84,125,148,204	89,217,239,248	27,57,159,227	29.3621	69,172,182,226	3,55,91,214	60,116,142,190	29.4102
7	30,46,60,73,123,148	29,46,107,132,170,217	29,75,131,152,159,183	35.1836	18,48,56,127,132,239	5,81,87,95,189,237	49,101,129,129,134,236	34.9284

Daisies

K	BSA				CoDE			
	θ_R	θ_G	θ_B	y_{best}	θ_R	θ_G	θ_B	y_{best}
3	68,135	72,120	94,143	24.6669	65,135	64,126	71,143	24.6587
5	49,100,162,220	72,101,123,221	40,80,124,143	34.1812	51,92,125,170	66,118,145,191	25,84,117,150	34.1605
7	38,53,87,97,141,183	57,100,114,151,186,216	46,57,102,129,154,186	40.8900	45,64,81,126,157,212	66,76,114,146,175,221	28,43,71,100,134,161	40.8545

Table 6.29 Best results obtained for QIGAMLTCI and QISAMLTCI for multi-level thresholding of Baboon, Bird, Elephant, Seaport, and Montreal for Huang's method [130].

Baboon

K	QIGAMLTCI θ_R	θ_G	θ_B	y_{best}	QISAMLTCI θ_R	θ_G	θ_B	y_{best}
3	102,250	15,201	79,129	2.5495	71,170	77,126	13,158	2.5493
5	8,45,164,211	13,182,204,215	9,32,84,236	2.5615	22,101,226,232	19,54,75,253	26,30,60,236	2.5576
7	26,35,36,72,205,248	15,56,192,200,203,252	100,101,109,139,169,227	2.6068	27,142,159,160,199,251	14,20,23,51,82,246	5,12,162,210,234,242	2.6096

Bird

K	QIGAMLTCI θ_R	θ_G	θ_B	y_{best}	QISAMLTCI θ_R	θ_G	θ_B	y_{best}
3	62,215	114,133	46,200	2.4089	50,74	65,232	117,206	2.4089
5	36,84,252,253	49,194,217,242	67,87,199,201	2.4089	142,175,175,186	39,185,217,244	45,69,226,248	2.4089
7	39,46,210,228,249,254	34,171,176,201,209,253	49,57,61,200,214,227	2.4825	102,104,120,218,242,251	11,156,157,171,198,245	37,60,61,203,233,236	2.4900

Elephant

K	QIGAMLTCI θ_R	θ_G	θ_B	y_{best}	QISAMLTCI θ_R	θ_G	θ_B	y_{best}
3	145,169	88,247	196,217	2.5404	158,211	111,134	93,175	2.5407
5	124,232,236,251	36,83,163,201	91,135,188,232	2.5489	6,48,116,234	58,100,128,220	8,134,182,233	2.5853
7	2,12,73,146,221,230	90,136,147,149,154,228	68,145,155,168,217,223	2.6299	35,62,101,116,122,250	6,114,153,155,197,253	50,51,55,69,189,246	2.6331

Seaport

K	QIGAMLTCI θ_R	θ_G	θ_B	y_{best}	QISAMLTCI θ_R	θ_G	θ_B	y_{best}
3	141,233	154,164	41,173	2.7252	124,172	153,163	159,201	2.7279
5	67,94,157,219	95,135,153,177	91,142,158,177	2.7333	82,84,241,243	4,136,188,199	60,158,162,226	2.7331
7	3,140,143,203,218,225	114,146,211,247,252,254	47,56,169,173,193,231	2.7381	28,115,136,137,162,222	191,196,199,228,237,243	22,27,107,118,151,204	2.7407

Montreal

K	QIGAMLTCI θ_R	θ_G	θ_B	y_{best}	QISAMLTCI θ_R	θ_G	θ_B	y_{best}
3	54,215	33,193	76,148	2.7459	116,225	193,234	65,132	2.7459
5	7,174,180,237	145,203,208,244	2,171,217,249	2.7474	2,173,195,201	110,137,197,222	114,197,211,230	2.7477
7	217,222,230,235,238,245	105,160,170,182,195,196	113,150,201,239,239,247	2.7483	158,169,206,229,229,233	18,99,142,156,214,247	134,136,158,162,164,239	2.7504

Table 6.30 Best results obtained for QIGAMLTCI and QISAMLTCI for multi-level thresholding of Monolake, Mona Lisa, Manhattan, Lighthouse, and Daisies for Huang's method [130].

Monolake

K	QIGAMLTCI				QISAMLTCI			
	θ_R	θ_G	θ_B	y_{best}	θ_R	θ_G	θ_B	y_{best}
3	6,152	42,230	18,226	2.6386	82,119	20,202	88,186	2.6487
5	39,155,158,201	33,133,150,239	49,145,173,225	2.6683	10,131,158,225	7,26,197,240	57,167,177,235	2.6587
7	66,93,104,125,163,217	13,57,66,111,134,231	22,32,72,81,136,224	2.6994	26,46,141,197,241,242	24,46,169,173,206,238	80,157,182,190,203,253	2.6951

Mona Lisa

K	QIGAMLTCI				QISAMLTCI			
	θ_R	θ_G	θ_B	y_{best}	θ_R	θ_G	θ_B	y_{best}
3	6,228	25,242	141,206	2.6048	27,227	172,189	220,252	2.6123
5	20,194,209,227	11,167,178,244	10,189,214,229	2.6101	10,166,231,237	75,78,96,136	71,142,213,217	2.6079
7	155,160,221,230,237,241	12,18,58,167,213,220	30,143,195,228,229,247	2.6270	14,34,194,199,207,208	36,149,185,199,224,234	135,137,170,195,236,243	2.6291

Manhattan

K	QIGAMLTCI				QISAMLTCI			
	θ_R	θ_G	θ_B	y_{best}	θ_R	θ_G	θ_B	y_{best}
3	171,236	94,133	58,208	2.6404	41,243	124,176	62,176	2.6404
5	29,53,135,238	5,151,191,217	28,138,141,237	2.6410	34,163,207,227	22,33,38,219	64,109,171,197	2.6408
7	10,147,165,168,239,248	30,33,163,222,225,247	16,71,76,138,153,210	2.6440	25,36,124,153,163,186	12,33,35,170,238,247	29,34,130,142,163,175	2.6416

Lighthouse

K	QIGAMLTCI				QISAMLTCI			
	θ_R	θ_G	θ_B	y_{best}	θ_R	θ_G	θ_B	y_{best}
3	97,194	59,204	213,225	2.4324	79,237	167,238	28,229	2.4313
5	162,177,226,250	35,39,61,208	30,97,235,236	2.4425	171,188,197,220	4,95,102,239	92,146,238,239	2.4421
7	106,187,204,221,228,246	65,67,75,202,212,214	43,72,72,213,214,239	2.4980	21,37,189,192,219,249	120,173,185,203,222,235	27,49,53,65,208,211	2.4834

Daisies

K	QIGAMLTCI				QISAMLTCI			
	θ_R	θ_G	θ_B	y_{best}	θ_R	θ_G	θ_B	y_{best}
3	41,82	68,116	13,206	2.6714	49,225	42,146	140,155	2.6714
5	62,163,202,239	28,129,142,186	11,52,210,218	2.6804	2,33,38,219	24,155,218,239	37,45,47,208	2.6839
7	10,13,193,222,241,253	19,22,177,178,216,252	5,97,213,218,237,252	2.7222	24,69,74,86,236,239	80,104,141,142,235,245	7,9,12,76,217,243	2.7406

Table 6.31 Best results obtained for QIPSOMLTCI and QIDEMLTCI for multi-level thresholding of Baboon, Bird, Elephant, Seaport, and Montreal for Huang's method [130].

Baboon

K	QIPSOMLTCI				QIDEMLTCI			
	θ_R	θ_G	θ_B	V_{best}	θ_R	θ_G	θ_B	V_{best}
3	107,239	139,174	56,240	2.5505	17,243	53,251	101,230	2.5487
5	5,9,99,233	79,81,203,249	30,117,118,237	2.5601	12,20,77,251	83,186,203,235	146,154,221,247	2.5611
7	10,66,178,184,214,226	151,169,189,192,236,238	133,143,148,188,228,229	2.4982	7,17,54,94,251,252	20,184,193,201,227,238	11,75,78,99,102,225	2.5775

Bird

K	QIPSOMLTCI				QIDEMLTCI			
	θ_R	θ_G	θ_B	V_{best}	θ_R	θ_G	θ_B	V_{best}
3	40,251	101,209	44,126	2.4089	185,199	23,215	46,204	2.4089
5	159,204,239,245	12,60,94,237	24,79,196,229	2.4089	44,57,203,223	3,9,58,207	24,164,227,234	2.4089
7	162,169,184,212,240,242	25,201,213,222,229,236	33,45,68,82,217,253	2.4770	4,35,40,171,241,245	149,167,191,212,222,247	19,142,183,216,217,235	2.4807

Elephant

K	QIPSOMLTCI				QIDEMLTCI			
	θ_R	θ_G	θ_B	V_{best}	θ_R	θ_G	θ_B	V_{best}
3	62,118	104,140	140,167	2.5406	76,250	133,180	19,110	2.5407
5	37,72,114,186	45,57,142,233	45,59,181,231	2.5861	72,109,116,189	27,62,113,182	14,72,183,223	2.5717
7	4,141,148,181,185,242	30,49,168,181,185,242	125,136,155,174,178,212	2.5935	88,93,106,180,224,251	144,150,155,157,183,242	17,48,176,202,216,226	2.6070

Seaport

K	QIPSOMLTCI				QIDEMLTCI			
	θ_R	θ_G	θ_B	V_{best}	θ_R	θ_G	θ_B	V_{best}
3	81,195	25,130	105,183	2.7268	180,211	98,220	148,157	2.7279
5	21,49,158,233	33,41,61,198	28,83,99,137	2.7332	42,54,61,255	45,124,235,254	43,93,102,133	2.7303
7	47,97,120,131,154,237	64,72,109,216,238,247	7,94,162,207,230,248	2.7373	18,22,76,148,220,253	25,39,55,104,112,141	117,127,142,170,190,251	2.7400

Montreal

K	QIPSOMLTCI				QIDEMLTCI			
	θ_R	θ_G	θ_B	V_{best}	θ_R	θ_G	θ_B	V_{best}
3	35,67	107,146	87,159	2.7459	33,186	42,174	66,170	2.7459
5	2,75,171,234	22,137,174,196	20,142,166,216	2.7463	214,225,236,245	219,219,226,244	144,157,219,230	2.7463
7	38,189,201,227,228,231	210,212,221,223,225,247	179,208,214,227,230,248	2.7496	3,15,57,162,174,194	15,21,129,176,205,211	25,37,40,156,228,229	2.7505

Table 6.32 Best results obtained for QIPSOMLTCI and QIDEMLTCI for multi-level thresholding of Monolake, Mona Lisa, Manhattan, Lighthouse, and Daisies for Huang's method [130].

Monolake

K	QIPSOMLTCI				QIDEMLTCI			
	θ_R	θ_G	θ_B	V_{best}	θ_R	θ_G	θ_B	V_{best}
3	221,254	76,198	19,164	2.6493	250,251	24,229	47,230	2.6386
5	47,50,85,134	11,212,228,241	34,57,134,219	2.6624	63,109,137,194	69,157,164,217	24,28,77,216	2.6689
7	68,159 216,225,225,242	86,99,117,127,184,227	4,75,120,174,237,246	2.6961	74,75,96,181,220,237	13,66,69,175,180,242	57,114,157,169,172,242	2.7030

Mona Lisa

K	QIPSOMLTCI				QIDEMLTCI			
	θ_R	θ_G	θ_B	V_{best}	θ_R	θ_G	θ_B	V_{best}
3	109,145	90,127	21,114	2.6052	89,139	32,167	51,167	2.6061
5	3,115,128,134	60,62,140,158	4,57,138,180	2.6185	47,172,202,212	158,170,177,188	23,140,151,246	2.6126
7	26,147,168,194,228,229	15,28,55,70,205,212	7,109,123,125,152,162	2.6238	19,40,51,225,248,249	24,153,179,198,231,246	28,94,140,163,176,203	2.6172

Manhattan

K	QIPSOMLTCI				QIDEMLTCI			
	θ_R	θ_G	θ_B	V_{best}	θ_R	θ_G	θ_B	V_{best}
3	78,232	75,159	117,185	2.6404	58,252	136,244	142,173	2.6404
5	6,123,153,163	60,72,131,215	74,83,139,188	2.6409	41,179,193,244	139,225,234,242	75,127,138,205	2.6415
7	122,127,147,185,204,215	17,98,200,215,232,241	10,26,148,199,201,238	2.6439	20,29,49,195,229,246	26,51,105,176,184,185	33,47,156,189,213,216	2.6459

Lighthouse

K	QIPSOMLTCI				QIDEMLTCI			
	θ_R	θ_G	θ_B	V_{best}	θ_R	θ_G	θ_B	V_{best}
3	41,179	141,183	251,254	2.4324	150,191	130,172	106,110	2.4336
5	132,161,197,200	1,23,43,180	96,115,242,245	2.4428	39,193,215,221	6,147,150,240	2,31,57,221	2.4430
7	23,50,107,202,203,254	41,43,177,208,231,233	15,169,179,181,204,221	2.4997	29,63,66,195,215,238	12,16,32,85,211,245	175,196,213,216,243,253	2.4899

Daisies

K	QIPSOMLTCI				QIDEMLTCI			
	θ_R	θ_G	θ_B	V_{best}	θ_R	θ_G	θ_B	V_{best}
3	44,133	65,221	70,127	2.6722	69,196	40,220	43,51	2.6726
5	72,101,105,219	63,86,228,250	4,194,229,247	2.6827	29,169,183,244	15,23,39,239	4,21,38,252	2.6838
7	18,29,152,154,213,226	61,68,71,103,124,231	37,42,144,150,155,234	2.7406	25,105,182,227,243,255	6,31,38,156,211,220	1,6,33,56,66,209	2.7177

Table 6.33 Best results obtained for GA and SA for multi-level thresholding of Baboon, Bird, Elephant, Seaport, and Montreal for Huang's method [130].

Baboon

K	GA				SA			
	θ_R	θ_G	θ_B	V_{best}	θ_R	θ_G	θ_B	V_{best}
3	169,208	168,244	41,59	2.5670	16,167	105,140	43,67	2.5705
5	27,199,211,240	40,64,137,172	22,47,124,151	2.6008	55,100,108,116	16,17,75,164	50,154,161,221	2.6017
7	26,38,39,48,100,146	25,39,130,177,213,250	92,114,120,162,163,223	2.6411	46,119,183,214,240,243	54,65,73,79,149,151	2,8,132,142,186,187	2.6428

Bird

K	GA				SA			
	θ_R	θ_G	θ_B	V_{best}	θ_R	θ_G	θ_B	V_{best}
3	108,205	11,75	26,179	2.4089	25,180	169,191	10,247	2.40894
5	15,25,33,145	64,105,174,239	34,47,109,195	2.4497	126,133,148,153	22,28,57,92	25,34,77,89	2.4596
7	23,28,41,42,167,229	41,59,78,91,185,190	199,201,207,222,238,247	2.5569	68,105,124,187,188,203	12,93,104,127,215,221	10,188,212,218,224,239	2.6077

Elephant

K	GA				SA			
	θ_R	θ_G	θ_B	V_{best}	θ_R	θ_G	θ_B	V_{best}
3	172,212	113,220	8,82	2.5415	33,72	92,171	41,129	2.5422
5	109,199,215,218	9,107,164,229	202,216,218,238	2.5943	80,106,112,149	36,73,106,171	27,58,190,232	2.6075
7	13,193,196,213,214,235	13,112,114,179,186,219	21,44,116,217,225,246	2.6410	33,65,83,103,222,233	29,76,215,219,240,251	116,132,174,205,233,236	2.6653

Seaport

K	GA				SA			
	θ_R	θ_G	θ_B	V_{best}	θ_R	θ_G	θ_B	V_{best}
3	39,69	38,44	99,250	2.7898	36,69	13,29	15,179	2.7959
5	174,190,224,244	28,59,141,173	104,118,178,213	2.7353	27,100,227,234	10,34,82,218	31,91,99,222	2.7352
7	179,194,208,223,250,255	34,81,89,120,136,211	76,124,148,218,239,246	2.7444	58,131,139,163,205,232	148,189,192,206,228,244	131,153,182,194,204,250	2.7416

Montreal

K	GA				SA			
	θ_R	θ_G	θ_B	V_{best}	θ_R	θ_G	θ_B	V_{best}
3	80,136	151,175	120,219	2.7459	135,178	3,33	61,231	2.7459
5	29,99,165,203	41,96,140,211	209,215,222,226	2.7509	102,147,160,224	10,48,69,89	8,71,206,210	2.7503
7	159,170,189,201,219,222	64,108,109,154,225,243	169,174,185,198,233,234	2.7607	48,121,141,162,167,223	54,110,147,157,187,191	61,94,148,149,161,174	2.7619

Table 6.34 Best results obtained for GA and SA for multi-level thresholding of Monolake, Mona Lisa, Manhattan, Lighthouse, and Daisies for Huang's method [130].

Monolake

K	GA				SA			
	θ_R	θ_G	θ_B	V_{best}	θ_R	θ_G	θ_B	V_{best}
3	35,240	90,244	125,240	2.6512	64,134	60,210	33,37	2.6506
5	150,235,235,245	18,44,211,249	5,9,153,192	2.6750	35,44,205,239	102,153,205,219	19,23,87,158	2.6718
7	123,126,169,171,177,242	30,33,113,128,172,190	23,25,218,226,240,255	2.7115	84,94,145,164,237,244	25,127,195,204,210,246	32,38,45,47,53,64	2.7123

Mona Lisa

K	GA				SA			
	θ_R	θ_G	θ_B	V_{best}	θ_R	θ_G	θ_B	V_{best}
3	138,166	66,235	24,39	2.6119	42,107	37,246	11,242	2.6050
5	20,31,73,77	8,18,154,241	28,112,125,184	2.6299	10,110,147,228	11,44,49,60	32,42,70,220	2.6282
7	19,22,26,38,47,245	15,67,87,128,164,241	26,147,170,186,199,250	2.6524	64,97,102,149,184,185	89,141,186,194,195,235	12,18,44,55,73,77	2.6599

Manhattan

K	GA				SA			
	θ_R	θ_G	θ_B	V_{best}	θ_R	θ_G	θ_B	V_{best}
3	112,221	19,195	57,240	2.6408	87,180	38,233	16,90	2.6408
5	25,77,108,121	40,66,72,156	43,120,158,208	2.6434	42,99,153,240	19,54,116,172	14,64,81,254	2.6436
7	4,40,160,166,195,238	63,82,132,148,166,202	3,17,176,199,206,234	2.6530	16,26,195,196,198,236	24,87,142,184,208,247	47,54,63,66,99,149	2.6637

Lighthouse

K	GA				SA			
	θ_R	θ_G	θ_B	V_{best}	θ_R	θ_G	θ_B	V_{best}
3	29,199	120,192	169,180	2.4331	106,219	5,35	40,111	2.4337
5	9,97,159,249	29,231,240,245	51,76,240,253	2.5000	212,216,226,227	42,80,187,204	80,175,180,249	2.4829
7	5,154,138,193,223,234	14,23,138,212,217,246	71,201,201,211,221,237	2.5713	130,156,163,210,215,242	153,172,189,220,222,231	8,42,68,74,211,249	2.5601

Daisies

K	GA				SA			
	θ_R	θ_G	θ_B	V_{best}	θ_R	θ_G	θ_B	V_{best}
3	24,51	85,139	166,253	2.6731	213,255	84,157	137,241	2.6736
5	7,36,96,124	64,71,219,232	28,30,53,184	2.7393	17,58,83,186	47,114,143,219	148,167,209,250	2.7382
7	32,49,104,125,152,243	7,26,118,148,220,226	41,42,131,133,190,239	2.7791	157,170,184,209,244,250	27,52,58,76,216,241	89,93,103,212,223,254	2.7390

Table 6.35 Best results obtained for PSO and DE for multi-level thresholding of Baboon, Bird, Elephant, Seaport, and Montreal for Huang's method [130].

Baboon

K	PSO θ_R	PSO θ_G	PSO θ_B	PSO y_{best}	DE θ_R	DE θ_G	DE θ_B	DE y_{best}
3	141,153	13,246	38,72	2.5504	54,150	23,61	53,234	2.5589
5	46,81,128,160	68,162,171,234	3,35,220,250	2.5800	32,54,59,78	66,159,211,243	49,75,221,246	2.5888
7	75,117,134,169,218,252	15,84,87,90,111,132	46,67,68,105,140,232	2.6510	28,31,79,102,106,107	93,123,154,166,227,245	8,35,60,130,156,181	2.6498

Bird

K	PSO θ_R	PSO θ_G	PSO θ_B	PSO y_{best}	DE θ_R	DE θ_G	DE θ_B	DE y_{best}
3	7,19	48,55	60,121	2.4089	15,116	18,113	175,240	2.4089
5	33,57,174,191	144,155,177,250	189,212,223,227	2.4738	18,49,84,175	15,80,137,181	28,56,188,213	2.4440
7	48,51,117,147,185,230	8,57,86,88,94,109	71,79,169,171,196,212	2.6099	53,75,167,201,218,254	218,220,224,230,236,245	69,156,183,198,202,243	2.5543

Elephant

K	PSO θ_R	PSO θ_G	PSO θ_B	PSO y_{best}	DE θ_R	DE θ_G	DE θ_B	DE y_{best}
3	61,222	235,240	18,231	2.5404	84,234	125,220	32,41	2.5406
5	186,216,224,232	6,94,193,243	182,203,221,229	2.5910	15,212,216,241	133,152,169,173	53,114,184,236	2.6109
7	152,158,166,200,202,237	19,149,150,195,212,236	122,160,177,210,229,247	2.6403	3,17,29,150,222,246	83,100,175,213,221,235	20,21,33,190,201,210	2.6406

Seaport

K	PSO θ_R	PSO θ_G	PSO θ_B	PSO y_{best}	DE θ_R	DE θ_G	DE θ_B	DE y_{best}
3	68,107	2,16	153,202	2.7836	36,69	13,29	15,179	2.7959
5	46,124,172,187	20,31,83,248	23,97,209,242	2.7345	126,131,142,183	81,188,208,222	85,129,173,206	2.7352
7	47,60,93,112,149,245	46,113,129,137,145,162	31,35,45,65,84,200	2.7423	115,149,170,238,238,243	122,151,154,166,178,218	6,17,45,184,216,237	2.7427

Montreal

K	PSO θ_R	PSO θ_G	PSO θ_B	PSO y_{best}	DE θ_R	DE θ_G	DE θ_B	DE y_{best}
3	187,244	51,174	111,190	2.7459	190,219	5,143	20,251	2.7459
5	13,42,61,71	158,164,171,202	34,63,113,165	2.7508	2,119,156,241	28,49,92,251	24,70,95,107	2.7510
7	108,170,174,196,236,244	15,109,126,128,135,207	32,59,91,145,172,192	2.7603	13,57,64,76,85,132	5,17,62,194,231,251	9,26,141,212,221,243	2.7604

Table 6.36 Best results obtained for PSO and DE for multi-level thresholding of Monolake, Mona Lisa, Manhattan, Lighthouse, and Daisies for Huang's method [130].

Monolake

K	PSO				DE			
	θ_R	θ_G	θ_B	ν_{best}	θ_R	θ_G	θ_B	ν_{best}
3	236,253	193,197	172,249	2.6440	206,220	220,242	191,244	2.6437
5	89,95,119,242	206,223,227,244	83,93,105,236	2.6762	150,235,235,245	18,44,211,249	5,9,153,192	2.6750
7	51,62,74,131,144,177	45,91,127,134,159,196	41,59,123,129,132,144	2.8759	1,46,80,108,157,230	12,22,52,86,105,107	4,47,59,98,148,149	2.8714

Mona Lisa

K	PSO				DE			
	θ_R	θ_G	θ_B	ν_{best}	θ_R	θ_G	θ_B	ν_{best}
3	4,139	4,42	67,101	2.6101	61,93	50,100	200,220	2.6524
5	12,67,186,234	17,35,242,249	28,38,146,203	2.6298	10,49,149,202	16,48,187,245	39,61,70,100	2.6233
7	7,36,55,82,104,144	80,123,139,185,199,252	144,170,220,223,227,238	2.6663	156,169,226,227,228,241	13,26,53,212,219,238	15,37,220,227,238,255	2.6332

Manhattan

K	PSO				DE			
	θ_R	θ_G	θ_B	ν_{best}	θ_R	θ_G	θ_B	ν_{best}
3	191,245	235,249	127,156	2.6408	134,188	46,227	149,243	2.6408
5	15,72,96,99	39,67,105,174	56,69,165,219	2.6443	15,16,31,220	47,124,125,147	18,43,170,215	2.6448
7	24,41,82,85,95,213	15,49,72,78,88,255	42,54,61,76,115,129	2.6660	16,26,195,196,198,236	24,87,142,184,208,247	47,54,63,66,99,149	2.6637

Lighthouse

K	PSO				DE			
	θ_R	θ_G	θ_B	ν_{best}	θ_R	θ_G	θ_B	ν_{best}
3	224,250	90,103	162,195	2.4327	11,14	182,228	88,163	2.4336
5	57,83,226,245	53,217,229,245	2,5,53,162	2.5242	22,130,145,182	47,55,97,104	56,83,106,238	2.4988
7	114,118,119,151,184,248	5,44,197,212,214,249	2,16,32,78,218,253	2.5347	111,120,189,191,223,230	26,56,147,210,213,227	23,54,81,85,241,248	2.5769

Daisies

K	PSO				DE			
	θ_R	θ_G	θ_B	ν_{best}	θ_R	θ_G	θ_B	ν_{best}
3	232,241	183,197	108,178	2.6731	15,235	154,250	163,254	2.6736
5	6,215,220,244	47,86,94,129	217,219,240,246	2.7134	59,126,187,227	43,120,172,180	136,140,210,220	2.7293
7	180,189,212,235,239,249	33,60,66,78,101,244	74,104,119,236,238,243	2.7474	51,84,85,88,243,254	26,52,167,223,250,254	130,139,159,208,219,241	2.7487

Table 6.37 Best results obtained for BSA and CoDE for multi-level thresholding of Baboon, Bird, Elephant, Seaport, and Montreal for Huang's method [130].

Baboon

	BSA				CoDE			
K	θ_R	θ_G	θ_B	V_{best}	θ_R	θ_G	θ_B	V_{best}
3	13,115	22,140	126,199	2.5567	54,112	193,229	13,233	2.5572
5	56,163,168,183	29,93,120,145	40,106,144,225	2.5850	125,134,226,251	6,59,71,116	46,66,107,254	2.5791
7	59,75,89,223,232,250	24,31,70,188,205,238	9,15,20,23,231,245	2.6147	52,65,71,83,86,221	6,12,36,46,211,219	172,176,189,206,225,248	2.6121

Bird

	BSA				CoDE			
K	θ_R	θ_G	θ_B	V_{best}	θ_R	θ_G	θ_B	V_{best}
3	129,235	217,227	221,235	2.4089	189,243	203,237	174,183	2.4089
5	169,179,232,248	132,133,209,217	5,31,174,225	2.4089	118,182,222,253	14,15,249,253	34,56,65,219	2.4089
7	5,12,22,37,180,201	21,91,206,209,211,213	10,12,26,48,217,249	2.4939	51,56,67,73,246,251	162,164,191,191,210,231	68,121,188,224,225,246	2.4993

Elephant

	BSA				CoDE			
K	θ_R	θ_G	θ_B	V_{best}	θ_R	θ_G	θ_B	V_{best}
3	3,255	121,154	78,132	2.5373	62,92	114,232	83,212	2.5373
5	10,30,39,213	50,93,207,230	2,85,120,137	2.5588	59,92,153,241	19,66,67,87	44,139,175,186	2.5858
7	12,100,105,157,189,226	3,28,38,42,61,173	21,22,48,71,75,128	2.6367	36,38,39,201,206,216	69,72,88,232,252,255	144,147,208,239,239,254	2.6400

Seaport

	BSA				CoDE			
K	θ_R	θ_G	θ_B	V_{best}	θ_R	θ_G	θ_B	V_{best}
3	138,201	157,250	152,156	2.7279	242,244	10,159	51,94	2.7279
5	112,167,217,237	61,78,210,231	59,179,239,244	2.7357	30,49,150,236	5,74,200,248	115,135,159,229	2.7352
7	8,178,236,238,246,248	8,43,62,79,138,172	127,185,198,214,242,255	2.7401	120,137,181,230,242,243	83,124,191,200,223,233	120,156,195,200,211,224	2.7408

Montreal

	BSA				CoDE			
K	θ_R	θ_G	θ_B	V_{best}	θ_R	θ_G	θ_B	V_{best}
3	40,233	65,148	69,152	2.7459	69,137	63,118	171,198	2.7459
5	116,159,183,211	60,87,135,137	171,189,194,252	2.7510	52,233,235,242	10,79,133,221	93,158,241,245	2.7497
7	72,76,105,118,177,187	13,56,80,97,114,152	31,67,97,103,204,245	2.7591	26,38,46,85,120,121	11,32,37,165,166,209	37,40,81,157,163,195	2.7532

Table 6.38 Best results obtained for BSA and CoDE for multi-level thresholding of Monolake, Mona Lisa, Manhattan, Lighthouse, and Daisies for Huang's method [130].

Monolake

K	BSA				CoDE			
	θ_R	θ_G	θ_B	V_{best}	θ_R	θ_G	θ_B	V_{best}
3	205,248	27,249	42,246	2.6415	21,38	14,106	169,203	2.6501
5	15,125,169,251	190,197,217,220	6,13,243,249	2.6637	8,190,197,210	26,36,182,237	43,83,194,224	2.6635
7	4,133,134,161,212,234	66,72,187,200,203,228	17,28,40,206,249,253	2.6883	13,23,135,192,233,235	36,153,175,201,203,215	1,27,40,236,243,248	2.6846

Mona Lisa

K	BSA				CoDE			
	θ_R	θ_G	θ_B	V_{best}	θ_R	θ_G	θ_B	V_{best}
3	136,247	23,98	141,229	2.6048	8,175	33,69	128,244	2.6048
5	3,65,95,180	45,65,126,136	11,37,44,186	2.6159	52,76,85,157	19,23,63,145	51,80,126,152	2.6308
7	1,169,170,191,195,209	8,153,184,203,230,234	14,153,173,199,201,208	2.6316	17,21,36,185,215,255	54,92,208,223,249,252	7,135,138,155,184,208	2.6309

Manhattan

K	BSA				CoDE			
	θ_R	θ_G	θ_B	V_{best}	θ_R	θ_G	θ_B	V_{best}
3	35,198	40,240	127,191	2.6408	178,197	227,246	22,24	2.6408
5	28,37,202,215	27,145,190,236	50,89,141,180	2.6420	124,67,134,241	7,17,24,160	187,193,208,238	2.6426
7	2,42,76,99,105,255	33,63,91,94,186,202	13,151,170,174,211,247	2.6619	9,144,176,202,231,255	17,40,43,163,225,232	11,18,20,45,195,207	2.6467

Lighthouse

K	BSA				CoDE			
	θ_R	θ_G	θ_B	V_{best}	θ_R	θ_G	θ_B	V_{best}
3	14,245	177,237	231,245	2.4324	80,95	28,164	56,228	2.4338
5	22,44,62,252	22,79,120,156	25,49,78,90	2.4714	19,109,136,215	43,216,219,231	12,142,207,246	2.5147
7	178,180,183,202,208,218	22,32,48,64,177,194	9,78,195,197,234,236	2.5616	65,215,208,224,228,251	10,210,222,233,234,241	13,37,78,89,220,230	2.5012

Daisies

K	BSA				CoDE			
	θ_R	θ_G	θ_B	V_{best}	θ_R	θ_G	θ_B	V_{best}
3	117,240	110,123	67,142	2.6726	34,38	30,188	15,71	2.6726
5	34,44,110,195	22,36,63,107	51,138,147,201	2.7251	186,207,230,250	20,54,141,250	27,61,201,219	2.7182
7	96,173,181,211,228,246	64,132,145,184,192,240	156,185,197,234,238,239	2.7545	28,30,128,135,217,223	15,20,25,27,37,97	37,70,107,148,176,209	2.7819

Table 6.39 Computational time (*t*) of QIGAMLTCI, QISAMLTCI,QIDEMLTCI, QIPSOMLTCI, GA, SA, DE, PSO, BSA, and CoDE for multi-level thresholding of Baboon, Bird, Elephant, Seaport, Montreal, Monolake, Mona Lisa, Manhattan, Lighthouse and Daisies for Kapur's method [140].

					Baboon					
K	(1)	(2)	(3)	(4)	(5)	(6)	(7)	(8)	(9)	(10)
3	05.12	06.13	01.18	05.12	21.18	32.10	15.12	16.13	11.18	11.10
5	08.18	08.32	04.13	07.22	30.13	44.12	28.18	28.32	30.13	24.12
7	10.22	10.51	09.12	12.22	48.52	62.27	46.13	46.51	38.52	35.27

					Bird					
K	(1)	(2)	(3)	(4)	(5)	(6)	(7)	(8)	(9)	(10)
3	04.14	05.30	01.17	04.26	22.17	29.25	14.14	15.30	12.17	19.25
5	07.24	09.20	03.57	08.18	32.27	44.16	25.24	26.18	22.27	24.16
7	11.22	12.05	10.02	10.15	52.15	62.15	42.16	44.05	32.15	33.15

					Elephant					
K	(1)	(2)	(3)	(4)	(5)	(6)	(7)	(8)	(9)	(10)
3	05.11	06.11	01.19	04.28	24.19	28.10	15.11	16.11	14.19	17.10
5	07.49	09.02	03.35	08.08	33.22	40.16	27.49	29.02	23.22	30.16
7	10.13	11.48	09.23	12.03	49.11	60.58	42.13	45.48	41.11	35.16

					Seaport					
K	(1)	(2)	(3)	(4)	(5)	(6)	(7)	(8)	(9)	(10)
3	05.04	06.10	01.10	04.17	24.10	30.11	15.04	17.10	16.12	14.21
5	08.37	08.39	04.03	08.12	34.03	43.18	28.37	28.39	24.16	22.15
7	11.23	12.01	10.14	11.04	52.14	62.15	41.23	42.01	39.24	33.19

					Montreal					
K	(1)	(2)	(3)	(4)	(5)	(6)	(7)	(8)	(9)	(10)
3	04.29	06.39	01.18	05.01	22.18	29.12	17.23	19.39	17.13	15.12
5	08.11	08.20	05.07	07.21	35.09	44.29	25.16	28.20	25.10	24.36
7	12.01	10.21	09.32	10.10	53.49	63.10	46.01	45.21	37.42	33.09

					Monolake					
K	(1)	(2)	(3)	(4)	(5)	(6)	(7)	(8)	(9)	(10)
3	05.18	05.30	01.13	05.10	23.13	31.09	12.18	12.30	13.13	11.21
5	08.28	09.49	05.10	07.24	32.38	43.14	22.28	27.49	27.10	22.17
7	10.39	10.52	09.12	09.22	53.11	65.18	42.39	42.52	35.11	34.05

					Mona Lisa					
K	(1)	(2)	(3)	(4)	(5)	(6)	(7)	(8)	(9)	(10)
3	04.49	06.10	01.11	04.50	22.11	31.20	12.49	14.10	14.25	13.22
5	07.22	09.03	05.28	07.11	32.22	41.30	27.22	26.03	26.20	21.32
7	10.05	10.50	09.13	09.48	55.13	59.42	47.05	47.50	38.13	35.12

					Manhattan					
K	(1)	(2)	(3)	(4)	(5)	(6)	(7)	(8)	(9)	(10)
3	05.30	05.49	01.32	05.12	20.32	28.12	13.30	14.49	17.18	17.10
5	08.10	08.18	04.22	08.31	34.11	40.30	24.10	22.18	28.11	22.31
7	09.48	11.14	09.07	10.09	52.06	63.11	39.48	40.14	37.06	34.16

					Lighthouse					
K	(1)	(2)	(3)	(4)	(5)	(6)	(7)	(8)	(9)	(10)
3	05.03	06.07	01.06	05.11	26.06	29.01	15.03	16.07	16.12	14.22
5	08.18	09.20	05.13	08.12	34.38	42.18	28.18	29.20	26.14	26.19
7	10.23	11.42	09.26	11.07	49.35	63.47	40.12	41.46	39.35	35.17

					Daisies					
K	(1)	(2)	(3)	(4)	(5)	(6)	(7)	(8)	(9)	(10)
3	05.12	06.12	01.18	04.22	21.18	30.12	15.12	16.12	18.26	14.16
5	08.37	09.13	04.48	08.14	33.08	41.11	28.37	29.13	23.20	21.16
7	11.07	12.10	10.11	11.21	50.10	60.10	42.07	42.15	35.10	32.15
(1)→ QIGAMLTCI (2)→ QISAMLTCI (3)→ QIPSOMLTCI (4)→ QIDEMLTCI										
(5)→ GA (6)→ SA (7)→ PSO (8)→ DE (9)→ BSA (10)→ CoDE										

Table 6.40 Computational time (*t*) of QIGAMLTCI, QISAMLTCI,QIDEMLTCI, QIPSOMLTCI, GA, SA, DE, PSO, BSA, and CoDE for multi-level thresholding of Baboon, Bird, Elephant, Seaport, Montreal, Monolake, Mona Lisa, Manhattan, Lighthouse, and Daisies for Huang's method [130].

K	(1)	(2)	(3)	(4)	(5)	(6)	(7)	(8)	(9)	(10)
Baboon										
3	05.12	06.13	01.23	06.11	21.28	30.12	17.18	18.15	11.18	12.10
5	08.18	08.32	08.16	08.42	31.16	40.19	27.15	28.15	20.23	24.12
7	10.22	10.51	10.02	10.31	49.28	59.05	44.22	48.27	38.54	32.27
Bird										
3	04.14	05.30	02.24	05.32	22.15	28.20	17.15	19.24	12.17	19.26
5	07.24	09.20	07.27	08.20	34.21	42.09	25.27	28.32	22.27	24.06
7	11.22	12.05	11.27	12.09	51.32	61.19	41.17	44.17	32.15	32.14
Elephant										
3	05.11	06.11	01.48	06.10	21.04	30.42	18.18	20.20	14.19	18.10
5	07.49	09.02	07.42	09.12	31.42	43.38	30.02	27.25	23.21	20.15
7	10.13	11.48	10.15	11.42	49.22	59.20	43.02	46.32	39.11	30.56
Seaport										
3	05.04	06.10	02.10	06.01	24.31	32.13	17.08	18.03	14.10	13.11
5	08.37	08.39	08.17	08.19	34.18	39.44	29.30	28.10	24.03	23.16
7	11.23	12.01	11.21	12.10	52.16	60.03	44.12	49.11	22.14	32.16
Montreal										
3	04.29	06.39	01.12	05.14	22.10	31.12	19.17	19.50	12.18	19.11
5	08.11	08.20	08.14	08.22	34.19	42.15	28.19	26.22	25.09	24.28
7	12.01	10.21	10.12	10.22	53.10	62.01	41.30	46.27	33.48	33.01
Monolake										
3	05.18	05.30	02.18	05.31	23.32	32.05	17.12	18.17	13.13	11.09
5	08.28	09.49	08.26	09.39	34.12	42.15	22.32	17.34	22.39	23.14
7	10.39	10.52	10.37	10.55	55.05	63.14	46.12	48.16	33.15	35.16
Mona Lisa										
3	04.49	06.10	02.13	06.09	22.14	29.20	17.38	18.15	12.11	11.20
5	07.22	09.20	07.20	09.06	33.22	41.17	31.22	28.24	22.23	21.30
7	10.05	10.50	10.07	10.53	54.17	62.44	44.23	49.17	35.13	29.48
Manhattan										
3	05.30	05.49	02.04	05.42	21.28	31.14	18.12	20.22	12.32	18.32
5	08.10	08.18	08.11	08.19	34.06	43.34	24.21	29.16	24.12	24.30
7	09.48	11.14	09.46	11.15	54.06	63.18	42.06	48.15	32.06	33.11
Lighthouse										
3	05.03	06.07	02.07	06.01	24.09	31.03	16.07	19.05	16.08	19.01
5	08.18	09.20	08.15	09.22	33.28	42.16	25.36	29.11	24.36	22.18
7	10.23	11.42	10.25	11.41	51.33	62.44	43.35	48.01	39.35	33.48
Daisies										
3	05.12	06.12	01.28	06.29	23.19	29.15	18.20	20.02	12.16	13.12
5	08.37	09.13	08.35	09.03	34.20	42.08	26.10	28.03	23.06	21.11
7	10.02	12.30	11.08	12.05	52.35	61.14	45.04	47.09	35.14	36.05

(1)→ QIGAMLTCI (2)→ QISAMLTCI (3)→ QIPSOMLTCI (4)→ QIDEMLTCI
(5)→ GA (6)→ SA (7)→ PSO (8)→ DE (9)→ BSA (10)→ CoDE

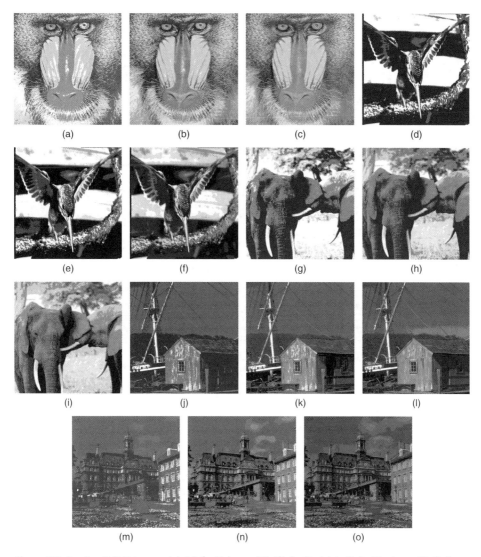

Figure 6.9 For $K = 3, 5, 7$, images (a)–(c), for Baboon, (d)–(f), for Bird, (g)–(i), for Elephant, (j)–(l), for Seaport, and (m)–(o), for Montreal, after using QIGAMLTCI, for multi-level thresholding for Kapur's method [140].

Later, the superiority among all algorithms has also been judged by performing another statistical test, called the median-based estimation. This procedure was first proposed by Doksum [75]. Basically, this test is conducted in two phases. At first, a measure, $D_p(i, j)$ is introduced for $K = 3, 5,$ and 7. $D_p(i, j)$ designates the difference between the fitness values of the algorithms i and j for input image p where, $1 \leq i < j \leq 10$. The computed values of each difference are given in Tables 6.45–6.50, respectively, for each fitness function. Afterwards, the medians for each $D(i, j)$, $i < j \leq 10$ are calculated. In the next part, another measure, $Y(i, j)$, called *unadjusted estimator* of $Q(i) - Q(j)$, is introduced for $K = 5$ and 7, where $i < j \leq 4$, $Y(i, j) = -Y(j, i)$, and $Y(i, i) = 0$. Moreover,

Figure 6.10 For $K = 3, 5, 7$, images (a)–(c), for Monolake, (d)–(f), for Mona Lisa, (g)–(i), for Manhattan, (j)–(l), for Lighthouse, and (m)–(o), for Daisies, after using QIGAMLTCI, for multi-level thresholding for Kapur's method [140].

the estimator of $Q(i) - Q(j)$ is calculated as $q(i) - q(j)$, where $q(i)$ is called the mean of unadjusted medians. $q(i) = \frac{\sum_{j=1}^{1} 0Y(i,j)}{10}$, where $i = 1, 2, \ldots, 10$. The results of the median-based estimation for each algorithm are reported in Tables 6.51 and 6.52 for $K = 3, 5$, and 7. We always get positive difference values for QIGAMLTCI from the remaining algorithms for each case. For any two algorithms, a positive difference value between each two of them indicates that the former algorithm outperforms the other. Hence, QIGAMLTCI is considered to be the best performing algorithm among all. Summarizing the above facts (based on two test results), it can be concluded that the

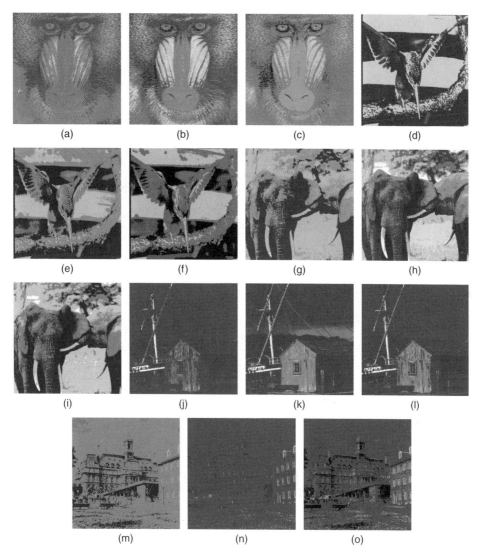

Figure 6.11 For $K = 3, 5, 7$, images (a)–(c), for Baboon, (d)–(f), for Bird, (g)–(i), for Elephant, (j)–(l), for Seaport, and (m)-(o), for Montreal, after using QIGAMLTCI, for multi-level thresholding for Huang's method [130].

proposed QIGAMLTCI outperforms the others for each level of thresholds. Apart from that, the computational times are found to be least for QIPSOMLTCI among others. Therefore, the proposed QIPSOMLTCI outperforms others with reference to computational time. For each $K = 5$ and 7, the convergence curves for the participating algorithms are presented in Figures 6.13–6.20, respectively, for each of the objective functions. These curves visually establish the efficiency of the proposed algorithms compared to others.

In another approach, the application of the proposed QITSMLTCI has been demonstrated on Tulips and Barge using a standard objective function offered by

Figure 6.12 For $K = 3, 5, 7$, images (a)–(c), for Monolake, (d)–(f), for Mona Lisa, (g)–(i), for Manhattan, (j)–(l), for Lighthouse, and (m)–(o), for Daisies, after using QIGAMLTCI, for multi-level thresholding for Huang's method [130].

Otsu's method [192]. The original test images are shown in Figures 6.1(k) and (l). The proposed QITSMLTCI algorithm has been run for 20 different runs. The best results of QITSMLTCI with reference to the number of thresholds (K), the component level optimal threshold values ($\theta_R, \theta_G, \theta_B$) of the test image, the fitness value ($\mathcal{F}_{best}$), and the execution time (t) (in seconds) is presented in Table 6.53 for $K = 2, 4$, and 6. The proposed algorithm has been compared with the conventional Tabu search algorithm for performance evaluation. To find the accuracy and stability of each algorithm, the value for mean fitness ($\mathcal{F}_{avg}$) and the standard deviation (σ) are reported in Table 6.54. Moreover, for each algorithm, a one-tailed t-test has been performed at the 5%

Table 6.41 PSNR of QIGAMLTCI, QISAMLTCI, QIDEMLTCI, QIPSOMLTCI, GA, SA, DE, PSO, BSA, and CoDE for multi-level thresholding of Baboon, Bird, Elephant, Seaport, Montreal, Monolake, Mona Lisa, Manhattan, Lighthouse, and Daisies for Kapur's method [140].

Baboon									
K	(1)	(2)	(3)	(4)	(5)	(6)	(7)	(8)	(9)	(10)
3	15.285	15.030	16.430	15.123	14.736	14.842	13.872	12.005	14.030	07.852
5	19.958	19.876	19.339	19.775	19.140	19.358	19.632	18.502	19.464	17.805
7	22.485	21.964	22.106	23.508	18.085	20.581	20.950	20.634	21.362	18.010

Bird									
K	(1)	(2)	(3)	(4)	(5)	(6)	(7)	(8)	(9)	(10)
3	14.052	13.966	12.212	13.111	10.566	10.205	13.368	12.808	11.801	12.296
5	16.035	15.809	16.697	17.272	16.432	15.788	15.463	15.639	15.833	16.271
7	20.599	20.484	23.482	20.748	18.788	17.114	18.644	16.373	18.062	18.010

Elephant									
K	(1)	(2)	(3)	(4)	(5)	(6)	(7)	(8)	(9)	(10)
3	15.309	16.999	15.830	15.353	15.296	14.904	9.780	11.351	13.114	15.031
5	20.924	21.514	20.677	20.919	20.445	22.204	19.953	20.069	20.385	20.564
7	25.080	23.159	23.022	23.625	24.387	23.104	20.578	17.928	22.994	23.136

Seaport									
K	(1)	(2)	(3)	(4)	(5)	(6)	(7)	(8)	(9)	(10)
3	20.171	19.593	19.628	19.624	13.475	18.392	16.870	17.395	16.686	16.911
5	22.935	22.133	22.761	22.692	18.392	22.373	20.053	18.655	21.449	19.783
7	25.466	25.922	25.281	25.622	25.014	24.681	21.946	22.389	24.975	25.340

Montreal									
K	(1)	(2)	(3)	(4)	(5)	(6)	(7)	(8)	(9)	(10)
3	17.655	18.593	15.214	18.056	9.173	9.506	13.074	11.258	11.471	13.265
5	23.711	23.814	23.910	24.390	13.376	23.230	23.369	22.491	23.702	22.010
7	25.780	25.169	24.371	26.093	21.735	19.455	17.296	20.626	20.280	21.483

Monolake									
K	(1)	(2)	(3)	(4)	(5)	(6)	(7)	(8)	(9)	(10)
3	15.141	14.105	14.936	16.590	8.599	13.291	12.557	8.291	13.882	7.827
5	21.387	22.423	21.610	22.502	23.266	18.633	21.373	18.720	19.030	16.391
7	25.994	23.595	23.527	24.755	22.154	21.492	19.663	24.354	20.446	21.200

Mona Lisa									
K	(1)	(2)	(3)	(4)	(5)	(6)	(7)	(8)	(9)	(10)
3	17.760	17.436	20.169	14.397	11.560	11.210	14.353	14.586	9.258	8.659
5	23.981	23.149	22.638	21.958	15.184	18.326	14.770	17.079	19.595	14.148
7	27.001	26.792	25.894	25.926	24.435	23.408	23.043	23.699	21.243	20.453

Manhattan									
K	(1)	(2)	(3)	(4)	(5)	(6)	(7)	(8)	(9)	(10)
3	16.702	14.295	16.062	14.683	13.878	14.890	7.969	11.867	13.532	13.629
5	17.505	18.138	16.321	19.525	14.657	15.296	11.955	17.587	17.016	16.031
7	26.349	23.544	26.217	23.715	21.780	21.163	24.527	21.149	20.715	20.985

Lighthouse									
K	(1)	(2)	(3)	(4)	(5)	(6)	(7)	(8)	(9)	(10)
3	16.531	16.307	16.680	16.665	15.605	13.777	7.999	12.882	9.600	13.986
5	20.084	21.736	20.198	19.943	18.324	17.749	19.322	19.749	19.940	20.595
7	20.871	20.063	22.038	23.382	22.887	23.161	20.921	23.165	19.610	22.931

Daisies									
K	(1)	(2)	(3)	(4)	(5)	(6)	(7)	(8)	(9)	(10)
3	14.458	15.174	14.650	16.015	14.091	14.317	13.669	14.512	13.853	14.494
5	19.697	20.009	19.768	20.052	17.788	17.186	19.382	19.466	18.867	18.761
7	23.154	23.868	23.019	23.678	24.152	20.080	21.837	23.238	21.169	20.894

(1)→ **QIGAMLTCI**　(2)→ **QISAMLTCI**　(3)→ **QIPSOMLTCI**　(4)→ **QIDEMLTCI**
(5)→ **GA**　(6)→ **SA**　(7)→ **PSO**　(8)→ **DE**　(9)→ **BSA**　(10)→ **CoDE**

Table 6.42 PSNR of QIGAMLTCI, QISAMLTCI, QIDEMLTCI, QIPSOMLTCI, GA, SA, DE, PSO, BSA, and CoDE for multi-level thresholding of Baboon, Bird, Elephant, Seaport, Montreal, Monolake, Mona Lisa, Manhattan, Lighthouse, and Daisies for Huang's method [130].

					Baboon					
K	(1)	(2)	(3)	(4)	(5)	(6)	(7)	(8)	(9)	(10)
3	14.798	15.236	15.389	14.161	10.608	10.563	11.052	9.938	13.003	12.357
5	17.483	17.265	17.961	17.110	16.724	14.070	16.589	11.823	16.831	14.337
7	18.500	18.112	18.624	18.489	18.267	17.400	16.648	14.342	15.791	16.645

					Bird					
K	(1)	(2)	(3)	(4)	(5)	(6)	(7)	(8)	(9)	(10)
3	15.286	13.967	15.568	13.823	13.451	12.839	7.693	13.013	12.509	12.182
5	18.805	18.598	18.025	17.017	17.479	12.236	15.384	17.794	15.898	15.321
7	17.114	18.420	17.857	17.540	15.370	17.237	17.002	15.179	17.162	18.236

					Elephant					
K	(1)	(2)	(3)	(4)	(5)	(6)	(7)	(8)	(9)	(10)
3	15.078	16.130	14.532	14.432	9.730	11.557	13.509	9.118	13.817	13.790
5	19.582	19.785	19.810	19.510	16.717	17.507	15.939	16.455	16.650	13.659
7	20.423	21.870	18.596	18.628	18.564	18.473	18.357	18.487	17.548	16.528

					Seaport					
K	(1)	(2)	(3)	(4)	(5)	(6)	(7)	(8)	(9)	(10)
3	17.772	17.875	18.664	16.059	8.143	7.634	7.584	14.481	17.277	10.923
5	21.578	20.503	18.976	19.624	17.471	18.291	17.568	19.606	19.524	17.931
7	19.749	18.773	22.300	20.589	18.259	18.422	21.599	19.679	17.230	18.728

					Montreal					
K	(1)	(2)	(3)	(4)	(5)	(6)	(7)	(8)	(9)	(10)
3	16.467	17.730	16.833	18.253	16.949	8.456	9.394	15.742	16.319	16.520
5	17.388	17.239	19.803	17.239	16.779	15.037	11.400	14.266	17.358	16.599
7	19.894	19.050	19.833	20.480	17.078	13.756	21.668	18.421	22.560	20.097

					Monolake					
K	(1)	(2)	(3)	(4)	(5)	(6)	(7)	(8)	(9)	(10)
3	15.717	17.177	15.462	15.691	14.296	10.853	10.100	13.291	14.690	6.954
5	20.037	19.099	19.438	19.828	17.766	15.916	18.709	16.564	15.811	17.982
7	22.737	20.472	22.457	21.745	17.570	12.638	16.317	15.657	16.833	17.965

					Mona Lisa					
K	(1)	(2)	(3)	(4)	(5)	(6)	(7)	(8)	(9)	(10)
3	14.510	14.438	13.417	15.536	8.394	11.661	8.029	13.531	13.848	11.710
5	17.098	16.516	16.252	16.950	12.475	11.176	15.531	14.397	16.301	17.209
7	18.097	17.043	20.028	17.160	16.367	13.266	17.435	15.223	18.101	17.948

					Manhattan					
K	(1)	(2)	(3)	(4)	(5)	(6)	(7)	(8)	(9)	(10)
3	16.490	16.748	18.919	16.342	16.376	12.435	15.312	16.500	16.212	07.514
5	17.709	17.436	20.380	18.305	14.813	12.303	12.903	16.534	18.130	16.430
7	19.217	18.798	18.752	19.225	18.428	18.311	16.920	18.903	17.757	17.655

					Lighthouse					
K	(1)	(2)	(3)	(4)	(5)	(6)	(7)	(8)	(9)	(10)
3	15.518	15.185	16.894	15.440	14.395	8.303	12.992	8.663	12.837	13.306
5	16.061	17.073	16.126	16.280	14.212	14.550	15.729	16.471	15.371	16.339
7	17.807	17.912	17.986	19.278	17.180	17.584	15.700	17.754	15.816	12.905

					Daisies					
K	(1)	(2)	(3)	(4)	(5)	(6)	(7)	(8)	(9)	(10)
3	13.688	14.561	15.501	14.738	12.046	11.998	11.040	12.802	12.627	13.882
5	19.998	18.458	17.797	17.791	16.890	16.094	11.514	16.380	16.705	14.813
7	18.040	20.271	21.221	21.994	23.204	15.247	14.679	16.583	15.009	16.725

(1)→ QIGAMLTCI (2)→ QISAMLTCI (3)→ QIPSOMLTCI (4)→ QIDEMLTCI
(5)→ GA (6)→ SA (7)→ PSO (8)→ DE (9)→ BSA (10)→ CoDE

Table 6.43 Data sets used in QIGAMLTCI, QISAMLTCI, QIPSOMLTCI, QIDEMLTCI, GA, SA, PSO, DE, BSA, and CoDE for Friedman test for K = 3, 5, and 7, respectively for Kapur's method [140]. The value in parentheses signifies the rank of the respective algorithm.

SN	Image	QIGAMLTCI	QISAMLTCI	QIPSOMLTCI	QIDEMLTCI	GA	SA	PSO	DE	BSA	CoDE
					For K=3						
1	Baboon	38.1106(3)	38.1053(4)	38.1191(1)	38.1170(2)	37.9784(7)	37.9686(8)	37.9606(10)	37.9664(9)	37.9892(6)	37.9983(5)
2	Bird	27.7920(1)	27.6512(4)	27.7700(2)	27.7692(3)	27.5126(6)	27.4627(10)	27.4770(9)	27.5439(5)	27.5058(8)	27.5091(7)
3	Elephant	19.4261(2.5)	19.4261(2.5)	19.4261(2.5)	19.4261(2.5)	19.0550(9)	19.0328(10)	19.3616(7)	19.3727(5)	19.3683(6)	19.0709(8)
4	Seaport	19.4944(1)	19.4453(4)	19.4934(2)	19.4704(3)	18.5216(9)	18.8618(7)	18.4979(10)	18.5438(8)	18.9487(5)	18.9448(6)
5	Montreal	21.3461(1)	20.9971(2)	20.4068(3)	20.3943(4)	20.3537(9)	20.3604(8)	20.3471(10)	20.3936(5)	20.3887(6)	20.3728(7)
6	Monolake	22.9667(1)	22.5170(4)	22.9651(2)	22.8912(3)	22.3014(9)	22.1184(10)	22.3318(8)	22.4737(5)	22.4351(6)	22.3835(7)
7	Mona Lisa	35.9646(3)	35.9594(4)	35.9679(2)	35.9708(1)	35.4958(10)	35.5675(9)	35.6474(8)	35.7945(7)	35.8757(6)	35.9014(5)
8	Manhattan	20.5450(3)	20.4452(4)	20.5766(1)	20.5543(2)	20.1754(8)	20.1180(10)	20.1655(9)	20.1957(7)	20.3019(5)	20.2649(6)
9	Lighthouse	21.8839(3)	21.7753(4)	21.9626(1)	21.9579(2)	21.3411(9)	21.3241(10)	21.4026(8)	21.4661(7)	21.6006(6)	21.6263(5)
10	Daisies	24.7691(3)	24.7433(4)	24.7745(2)	24.7852(1)	24.6400(7)	24.6209(9)	24.6201(10)	24.6329(8)	24.6669(5)	24.6587(6)
	Average rank	2.15	3.65	1.85	2.35	8.30	9.10	8.90	6.60	5.90	6.20
					For K=5						
1	Baboon	56.4427(1)	55.7638(4)	56.4036(2)	56.3867(3)	55.5483(7)	55.5499(6)	55.5318(9)	55.5391(8)	55.5556(5)	55.4125(10)
2	Bird	39.7076(1)	39.3170(4)	39.4554(2)	39.4510(3)	38.6345(6)	38.0177(10)	38.4960(8)	38.4133(9)	38.5333(7)	38.8943(5)
3	Elephant	24.9208(2)	24.2754(6)	24.9244(1)	24.2858(5)	24.1188(7)	24.1444(9)	24.1186(8)	24.1182(10)	24.4105(4)	24.5744(3)
4	Seaport	26.3884(1)	26.0978(4)	26.2047(3)	26.2414(2)	24.9379(5)	23.4984(10)	24.3427(8)	24.7252(6)	24.1767(9)	24.3885(7)
5	Montreal	29.2005(1)	28.9368(4)	29.0734(2)	28.8920(3)	26.7768(6)	26.7635(7)	26.6425(9)	26.5740(10)	26.9463(5)	26.7383(8)
6	Monolake	31.7998(1)	31.1351(3)	31.2537(2)	31.1196(4)	30.1623(7)	29.2394(10)	30.0592(9)	30.0847(8)	30.2282(5)	30.1864(6)
7	Mona Lisa	53.2803(1)	52.3239(5)	53.1221(2)	53.1210(3)	52.3380(4)	51.9529(10)	52.1727(7)	52.1185(9)	52.2325(6)	52.1575(8)
8	Manhattan	28.2676(1)	27.5575(4)	27.9636(2)	27.8503(3)	26.1379(9)	26.2211(7)	26.0354(10)	26.1850(8)	26.4056(6)	26.4182(5)
9	Lighthouse	30.6146(1)	30.5153(2)	30.5039(3)	30.4884(4)	29.1642(8)	29.1657(7)	29.1259(10)	29.1518(9)	29.3621(6)	29.4102(5)
10	Daisies	34.8079(1)	34.4185(4)	34.5995(2)	34.5201(3)	34.1841(6)	34.2137(5)	34.1360(9)	34.1273(10)	34.1812(7)	34.1605(8)
	Average rank	1.10	4.00	2.10	3.30	6.50	8.10	8.70	8.70	6.00	6.70
					For K=7						
1	Baboon	72.5895(1)	71.9987(4)	72.2779(2)	72.1471(3)	71.1919(6)	70.5951(10)	70.7130(9)	70.7642(8)	71.2335(5)	71.1250(7)
2	Bird	48.1137(3)	48.0243(4)	48.2604(1)	48.1517(2)	45.2848(6)	44.2234(10)	44.6108(8)	44.4637(9)	45.4073(7)	45.3422(5)
3	Elephant	27.4002(3)	27.3890(4)	27.6376(1)	27.6274(2)	26.8600(5)	24.4608(10)	26.3680(8)	26.3799(7)	26.4115(6)	26.1802(9)
4	Seaport	30.8059(1)	29.9389(3)	29.3007(4)	30.6428(2)	28.2911(7)	27.3997(10)	27.5768(8)	27.4480(9)	28.8505(5)	28.8068(6)
5	Montreal	31.5739(3)	31.5541(4)	32.0464(1)	31.7646(2)	30.1747(8)	30.3882(6)	30.3928(5)	30.1752(7)	30.1502(9)	30.1179(10)
6	Monolake	37.8726(1)	35.6220(4)	36.4533(2)	36.3196(3)	34.7246(7)	34.4819(10)	34.5598(8)	34.4870(9)	35.1292(6)	35.2227(5)
7	Mona Lisa	68.3283(1)	67.5836(4)	67.7713(2)	67.7204(3)	65.4327(9)	65.7589(6)	65.4037(10)	65.5240(8)	65.8547(5)	65.6869(7)
8	Manhattan	31.5377(2)	31.4702(4)	31.4277(3)	31.8317(1)	29.9079(7)	29.2814(8)	29.2113(10)	29.2357(9)	30.2522(5)	29.9357(6)
9	Lighthouse	37.6379(1)	36.0877(4)	36.8774(2)	36.3339(3)	34.1485(8)	34.0885(9)	34.2636(7)	34.0844(10)	35.1836(5)	34.9284(6)
10	Daisies	42.8365(1)	41.0997(4)	41.8475(2)	41.5488(3)	40.7998(7)	40.3464(9)	40.3380(10)	40.5699(8)	40.8300(6)	40.8545(5)
	Average rank	1.70	3.90	2.00	2.40	6.30	8.80	8.30	8.40	5.90	6.60

Table 6.44 Data sets used in QIGAMLTCI, QISAMLTCI, QIPSOMLTCI, QIDEMLTCI, GA, SA, PSO, DE, BSA, and CoDE for the Friedman test for K = 3, 5, and 7, respectively for Huang's method [130]. The value in parentheses signifies the rank of the respective algorithm.

SN	Image	QIGAMLTCI	QISAMLTCI	QIPSOMLTCI	QIDEMLTCI	GA	SA	PSO	DE	BSA	CoDE
					For K=3						
1	Baboon	2.5495 (3)	2.5491 (1)	2.5505 (5)	2.5493 (2)	2.5670 (9)	2.5705 (10)	2.5504 (4)	2.5589 (8)	2.5567 (6)	2.5572 (7)
2	Bird	2.4089 (5.5)	2.4089 (5.5)	2.4089 (5.5)	2.4089 (5.5)	2.4089 (5.5)	2.4089 (5.5)	2.4089 (5.5)	2.4089 (5.5)	2.4089 (5.5)	2.4089 (5.5)
3	Elephant	2.5404 (3.5)	2.5407 (7.5)	2.5406 (5.5)	2.5407 (7.5)	2.5415 (9)	2.5422 (10)	2.5404 (3.5)	2.5406 (5.5)	2.5373 (1.5)	2.5373 (1.5)
4	Seaport	2.7252 (1)	2.7279 (4.5)	2.7268 (2)	2.7279 (4.5)	2.7898 (8)	2.7959 (9.5)	2.7836 (7)	2.7959 (9.5)	2.7279 (4.5)	2.7279 (4.5)
5	Montreal	2.7459 (5.5)	2.7459 (5.5)	2.7459 (5.5)	2.7459 (5.5)	2.7459 (5.5)	2.7459 (5.5)	2.7459 (5.5)	2.7459 (5.5)	2.7459 (5.5)	2.7459 (5.5)
6	Monolake	2.6386 (1.5)	2.6487 (6)	2.6493 (7)	2.6386 (1.5)	2.6512 (10)	2.6506 (9)	2.6440 (5)	2.6437 (4)	2.6415 (3)	2.6501 (8)
7	Mona Lisa	2.6048 (2)	2.6123 (9)	2.6052 (5)	2.6061 (6)	2.6119 (8)	2.6050 (4)	2.6101 (7)	2.6524 (10)	2.6048 (2)	2.6048 (2)
8	Manhattan	2.6404 (2.5)	2.6404 (2.5)	2.6404 (2.5)	2.6404 (2.5)	2.6408 (7.5)	2.6408 (7.5)	2.6408 (7.5)	2.6408 (7.5)	2.6408 (7.5)	2.6408 (7.5)
9	Lighthouse	2.4324 (3)	2.4313 (1)	2.4324 (3)	2.4336 (7.5)	2.4331 (6)	2.4337 (9)	2.4327 (5)	2.4336 (7.5)	2.4324 (3)	2.4338 (10)
10	Daisies	2.6714 (1.5)	2.6714 (1.5)	2.6722 (3)	2.6726 (5)	2.6731 (7.5)	2.6736 (9.5)	2.6731 (7.5)	2.6736 (9.5)	2.6726 (5)	2.6726 (5)
	Average rank	2.80	4.40	4.20	4.75	7.60	7.95	5.75	7.25	4.35	5.65
					For K=5						
1	Baboon	2.5615 (4)	2.5576 (1)	2.5601 (2)	2.5611 (3)	2.6008 (9)	2.6017 (10)	2.5800 (6)	2.5888 (8)	2.5850 (7)	2.5791 (5)
2	Bird	2.4089 (3.5)	2.4089 (3.5)	2.4089 (3.5)	2.4089 (3.5)	2.4497 (8)	2.4596 (9)	2.4738 (10)	2.4440 (7)	2.4089 (3.5)	2.4089 (3.5)
3	Elephant	2.5489 (1)	2.5853 (4)	2.5861 (6)	2.5717 (3)	2.5943 (8)	2.6075 (9)	2.5910 (7)	2.6109 (10)	2.5588 (2)	2.5858 (5)
4	Seaport	2.7333 (4)	2.7331 (2)	2.7332 (3)	2.7303 (1)	2.7353 (9)	2.7352 (7)	2.7345 (5)	2.7352 (7)	2.7357 (10)	2.7352 (7)
5	Montreal	2.7474 (3)	2.7477 (4)	2.7463 (1.5)	2.7463 (1.5)	2.7509 (8)	2.7503 (6)	2.7508 (7)	2.7510 (9.5)	2.7510(9.5)	2.7497 (5)
6	Monolake	2.6683 (5)	2.6587 (1)	2.6624 (2)	2.6689 (6)	2.6750 (8.5)	2.6718 (7)	2.6762 (10)	2.6750 (8.5)	2.6637 (4)	2.6635 (3)
7	Mona Lisa	2.6101 (2)	2.6079 (1)	2.6185 (5)	2.6126 (3)	2.6299 (9)	2.6282 (7)	2.6298 (8)	2.6233 (6)	2.6159 (4)	2.6308 (10)
8	Manhattan	2.6410 (3)	2.6408 (1)	2.6409 (2)	2.6415 (4)	2.6434 (7)	2.6436 (8)	2.6443 (9)	2.6448 (10)	2.6420 (5)	2.6426 (6)
9	Lighthouse	2.4425 (2)	2.4421 (1)	2.4428 (3)	2.4430 (4)	2.5000 (8)	2.4829 (6)	2.5242 (10)	2.4988 (7)	2.4714 (5)	2.5147 (9)
10	Daisies	2.6804 (1)	2.6839 (4)	2.6827 (2)	2.6838 (3)	2.7393 (10)	2.7382 (9)	2.7134 (5)	2.7293 (8)	2.7251 (7)	2.7182 (6)
	Average rank	2.95	2.25	3.00	3.20	8.45	7.80	7.70	8.10	5.70	5.95
					For K=7						
1	Baboon	2.6068 (3)	2.6096 (4)	2.4982 (1)	2.5775 (2)	2.6411 (7)	2.6428(8)	2.6510 (10)	2.6499 (9)	2.6147 (5)	2.6121 (5)
2	Bird	2.4825 (3)	2.4900 (4)	2.4770 (1)	2.4807 (2)	2.5569 (8)	2.6077 (9)	2.6099 (10)	2.5543 (7)	2.4939 (5)	2.4993 (6)
3	Elephant	2.6299 (3)	2.6331 (4)	2.5935 (1)	2.6070 (2)	2.6410 (9)	2.6653 (10)	2.6403 (7)	2.6406 (8)	2.6367 (5)	2.6400 (6)
4	Seaport	2.7381 (2)	2.7407 (5)	2.7373 (2)	2.7400 (3)	2.7444 (10)	2.7416 (7)	2.7423 (8)	2.7427 (9)	2.7401 (4)	2.7408 (6)
5	Montreal	2.7483 (1)	2.7504 (5)	2.7496 (2)	2.7505 (4)	2.7607 (9)	2.7619 (10)	2.7603(7)	2.7604 (8)	2.7591(6)	2.7532 (5)
6	Monolake	2.6994 (5)	2.6951 (3)	2.6961 (4)	2.7030 (6)	2.7115 (7)	2.7123 (8)	2.8759 (10)	2.8714 (9)	2.6883 (2)	2.6846 (1)
7	Mona Lisa	2.6270 (3)	2.6291 (4)	2.6238 (2)	2.6172 (1)	2.6524 (8)	2.6599 (9)	2.6663 (10)	2.6332 (7)	2.6316 (6)	2.6309 (5)
8	Manhattan	2.6440 (3)	2.6416 (1)	2.6439 (2)	2.6459 (4)	2.6530 (6)	2.6637 (8.5)	2.6660 (9)	2.6637 (8.5)	2.6619 (7)	2.6467 (5)
9	Lighthouse	2.4980 (3)	2.4834 (1)	2.4997 (4)	2.4899 (2)	2.5713 (9)	2.5601 (7)	2.5347 (6)	2.5769 (10)	2.5616 (8)	2.5012 (5)
10	Daisies	2.7222 (2)	2.7406 (4.5)	2.7406 (4.5)	2.7177 (1)	2.7791 (9)	2.7390 (3)	2.7474 (6)	2.7487 (7)	2.7545 (8)	2.7819 (10)
	Average rank	2.80	3.35	2.35	2.70	8.20	7.95	8.30	8.25	5.70	5.40

Table 6.45 Difference between pairwise fitness values of the different techniques for $K = 3$ for Kapur's method [140]. Columnwise median values are given at the bottom of the table.

For K=3

SN	Image	D(1,2)	D(1,3)	D(1,4)	D(1,5)	D(1,6)	D(1,7)	D(1,8)	D(1,9)	D(1,10)	D(2,3)	D(2,4)	D(2,5)	D(2,6)	D(2,7)	D(2,8)	D(2,9)	D(2,10)	D(3,4)	D(3,5)	D(3,6)	D(3,7)	D(3,8)	D(3,9)
1	Baboon	0.0053	-0.0085	-0.0064	0.1322	0.142	0.15	0.1442	0.1214	0.1123	-0.0138	-0.0117	0.1269	0.1367	0.1447	0.1389	0.1161	0.107	0.0021	0.1407	0.1505	0.1585	0.1527	0.1299
2	Bird	0.1408	0.022	0.0228	0.2794	0.3293	0.315	0.2481	0.2862	0.2829	-0.1188	-0.118	0.1386	0.1885	0.1742	0.1073	0.1454	0.1421	0.0008	0.2574	0.3073	0.293	0.2261	0.2642
3	Elephant	0	0	0	0.3711	0.3933	0.0645	0.0534	0.0578	0.3552	0	0	0.3711	0.3933	0.0645	0.0534	0.0578	0.3552	0	0.3711	0.3933	0.0645	0.0534	0.0578
4	Seaport	0.0491	0.001	0.024	0.9728	0.6326	0.9965	0.9506	0.5457	0.5496	-0.0481	-0.0251	0.9237	0.5835	0.9474	0.9015	0.4966	0.5005	0.023	0.9718	0.6316	0.9955	0.9496	0.5447
5	Montreal	0.349	0.9393	0.9518	0.9924	0.9857	0.999	0.9525	0.9574	0.9733	0.5903	0.6028	0.6434	0.6367	0.65	0.6035	0.6084	0.6243	0.0125	0.0531	0.0464	0.0597	0.0132	0.0181
6	Monolake	0.4497	0.0016	0.0755	0.6653	0.8483	0.6349	0.493	0.5316	0.5832	-0.4481	-0.3742	0.2156	0.3986	0.1852	0.0433	0.0819	0.1335	0.0739	0.6637	0.8467	0.6333	0.4914	0.53
7	Mona Lisa	0.0052	-0.0033	-0.0062	0.4688	0.3971	0.3172	0.1701	0.0889	0.0632	-0.0085	-0.0114	0.4636	0.3919	0.312	0.1649	0.0837	0.058	-0.0029	0.4721	0.4004	0.3205	0.1734	0.0922
8	Manhattan	0.0998	-0.0316	-0.0093	0.3696	0.427	0.3795	0.3493	0.2431	0.2801	-0.1314	-0.1091	0.2698	0.3272	0.2797	0.2495	0.1433	0.1803	0.0223	0.4012	0.4586	0.4111	0.3809	0.2747
9	Lighthouse	0.1086	-0.0787	-0.074	0.5428	0.5598	0.4813	0.4178	0.2833	0.2576	-0.1873	-0.1826	0.4342	0.4512	0.3727	0.3092	0.1747	0.149	0.0047	0.6215	0.6385	0.56	0.4965	0.362
10	Daisies	0.0258	-0.0054	-0.0161	0.1291	0.1482	0.149	0.1362	0.1022	0.1104	-0.0312	-0.0419	0.1033	0.1224	0.1232	0.1104	0.0764	0.0846	-0.0107	0.1345	0.1536	0.1544	0.1416	0.1076
	Median	0.0745	-0.0017	-0.0031	0.4199	0.412	0.3484	0.2987	0.2632	0.2815	-0.0397	-0.0335	0.3205	0.3926	0.2324	0.1519	0.1297	0.1456	0.0034	0.3862	0.3968	0.3068	0.1998	0.1971

SN	Image	D(3,10)	D(4,5)	D(4,6)	D(4,7)	D(4,8)	D(4,9)	D(4,10)	D(5,6)	D(5,7)	D(5,8)	D(5,9)	D(5,10)	D(6,7)	D(6,8)	D(6,9)	D(6,10)	D(7,8)	D(7,9)	D(7,10)	D(8,9)	D(8,10)	D(9,10)
1	Baboon	0.1208	0.1386	0.1484	0.1564	0.1506	0.1278	0.1187	0.0098	0.0178	0.012	-0.0108	-0.0199	0.008	0.0022	-0.0206	-0.0297	-0.0058	-0.0286	-0.0377	-0.0228	-0.0319	-0.0091
2	Bird	0.2609	0.2566	0.3065	0.2922	0.2253	0.2634	0.2601	0.0499	0.0356	-0.0313	0.0068	0.0035	-0.0143	-0.0431	-0.0464	-0.0669	-0.0288	-0.0321	-0.0354	0.0381	0.0348	-0.0033
3	Elephant	0.3552	0.3711	0.3933	0.0645	0.0534	0.0578	0.3552	0.0222	-0.3066	-0.3177	-0.3133	-0.0159	-0.3288	-0.3399	-0.0381	-0.0111	-0.0067	-0.4508	0.2907	0.0044	0.3018	0.2974
4	Seaport	0.5486	0.9488	0.6086	0.9725	0.9266	0.5217	0.5256	-0.3402	0.0237	-0.0222	-0.4271	-0.4232	0.3639	0.318	-0.083	-0.0459	-0.0416	-0.4508	-0.4469	-0.4049	-0.401	0.0039
5	Montreal	0.034	0.0406	0.0339	0.0472	0.0007	0.0056	0.0215	-0.0067	0.0066	-0.0399	-0.035	-0.0191	0.0133	-0.0332	-0.0124	-0.0465	-0.0416	-0.1033	-0.0257	0.0049	0.0208	0.0039
6	Monolake	0.5816	0.5898	0.7728	0.5594	0.4175	0.4561	0.5077	0.183	-0.0304	-0.1723	-0.1337	-0.0821	-0.2134	-0.3553	-0.3167	-0.1419	-0.2651	-0.3339	-0.0517	0.0386	0.0902	0.0516
7	Mona Lisa	0.0665	0.475	0.4033	0.3234	0.1763	0.0951	0.0694	-0.0717	-0.1516	-0.2987	-0.3799	-0.4056	-0.0799	-0.227	-0.3082	-0.3339	-0.1471	-0.2283	-0.254	-0.0812	-0.1069	-0.0257
8	Manhattan	0.3117	0.3789	0.4363	0.3888	0.3586	0.2524	0.2894	0.0574	0.0099	-0.0203	-0.1265	-0.0895	-0.0475	-0.0777	-0.1839	-0.1469	-0.0302	-0.1364	-0.0994	-0.1062	-0.0692	0.037
9	Lighthouse	0.3363	0.6168	0.6338	0.5553	0.4918	0.3573	0.3316	0.017	-0.0615	-0.125	-0.2595	-0.2852	-0.0785	-0.142	-0.2765	-0.3022	-0.0635	-0.198	-0.2237	-0.1345	-0.1602	-0.0257
10	Daisies	0.1158	0.1452	0.1643	0.1651	0.1523	0.1183	0.1265	0.0191	0.0199	0.0071	-0.0269	-0.0187	0.0008	-0.012	-0.046	-0.0378	-0.0128	-0.0468	-0.0386	-0.034	-0.0258	0.0082
	Median	0.2863	0.375	0.3983	0.3078	0.2008	0.1901	0.2748	0.0181	0.0082	-0.0356	-0.1301	-0.051	-0.0309	-0.0794	-0.1354	-0.0647	-0.0462	-0.075	-0.0451	-0.0284	-0.0289	0.006

Table 6.46 Difference between pairwise fitness values of the different techniques for $K = 5$ for Kapur's method [140]. Columnwise median values are given at the bottom of the table.

For K=5

SN	Image	D(1,2)	D(1,3)	D(1,4)	D(1,5)	D(1,6)	D(1,7)	D(1,8)	D(1,9)	D(1,10)	D(2,3)	D(2,4)	D(2,5)	D(2,6)	D(2,7)	D(2,8)	D(2,9)	D(2,10)	D(3,4)	D(3,5)	D(3,6)	D(3,7)	D(3,8)	D(3,9)
1	Baboon	0.6789	0.0391	0.056	0.172	0.8928	0.9109	0.9036	0.8871	1.0302	-0.6398	-0.6229	-0.5069	0.2139	0.232	0.2247	0.2082	0.3513	0.0169	.1329	0.8537	0.8718	0.8645	0.848
2	Bird	0.3906	0.2522	0.2566	1.0731	1.6899	1.2116	1.2943	1.1743	0.8133	-0.1384	-0.134	0.6825	1.2993	0.821	0.9037	0.7837	0.4227	0.0044	.8209	1.4377	0.9594	1.0421	0.9221
3	Elephant	0.6454	-0.0036	0.635	0.802	0.7764	0.8022	0.8026	0.5103	0.3464	-0.649	-0.0104	0.1566	0.131	0.1568	0.1572	-0.1351	-0.299	0.6386	.8056	0.78	0.8058	0.8062	0.5139
4	Seaport	0.2906	0.1837	0.147	1.4505	2.89	2.0457	1.6632	2.2117	1.9999	-0.1069	-0.1436	1.1599	2.5994	1.7551	1.3726	1.9211	1.7093	-0.0367	.2668	2.7063	1.862	1.4795	2.028
5	Montreal	0.2637	0.1271	0.2085	2.4237	2.437	2.558	2.6265	2.2542	2.4622	-0.1366	-0.0552	2.16	2.1733	2.2943	2.3628	1.9905	2.1985	0.0814	-.2966	2.3099	2.4309	2.4994	2.1271
6	Monolake	0.6647	0.5461	0.6802	1.6375	2.5604	1.7406	1.7151	1.5716	1.6134	-0.1186	0.0155	0.9728	1.8957	1.0759	1.0504	0.9069	0.9487	0.1341	.0914	2.0143	1.1945	1.169	1.0255
7	Mona Lisa	0.9564	0.1582	0.1593	0.9423	1.3274	1.1076	1.1618	1.0478	1.1228	-0.7982	-0.7971	-0.0141	0.371	0.1512	0.2054	0.0914	0.1664	0.0011	.7841	1.1692	0.9494	1.0036	0.8896
8	Manhattan	0.7101	0.304	0.4173	2.1297	2.0465	2.2322	2.0826	1.862	1.8494	-0.4061	-0.2928	1.4196	1.3364	1.5221	1.3725	1.1519	1.1393	0.1133	1.8257	1.7425	1.9282	1.7786	1.558
9	Lighthouse	0.0993	0.1107	0.1107	1.4504	1.4489	1.4887	1.4628	1.2525	1.2044	0.0114	0.0114	1.3511	1.3496	1.3894	1.3635	1.1532	1.1051	0	1.3397	1.3382	1.378	1.3521	1.1418
10	Daisies	0.3894	0.2084	0.2878	0.6238	0.5942	0.6719	0.6806	0.6267	0.6474	-0.181	-0.1016	0.2344	0.2048	0.2825	0.2912	0.2373	0.258	0.0794	0.4154	0.3858	0.4635	0.4722	0.4183
	Median	0.518	0.1709	0.2326	1.2618	1.5694	1.3502	1.3786	1.2134	1.1636	-0.1597	-0.1178	0.8277	1.3179	0.9484	0.9771	0.8453	0.6857	0.0481	0.9561	1.388	1.077	1.1056	0.9738

SN	Image	D(3,10)	D(4,5)	D(4,6)	D(4,7)	D(4,8)	D(4,9)	D(4,10)	D(5,6)	D(5,7)	D(5,8)	D(5,9)	D(5,10)	D(6,7)	D(6,8)	D(6,9)	D(6,10)	D(7,8)	D(7,9)	D(7,10)	D(8,9)	D(8,10)	D(9,10)
1	Baboon	0.9911	0.116	0.8368	0.8549	0.8476	0.8311	0.9742	0.7208	0.7389	0.7316	0.7151	0.8582	0.0181	-0.0057	0.1374	-0.0073	-0.0238	0.1193	-0.0165	0.1266	0.1431	
2	Bird	0.5611	0.8165	1.4333	0.955	1.0377	0.9177	0.5567	0.6168	0.1385	0.2212	0.1012	-0.2598	-0.4783	-0.5156	-0.8766	0.0827	-0.0373	-0.3983	-0.12	-0.481	-0.361	
3	Elephant	0.35	0.167	0.1414	0.1672	0.1676	-0.1247	-0.2886	-0.0256	0.0002	0.0006	-0.2917	-0.4556	0.0258	0.0262	-0.43	0.0004	-0.2919	-0.4558	-0.2923	-0.4562	-0.1639	
4	Seaport	1.8162	1.3035	2.743	1.8987	1.5162	2.0647	1.8529	1.4395	0.5952	0.2127	0.7612	0.5494	-0.8443	-1.2268	-0.8901	-0.3825	0.166			0.5485	0.3367	-0.2118
5	Montreal	2.3351	2.2152	2.2285	2.3495	2.418	2.0457	2.2537	0.0133	0.1343	0.2028	-0.1695	0.0385	0.121	0.1895	0.0252	0.0685	-0.3035	-0.0958	-0.1272	-0.3723	-0.1643	0.208
6	Monolake	1.0673	0.9573	1.8802	1.0604	1.0349	0.8914	0.9332	0.9229	0.1031	0.0776	-0.0659	-0.0241	-0.8198	-0.8453	-0.947	-0.0255	-0.169	0.0152		-0.1435	-0.1017	0.0418
7	Mona Lisa	0.9646	0.783	1.1681	0.9483	1.0025	0.8885	0.9635	0.3851	0.1653	0.2195	0.1055	0.1805	-0.2198	-0.1656	-0.2046	0.0542	-0.059	-0.3828	0.0152	-0.114	-0.039	0.075
8	Manhattan	1.5454	1.7124	1.6292	1.8149	1.6653	1.4447	1.4321	-0.0832	0.1025	-0.0471	-0.2677	-0.2803	0.1857	0.0361	-0.1971	-0.1496	-0.3702	-0.2843		-0.2206	-0.2332	-0.0126
9	Lighthouse	1.0937	1.3397	1.3382	1.378	1.3521	1.1418	1.0937	-0.0015	0.0383	0.0124	-0.1979	-0.246	0.0398	0.0139	-0.2445	-0.0259	-0.2362	-0.0245		-0.2103	-0.2584	-0.0481
10	Daisies	0.439	0.336	0.3064	0.3841	0.3928	0.3389	0.3596	-0.0296	0.0481	0.0568	0.0029	0.0236	0.0777	0.0864	0.0532	0.0087	-0.0452	-0.1115		-0.0539	-0.0332	0.0207
	Median	1.0292	0.8869	1.3858	1.0077	1.0363	0.9046	0.9689	0.1992	0.1187	0.1402	-0.0315	-0.0002	0.022	0.0124	-0.2245	-0.0035	-0.0114	-0.1115		-0.1317	-0.133	0.004

Table 6.47 Difference between pairwise fitness values of the different techniques for $K = 7$ for Kapur's method [140]. Columnwise median values are given at the bottom of the table.

		For $K=7$																					
SN	Image	D(1,2)	D(1,3)	D(1,4)	D(1,5)	D(1,6)	D(1,7)	D(1,8)	D(1,9)	D(1,10)	D(2,3)	D(2,4)	D(2,5)	D(2,6)	D(2,7)	D(2,8)	D(2,9)	D(2,10)	D(3,4)	D(3,5)	D(3,6)	D(3,7)	D(3,8)	D(3,9)
1	Baboon	0.5908	0.3116	0.4424	0.8001	2.2222	1.3976	1.9944	1.8765	1.8253	-0.2792	-0.1484	0.2093	1.6314	0.8068	1.4036	1.2857	1.2345	0.1308	0.4885	1.9106	1.086	1.6828	1.5649
2	Bird	0.0894	-0.1467	-0.038	1.3796	3.8903	2.8289	3.8903	3.5029	3.65	-0.2361	-0.1274	1.2902	3.8009	2.7395	3.8009	3.4135	3.5606	0.1087	1.5263	4.037	2.9756	4.037	3.6496
3	Elephant	0.0112	-0.2374	-0.2272	0.5402	2.9394	0.5402	2.9394	1.0322	1.0203	-0.2486	-0.2384	0.529	2.9282	0.529	2.9282	1.021	1.0091	0.0102	0.7776	3.1768	0.7776	3.1768	1.2696
4	Seaport	0.867	1.5052	0.1631	2.5148	4.1031	2.5148	3.4062	3.2291	3.3579	0.6382	-0.7039	1.6478	3.2361	1.6478	2.5392	2.3621	2.4909	-1.3421	1.0096	2.5979	1.0096	1.901	1.7239
5	Montreal	0.0198	-0.4725	-0.1907	1.224	1.1857	1.3992	1.1857	1.1811	1.3987	-0.4923	-0.2105	1.2042	1.1659	1.3794	1.1659	1.1613	1.3789	0.2818	1.6965	.6582	1.8717	1.6582	1.6536
6	Monolake	2.2506	1.4193	1.553	2.2579	4.6186	3.148	3.3907	3.3128	3.3856	-0.8313	-0.6976	0.0073	2.368	0.8974	1.1401	1.0622	1.135	0.1337	0.8386	.1993	1.7287	1.9714	1.8935
7	Mona Lisa	0.7447	0.557	0.6079	1.4593	3.8153	2.8956	2.5694	2.9246	2.8043	-0.1877	-0.1368	0.7146	3.0706	2.1509	1.8247	2.1799	2.0596	0.0509	0.9023	3.2583	2.3386	2.0124	2.3676
8	Manhattan	0.0675	0.11	-0.294	3.0382	3.0382	1.6298	2.2563	2.3264	2.302	0.0425	-0.3615	-0.3615	2.9707	1.5623	2.1888	2.2589	2.2345	-0.404	-0.404	2.9282	1.5198	2.1463	2.2164
9	Lighthouse	1.5502	0.7605	1.304	1.8633	3.8768	3.4894	3.5494	3.3743	3.5535	-0.7897	-0.2462	0.3131	2.3266	1.9392	1.9992	1.8241	2.0033	0.5435	1.1028	3.1163	2.7289	2.7889	2.6138
10	Daisies	1.7368	0.989	1.2877	2.0367	3.2383	2.0367	2.4901	2.4985	2.2666	-0.7478	-0.4491	0.2999	1.5015	0.2999	0.7533	0.7617	0.5298	0.2987	1.0477	2.2493	1.0477	1.5011	1.5095
	Median	0.6677	0.4343	0.3028	1.4195	3.5268	2.2758	2.7544	2.7116	2.5532	-0.2639	-0.2423	0.421	2.6481	1.4709	1.912	1.5549	1.6911	0.1198	0.9559	3.0223	1.6243	1.9919	1.8087

SN	Image	D(3,10)	D(4,5)	D(4,6)	D(4,7)	D(4,8)	D(4,9)	D(4,10)	D(5,6)	D(5,7)	D(5,8)	D(5,9)	D(5,10)	D(6,7)	D(6,8)	D(6,9)	D(6,10)	D(7,8)	D(7,9)	D(7,10)	D(8,9)	D(8,10)	D(9,10)
1	Baboon	1.5137	0.3577	1.7798	0.9552	1.552	1.4341	1.3829	1.4221	0.5975	1.1943	1.0764	1.0252	-0.8246	-0.2278	-0.3969	-0.3457	0.5968	0.4789	0.4277	-0.1179	-0.1691	-0.0512
2	Bird	3.7967	4.1176	3.9283	2.8669	3.9283	3.5409	3.688	2.5107	1.4493	2.5107	2.1233	2.2704	-1.0614	0	-0.2403	-0.2403	0.674	0.8211	-0.3874	0.1471	-0.3874	-0.0512
3	Elephant	1.2577	0.7674	3.1666	0.7674	3.1666	1.2594	1.2475	2.3992	0	2.3992	0.492	0.4801	-2.3992	0	-1.9072	-1.9191	0.492	0.4801	-1.9072	-0.0119	-1.9191	-0.0119
4	Seaport	1.8527	2.3517	3.94	2.3517	3.2431	3.066	3.1948	1.5883	0	0.8914	0.7143	0.8431	-1.5883	-0.6969	-0.7452	-0.874	0.7143	0.8431	-0.771	0.1288	-0.874	0.1288
5	Montreal	1.8712	1.4147	1.3764	1.5899	1.3764	1.3718	1.5894	-0.0383	0.1752	-0.0383	-0.0429	0.1747	0.2135	0	0.213	0.2427	-0.2181	-0.0005	0.2376	0.2176	0.046	0.2176
6	Monolake	1.9663	0.7049	3.0656	1.595	1.3377	1.7598	1.8326	2.3607	0.8901	1.1328	1.0549	1.1277	-1.4706	-1.3058	-1.233	-1.233	0.1648	0.2376	-0.0779	0.0728	-0.0779	0.0728
7	Mona Lisa	2.2473	3.2074	2.2877	2.2877	1.9615	2.3167	2.1964	2.356	1.4363	1.1101	1.4653	1.345	-0.9197	-1.2459	-1.011	-1.011	0.029	-0.0913	0.3552	-0.1203	0.3552	-0.1203
8	Manhattan	2.192	0	3.3322	1.9238	2.5503	2.6204	2.596	3.3322	1.9238	2.5503	2.6204	2.596	-1.4084	-0.7819	-0.7362	-0.7118	0.6966	0.06	0.6722	-0.0244	0.0731	-0.0244
9	Lighthouse	2.793	0.5593	2.5728	2.1854	2.2454	2.0703	2.2495	2.0135	1.6261	1.6861	1.511	1.6902	-0.3874	-0.3274	-0.3233	-0.5025	-0.1151	0.0457	0.0641	0.1792	-0.1751	0.1792
10	Daisies	1.2776	0.749	1.9506	0.749	1.2024	1.2108	0.9789	1.2016	0	0.4534	0.4618	0.2299	-1.2016	-0.7482	-0.9717	-0.9717	0.4618	0.4534	0.2299	-0.2319	0.0054	-0.2319
	Median	1.9188	0.7582	3.1161	1.7594	2.1035	1.9151	2.0145	2.1848	0.7438	1.1636	1.0657	1.0765	-1.1315	-0.5122	-0.7407	-0.7258	0.4704	0.3327	0.3327	-0.0267	-0.0579	0.0305

Table 6.48 Difference between pairwise fitness values of the different techniques for $K = 3$ for Huang's method [130]. Columnwise median values are given at the bottom of the table.

For K=3

SN	Image	D(1,2)	D(1,3)	D(1,4)	D(1,5)	D(1,6)	D(1,7)	D(1,8)	D(1,9)	D(1,10)	D(2,3)	D(2,4)	D(2,5)	D(2,6)	D(2,7)	D(2,8)	D(2,9)	D(2,10)	D(3,4)	D(3,5)	D(3,6)	D(3,7)	D(3,8)	D(3,9)
1	Baboon	0.0006	0.0011	0.001	-0.0173	-0.0208	-0.0007	-0.0092	-0.007	-0.0075	0.0005	0.0004	-0.0179	-0.0214	-0.0013	-0.0098	-0.0076	-0.0081	-0.0001	-0.0184	-0.0219	-0.0018	-0.0103	-0.0081
2	Bird	0	0	0	0	0	0	0	0	0	0	0	0	0	0	0	0	0	0	0	0	0	0	0
3	Elephant	0.0001	0	0.0001	-0.0044	-0.0051	-0.0033	-0.0035	-0.0002	-0.0002	-0.0001	0	-0.0045	-0.0052	-0.0034	-0.0036	-0.0003	-0.0003	0.0001	-0.0044	-0.0051	-0.0033	-0.0035	-0.0002
4	Seaport	-0.0027	-0.0016	-0.0027	-0.0646	-0.0707	-0.0584	-0.0707	-0.0027	-0.0027	0.0011	0	-0.0619	-0.068	-0.0557	-0.068	0	0	-0.0011	-0.063	-0.0691	-0.0568	-0.0691	-0.0011
5	Montreal	0	0	0	0	0	0	0	0	0	0	0	0	0	0	0	0	0	0	0	0	0	0	0
6	Monolake	-0.0013	-0.0012	0	-0.0126	-0.012	-0.0054	-0.0051	-0.0029	-0.0115	0.0001	0.0013	-0.0113	-0.0107	-0.0041	-0.0038	-0.0016	-0.0102	0.0012	-0.0114	-0.0138	-0.0042	-0.0039	-0.0017
7	Mona Lisa	-0.0075	0	-0.0053	-0.0071	-0.0002	-0.0053	-0.0476	0.0006	0	0.0075	0.0022	0.0004	0.0073	0.0022	-0.0401	0.0075	0.0075	-0.0053	-0.0071	-0.0002	-0.0053	-0.0476	0
8	Manhattan	-0.036	0.0356	0.0303	-0.0004	-0.0004	-0.0004	-0.0004	-0.0004	-0.0004	0.0356	0.0303	-0.0004	-0.0004	-0.0004	-0.0004	-0.0004	-0.0004	-0.0053	-0.036	-0.056	-0.036	-0.036	-0.036
9	Lighthouse	0.0011	-0.0008	-0.0012	-0.0017	-0.0022	-0.0017	-0.0022	-0.0012	-0.0014	-0.0011	-0.0023	-0.0018	-0.0024	-0.0014	-0.0023	-0.0011	-0.0025	-0.0012	-0.0007	-0.0013	-0.0003	-0.0012	0
10	Daisies	0	-0.0008	-0.0012	-0.0017	-0.0022	-0.0017	-0.0022	-0.0012	-0.0008	0	-0.0012	-0.0017	-0.0022	-0.0017	-0.003	-0.0004	-0.0004	-0.0004	-0.0009	-0.0014	-0.0009	-0.0014	-0.0004
	Median	0	0	0	-0.003	-0.0017	-0.0012	-0.0028	-0.0003	-0.0008	0	0	-0.0017	-0.0023	-0.0013	-0.003	-0.0004	-0.0004	-0.0003	-0.0057	-0.0032	-0.0025	-0.0037	-0.0003

SN	Image	D(3,10)	D(4,5)	D(4,6)	D(4,7)	D(4,8)	D(4,9)	D(4,10)	D(5,6)	D(5,7)	D(5,8)	D(5,9)	D(5,10)	D(6,7)	D(6,8)	D(6,9)	D(6,10)	D(7,8)	D(7,9)	D(7,10)	D(8,9)	D(8,10)	D(9,10)
1	Baboon	-0.0086	-0.0183	-0.0218	-0.0017	-0.0102	-0.008	-0.0085	-0.0035	0.0166	0.0081	0.0103	0.0098	0.0201	0.0116	0.0138	0.0133	-0.0085	-0.0063	-0.0068	0.0022	0.0017	-0.0005
2	Bird	0	0	0	0	0	0	0	0	0	0	0	0	0	0	0	0	0	0	0	0	0	0
3	Elephant	-0.0002	-0.0045	-0.0052	-0.0034	-0.0036	-0.0003	-0.0003	-0.0007	0.0011	0.0009	0.0042	0.0042	0.0018	0.0016	0.0049	0.0049	-0.0002	0.0031	0.0031	0.0033	0.0033	0
4	Seaport	-0.0011	-0.0619	-0.068	-0.0557	-0.068	0	0	-0.0061	0.0062	-0.0061	0.0619	0.0619	0.0123	0.068	0.068	0.068	-0.0123	0.0557	0.0557	0.068	0.068	0
5	Montreal	0	0	0	0	0	0	0	0	0	0	0	0	0	0	0	0	0	0	0	0	0	0
6	Monolake	-0.0103	-0.0126	-0.012	-0.0054	-0.0051	-0.0029	-0.0115	0.0006	0.0072	0.0075	0.0097	0.0011	0.0066	0.0069	0.0091	0.0005	0.0003	0.0025	-0.0061	0.0022	-0.0064	-0.0086
7	Mona Lisa	0	-0.0018	0.0051	0.0053	0.0053	0.0053	0.0053	0.0069	0.0018	-0.0405	0.0071	0.0071	-0.0051	-0.0474	0.0002	0.0002	-0.0423	0.0053	0.0053	0.0053	0.0476	0.0476	0
8	Manhattan	-0.036	-0.0307	-0.0307	-0.0307	-0.0307	-0.0307	-0.0307	0	0	0	0	0	0	0	0	0	0	0	0	0	0	0
9	Lighthouse	-0.0014	0.0005	-0.0001	0.0009	0	0.0012	-0.0002	-0.0006	0.0004	-0.0005	0.0007	-0.0007	0.001	0.0001	0.0013	-0.0001	-0.0009	0.0003	-0.0011	0.0012	-0.0002	-0.0014
10	Daisies	-0.0004	-0.0005	-0.001	-0.0005	-0.001	-0.0005	-0.0005	-0.0005	0.0004	-0.0005	0.0005	0.0005	0.0005	0.0005	0.001	0.001	-0.0005	0.0005	0.0005	0.001	0.001	0.0005
	Median	-0.0008	-0.0031	-0.0031	-0.0011	-0.0044	0	0	-0.0003	0.0008	0	0.0025	0.0008	0.0008	0	0.0011	0.0003	-0.0004	0.0004	0	0.0017	0.0005	0

Table 6.49 Difference between pairwise fitness values of the different techniques for K = 5 for Huang's method [130]. Columnwise median values are given at the bottom of the table.

For K=5

SN	Image	D(1,2)	D(1,3)	D(1,4)	D(1,5)	D(1,6)	D(1,7)	D(1,8)	D(1,9)	D(1,10)	D(2,3)	D(2,4)	D(2,5)	D(2,6)	D(2,7)	D(2,8)	D(2,9)	D(2,10)	D(3,4)	D(3,5)	D(3,6)	D(3,7)	D(3,8)	D(3,9)
1	Baboon	0.0039	0.0014	0.0004	-0.0393	-0.0402	-0.0185	-0.0273	-0.0235	-0.0176	-0.0025	-0.0035	-0.0432	-0.0441	-0.0224	-0.0312	-0.0274	-0.0215	-0.001	-0.0407	-0.0416	-0.0199	-0.0287	-0.0249
2	Bird	0	0	0	-0.0408	-0.0507	-0.0649	-0.0351	-0.0298	-0.0298	0	0	-0.0408	-0.0507	-0.0649	-0.0351	-0.0298	-0.0298	0	-0.0408	-0.0507	-0.0649	-0.0351	-0.0298
3	Elephant	-0.0364	-0.0372	-0.0228	-0.0454	-0.0586	-0.0421	-0.062	-0.0099	-0.0369	-0.0008	0.0136	-0.009	-0.0222	-0.0057	-0.0256	0.0265	-0.0005	0.0144	-0.0082	-0.0214	-0.0049	-0.0248	0.0273
4	Seaport	0.0002	0.0001	0.003	-0.1161	-0.1168	-0.0827	-0.1257	-0.0024	-0.0019	-0.0001	0.0028	-0.1163	-0.117	-0.0829	-0.1259	-0.0026	-0.0021	0.0029	-0.1162	-0.1169	-0.0828	-0.1258	-0.0025
5	Montreal	-0.0003	0.0011	0.0011	-0.0035	-0.0029	-0.0034	-0.0036	-0.0117	-0.0058	0.0014	0.0014	-0.0032	-0.0026	-0.0031	-0.0033	-0.0114	-0.0055	0	-0.0046	-0.004	-0.0045	-0.0047	-0.0128
6	Monolake	0.0096	0.0059	-0.0006	-0.0067	-0.0035	-0.0079	-0.0067	0.0046	0.0068	-0.0037	-0.0102	-0.0163	-0.0131	-0.0175	-0.0163	-0.005	-0.0048	-0.0065	-0.0126	-0.0094	-0.0138	-0.0126	-0.0013
7	Mona Lisa	0.0051	-0.0055	0.0029	-0.0169	-0.0152	-0.0168	-0.0103	-0.0029	-0.0178	-0.0106	-0.0022	-0.022	-0.0203	-0.0219	-0.0154	-0.008	-0.0229	0.0084	-0.0114	-0.0097	-0.0113	-0.0048	0.0026
8	Manhattan	0.0002	0.0001	0.0005	-0.0024	-0.0026	-0.0033	-0.0038	-0.001	-0.0016	-0.0001	-0.0007	-0.0026	-0.0028	-0.0035	-0.004	-0.0012	-0.0018	-0.0006	-0.0025	-0.0027	-0.0034	-0.0039	-0.0011
9	Lighthouse	0.0004	-0.0003	-0.0005	-0.0575	-0.0404	-0.0817	-0.0563	-0.0289	-0.0722	-0.0007	-0.0009	-0.0579	-0.0408	-0.0821	-0.0567	-0.0293	-0.0726	-0.0002	-0.0572	-0.0401	-0.0814	-0.056	-0.0286
10	Daisies	-0.0035	-0.0023	-0.0034	-0.0589	-0.0578	-0.033	-0.0489	-0.0447	-0.0378	0.0012	0.0001	-0.0554	-0.0543	-0.0295	-0.0454	-0.0412	-0.0343	-0.0011	-0.0566	-0.0555	-0.0307	-0.0466	-0.0424
	Median	0.0002	0	-0.0003	-0.04	-0.0403	-0.0257	-0.0312	-0.0108	-0.0177	-0.0004	-0.0004	-0.0314	-0.0315	-0.0222	-0.0284	-0.0097	-0.0135	-0.0001	-0.0267	-0.0307	-0.0169	-0.0268	-0.0076

SN	Image	D(3,10)	D(4,5)	D(4,6)	D(4,7)	D(4,8)	D(4,9)	D(4,10)	D(5,6)	D(5,7)	D(5,8)	D(5,9)	D(5,10)	D(6,7)	D(6,8)	D(6,9)	D(6,10)	D(7,8)	D(7,9)	D(7,10)	D(8,9)	D(8,10)	D(9,10)
1	Baboon	-0.019	-0.0397	-0.0406	-0.0189	-0.0277	-0.0239	-0.018	-0.0009	0.0208	0.012	0.0158	0.0217	0.0217	0.0167	0.0209	0.0226	-0.0088	-0.005	0.0009	0.0038	0.0097	0.0059
2	Bird	-0.0298	-0.0408	-0.0507	-0.0649	-0.0351	-0.0298	-0.0298	-0.0099	-0.0241	0.0057	0.011	0.011	-0.0142	0.0209	0.0209	0.0209	0.0298	0.0351	0.0351	0.0053	0.0053	0
3	Elephant	0.0003	-0.0226	-0.0358	-0.0193	-0.0392	0.0129	-0.0141	-0.0132	0.0033	-0.0166	0.0355	0.0085	0.0165	-0.0034	0.0487	0.0217	-0.0199	0.0322	0.0052	0.0521	0.0251	-0.027
4	Seaport	-0.002	-0.1191	-0.1198	-0.0857	-0.1287	-0.0054	-0.0049	-0.0007	0.0334	-0.0096	0.1137	0.1142	0.0341	0.1144	0.1149	-0.043	0.0803	0.0808	0.1233	0.1238	-0.0022	0.0005
5	Montreal	-0.0069	-0.0046	-0.004	-0.0045	-0.0047	-0.0128	-0.0069	0.0006	0.0001	-0.0001	0.0113	-0.0023	-0.0005	-0.0007	-0.0023	-0.0029	0.0125	0.0139	0.0127	0.0113	0.0115	0.0059
6	Monolake	-0.0011	-0.0061	-0.0029	-0.0073	-0.0061	0.0052	0.0054	0.0032	0.0006	-0.0012	0	0.0115	-0.0044	-0.0032	-0.0044	-0.0026	0.0012	0.0125	-0.001	0.0074	-0.0075	0.0002
7	Mona Lisa	-0.0123	-0.0198	-0.0181	-0.0197	-0.0132	-0.0058	-0.0207	0.0017	0.0001	0.0066	0.014	-0.0009	-0.0016	0.0049	0.0123	0.001	0.0065	0.0023	0.0017	0.0028	0.0022	-0.0149
8	Manhattan	-0.0017	-0.0019	-0.0021	-0.0028	-0.0033	-0.0005	-0.0005	-0.0002	-0.0009	-0.0014	0.0014	0.0008	-0.0007	-0.0012	0.0016	0.001	-0.0005	0.0023	0.0095	0.0023	-0.0159	-0.0006
9	Lighthouse	-0.0719	-0.057	-0.0399	-0.0812	-0.0558	-0.0284	-0.0717	0.0171	-0.0242	0.0012	0.0286	-0.0147	-0.0413	-0.0159	0.0115	-0.0318	0.0254	0.0528	-0.0048	0.0159	0.0159	-0.0433
10	Daisies	-0.0355	-0.0555	-0.0544	-0.0296	-0.0455	-0.0413	-0.0344	0.0011	0.0259	0.01	0.0142	0.0211	0.0248	0.0089	0.0131	0.02	-0.0159	-0.0117	0.0035	0.0111	0.0111	0.0069
	Median	-0.0096	-0.0312	-0.0378	-0.0195	-0.0314	-0.0093	-0.016	0.0002	0.0001	0.0006	0.0141	0.0098	-0.0006	-0.0009	0.0127	0.0142	-0.0004	0.0132	0.0035	0.0064	0.0075	0.0001

Table 6.50 Difference between pairwise fitness values of the different techniques for K = 7 for Huang's method [130]. Columnwise median values are given at the bottom of the table.

For K=7

SN	Image	D(1,2)	D(1,3)	D(1,4)	D(1,5)	D(1,6)	D(1,7)	D(1,8)	D(1,9)	D(1,10)	D(2,3)	D(2,4)	D(2,5)	D(2,6)	D(2,7)	D(2,8)	D(2,9)	D(2,10)	D(3,4)	D(3,5)	D(3,6)	D(3,7)	D(3,8)	D(3,9)
1	Baboon	-0.0028	0.1086	0.0293	-0.0343	-0.036	-0.0442	-0.043	-0.0079	-0.0053	0.1114	0.0321	-0.0315	-0.0332	-0.0414	-0.0402	-0.0051	-0.0025	-0.0793	-0.1129	-0.1446	-0.1528	-0.1516	-0.1165
2	Bird	-0.0075	0.0055	0.0018	-0.0744	-0.1252	-0.1274	-0.0718	-0.0114	-0.0168	0.013	0.0093	-0.0669	-0.1177	-0.1199	-0.0643	-0.0039	-0.0093	-0.0037	-0.0799	-0.1307	-0.1329	-0.0773	-0.0169
3	Elephant	-0.0032	0.0364	0.0229	-0.0111	-0.0354	-0.0104	-0.0107	-0.0068	-0.0101	0.0396	0.0261	-0.0079	-0.0322	-0.0072	-0.0075	-0.0036	-0.0069	-0.0135	-0.0475	-0.0718	-0.0468	-0.0471	-0.0432
4	Seaport	-0.0026	0.0008	-0.0019	-0.0063	0.0035	-0.0042	-0.0046	-0.002	-0.0027	0.0034	0.0007	-0.0037	-0.0009	-0.0016	-0.002	0.0006	-0.0001	-0.0027	-0.0071	-0.0043	-0.005	-0.0054	-0.0028
5	Montreal	-0.0021	-0.0013	-0.0022	-0.0019	-0.0136	-0.012	-0.0121	-0.0108	-0.0049	0.0008	-0.0001	-0.0103	-0.0115	-0.0099	-0.01	-0.0087	-0.0028	-0.0009	-0.0111	-0.0123	-0.0108	-0.0108	-0.0095
6	Monolake	0.0043	0.0033	-0.0036	-0.0121	-0.0129	-0.1765	-0.172	-0.1697	-0.1564	-0.001	-0.0079	-0.0164	-0.0172	-0.1808	-0.1763	-0.174	-0.1607	-0.0069	-0.3154	-0.0162	-0.1798	-0.1753	-0.173
7	Mona Lisa	-0.0021	0.0032	0.0098	-0.0254	-0.0329	-0.0393	-0.0062	-0.0314	-0.0252	0.0053	0.0119	-0.0233	-0.0308	-0.0372	-0.0041	-0.0293	-0.0231	0.0066	-0.3286	-0.0361	-0.0425	-0.0094	-0.0346
8	Manhattan	0.0024	0.0001	-0.0019	-0.009	-0.0197	-0.022	-0.0197	-0.0179	-0.0027	-0.0023	-0.0043	-0.0114	-0.0221	-0.0244	-0.0221	-0.0203	-0.0051	-0.002	-0.0091	-0.0198	-0.0221	-0.0198	-0.018
9	Lighthouse	0.0146	-0.0017	0.0081	-0.0733	-0.0621	-0.0367	-0.0789	-0.0636	-0.0032	-0.0163	-0.0065	-0.0879	-0.0767	-0.0513	-0.0935	-0.0782	-0.0178	0.0098	-0.0716	-0.0604	-0.035	-0.0772	-0.0619
10	Daisies	-0.0184	-0.0184	0.0045	-0.0569	-0.0168	-0.0252	-0.0265	-0.0323	-0.0597	0	0.0229	-0.0385	0.0016	-0.0068	-0.0081	-0.0139	-0.0413	0.0229	-0.0385	0.0016	-0.0068	-0.0081	-0.0139
	Median	-0.0024	0.002	0.0031	-0.0189	-0.0263	-0.0309	-0.0231	-0.0147	-0.0077	0.0021	0.005	-0.0198	-0.0264	-0.0308	-0.0161	-0.0113	-0.0081	-0.0024	-0.0336	-0.0279	-0.0388	-0.0335	-0.0263

SN	Image	D(3,10)	D(4,5)	D(4,6)	D(4,7)	D(4,8)	D(4,9)	D(4,10)	D(5,6)	D(5,7)	D(5,8)	D(5,9)	D(5,10)	D(6,7)	D(6,8)	D(6,9)	D(6,10)	D(7,8)	D(7,9)	D(7,10)	D(8,9)	D(8,10)	D(9,10)
1	Baboon	-0.1139	-0.0636	-0.0653	-0.0735	-0.0723	-0.0372	-0.0346	-0.0017	-0.0099	-0.0087	0.0264	0.029	-0.0082	-0.007	0.0281	0.0307	0.0012	0.0363	0.0389	0.0351	0.0377	0.0026
2	Bird	-0.0223	-0.0762	-0.127	-0.1292	-0.0736	-0.0132	-0.0186	-0.0508	-0.053	0.0026	0.063	0.0576	-0.0022	0.0534	0.1138	0.1084	0.0556	0.116	0.1106	0.0604	0.055	-0.0054
3	Elephant	-0.0465	-0.034	-0.0583	-0.0333	-0.0336	-0.0297	-0.033	-0.0243	0.0007	0.0004	0.0043	0.001	0.025	0.0247	0.0286	0.0253	-0.0003	0.0036	0.0003	0.0039	0.0006	-0.0033
4	Seaport	-0.0035	-0.0044	-0.0016	-0.0023	-0.0027	-0.0001	-0.0008	0.0028	0.0021	0.0017	0.0043	0.0036	0.0016	0.0015	0.0015	0.0008	-0.0004	0.0022	0.0015	0.0026	0.0019	-0.0007
5	Montreal	-0.0036	-0.0102	-0.0114	-0.0098	-0.0099	-0.0086	-0.0027	-0.0012	0.0004	0.0003	0.0016	0.0075	0.0016	0.0015	0.0028	0.0087	-0.0001	0.0012	0.0071	0.0013	0.0072	0.0059
6	Monolake	-0.1597	-0.0085	-0.0093	-0.1729	-0.1684	-0.1661	-0.1528	-0.0008	-0.1644	-0.1599	-0.1576	-0.1443	-0.1636	-0.1591	-0.1568	-0.1435	0.0045	0.0068	-0.0201	0.0023	0.0156	0.0133
7	Mona Lisa	-0.0284	-0.0352	-0.0427	-0.0491	-0.016	-0.0412	-0.035	-0.0075	-0.0139	0.0192	-0.006	0.0002	-0.0064	0.0267	0.0015	0.0077	0.0331	0.0079	0.0141	-0.0252	-0.019	0.0062
8	Manhattan	-0.0028	-0.0071	-0.0178	-0.0201	-0.0178	-0.016	-0.0008	-0.0107	-0.013	-0.0107	-0.0089	0.0063	-0.0023	0	0.0018	0.017	0.0023	0.0041	0.0193	0.0018	0.017	0.0152
9	Lighthouse	-0.0015	-0.0814	-0.0702	-0.0448	-0.087	-0.0717	-0.0113	0.0112	0.0366	-0.0556	0.0097	0.0701	0.0254	-0.0168	-0.0015	0.0589	-0.0422	-0.0269	0.0335	0.0153	0.0757	0.0604
10	Daisies	-0.0413	-0.0614	-0.0213	-0.0297	-0.031	-0.0368	-0.0642	0.0401	0.0317	0.0304	0.0246	-0.0028	-0.0084	-0.0097	-0.0155	-0.0429	-0.0013	-0.0071	-0.0345	-0.3058	-0.0332	-0.0274
	Median	-0.0254	-0.0346	-0.032	-0.039	-0.0323	-0.0333	-0.0258	-0.0014	-0.0048	0.0003	0.0043	0.005	-0.0023	-0.0006	0.0016	0.0128	0.0005	0.0038	0.0167	0.0025	0.0114	0.0043

Table 6.51 The median-based estimation among the participating techniques for $K = 3, 5,$ and 7, respectively for Kapur's method [140].

	For K=3									
	(1)	(2)	(3)	(4)	(5)	(6)	(7)	(8)	(9)	(10)
(1)	0.000	0.0868	0.0275	0.1063	0.3785	0.4021	0.3432	0.2840	0.2498	0.2897
(2)	−0.0868	0.000	−0.0592	0.0195	0.2917	0.3153	0.2564	0.1972	0.1630	0.2029
(3)	−0.0275	0.05928	0.000	0.0788	0.35098	0.3746	0.3156	0.2565	0.2223	0.2622
(4)	−0.1063	−0.0195	−0.0788	0.000	0.2722	0.2958	0.2369	0.1777	0.1435	0.1834
(5)	−0.3785	−0.2917	−0.3509	−0.2722	0.000	0.0236	−0.035	−0.0944	−0.1286	−0.0887
(6)	−0.4021	−0.3153	−0.3746	−0.2958	−0.0236	0.000	−0.0589	−0.1180	−0.1523	−0.1123
(7)	−0.3432	−0.2564	−0.3156	−0.2369	0.0353	0.0589	0.000	−0.0591	−0.0933	−0.0534
(8)	−0.2840	−0.1972	−0.2565	−0.1777	0.0944	0.1180	0.0591	0.000	−0.0342	0.0057
(9)	−0.2498	−0.1630	−0.2223	−0.1435	0.1290	0.1523	0.0933	0.0342	0.000	−0.0804
(10)	−0.2897	−0.2029	−0.2622	−0.1834	0.0887	0.1123	0.0534	−0.0057	0.0804	0.000
	For K=5									
	(1)	(2)	(3)	(4)	(5)	(6)	(7)	(8)	(9)	(10)
(1)	0.000	0.4051	0.2291	0.4605	1.2364	1.5140	1.3611	1.3769	1.2282	1.2240
(2)	−0.4051	0.000	−0.176	0.0553	0.8312	1.1088	0.9560	0.9718	0.8230	0.8188
(3)	−0.2291	0.176	0.000	0.2313	1.0072	1.2848	1.1320	1.1478	0.9990	0.9948
(4)	−0.4605	−0.0553	−0.2313	0.000	0.000	0.7759	1.0535	0.9006	0.9164	0.7677
(5)	−1.2364	−0.8312	−1.0072	−0.7759	0.000	0.2775	0.1247	0.1405	−0.0081	−0.0123
(6)	−1.5140	−1.1088	−1.2848	−1.0535	−0.2775	0.000	−0.1528	−0.1370	−0.2857	−0.2899
(7)	−1.3611	−0.9560	−1.132	−0.9006	−0.1247	0.1528	0.000	0.0158	−0.1329	−0.1371
(8)	−1.3769	−0.9718	−1.1478	−0.9164	−0.1405	0.1370	−0.015	0.000	−0.1487	−0.1529
(9)	−1.2282	−0.8230	−0.9990	−0.7677	0.0081	0.2857	0.1329	0.1487	0.000	−0.3382
(10)	−1.2240	−0.8188	−0.9948	−0.7635	0.0123	0.2899	0.1371	0.1529	0.3382	0.000
	For K=7									
	(1)	(2)	(3)	(4)	(5)	(6)	(7)	(8)	(9)	(10)
(1)	0.000	5.3375	0.5374	0.6676	1.3966	3.4254	2.2060	2.6709	2.5318	2.5496
(2)	−5.3375	0.000	5.8749	6.0051	6.7341	8.7629	7.5435	8.0084	7.8693	7.8871
(3)	−0.5374	−5.8749	0.000	0.1301	0.8591	2.8879	1.6685	2.1334	1.9943	2.0121
(4)	−0.6676	−6.0051	−0.1301	0.000	0.7290	2.7578	1.5384	2.0033	1.8642	1.882
(5)	−1.3966	−6.7341	−0.8591	−0.7290	0.000	2.0288	0.8094	1.2742	1.1352	1.1529
(6)	−3.4254	−8.7629	−2.8879	−2.7578	−2.0288	0.000	−1.2193	−0.7545	−0.8936	−0.8758
(7)	−2.2060	−7.5435	−1.6685	−1.5384	−0.8094	1.2193	0.000	0.4648	0.3257	0.3435
(8)	−2.6709	−8.0084	−2.1334	−2.0033	−1.2742	0.7545	−0.4648	0.000	−0.13907	−0.1213
(9)	−2.5318	−7.8693	−1.9943	−1.8642	−1.1352	0.8936	−0.3257	0.1390	0.000	−0.8849
(10)	−2.5496	−7.8871	−2.0121	−1.882	−1.1529	0.8758	−0.3435	0.1213	0.8849	0.000
(1)→ QIGAMLTCI (2)→ QISAMLTCI (3)→ QIPSOMLTCI (4)→ QIDEMLTCI										
(5)→ GA (6)→ SA (7)→ PSO (8)→ DE (9)→ BSA (10)→ CoDE										

Table 6.52 The median-based estimation among the participating techniques for $K = 3, 5,$ and 7, respectively, for Huang's method [130].

	(1)	(2)	(3)	(4)	(5)	(6)	(7)	(8)	(9)	(10)
					For K=3					
(1)	0.000	0.0073	1E−04	0.0020	−0.0020	−0.0019	−0.0012	−0.0019	−0.0024	−0.0019
(2)	−0.0073	0.000	−0.0072	−0.0053	−0.0093	−0.0092	−0.0085	−0.0092	−0.0097	−0.00921
(3)	−1E−04	0.007	0.000	0.0019	−0.0021	−0.002	−0.0013	−0.0020	−0.0025	−0.0020
(4)	−0.0020	0.0053	−0.0019	0.000	−0.0040	−0.0039	−0.0032	−0.0039	−0.0044	−0.0039
(5)	0.0020	0.0093	0.0021	0.0040	0.000	0.0001	0.0008	0.0001	−0.0004	0.0001
(6)	0.0019	0.0092	0.0020	0.0039	−0.0001	0.000	0.0007	0	−0.0005	0
(7)	0.0012	0.0085	0.0013	0.0032	0.0008	−0.0007	0.000	−0.0007	−0.0012	−0.0007
(8)	0.0019	0.0092	0.0020	0.0039	−0.0001	0	0.0007	0.000	−0.0005	0
(9)	0.0024	0.0097	−0.0025	0.0044	0.0004	0.0005	0.0012	0.0005	0.000	0.0011
(10)	0.0019	0.0092	0.0020	0.0039	−0.0001	0	0.0007	0	−0.0011	0.000
					For K=5					
	(1)	(2)	(3)	(4)	(5)	(6)	(7)	(8)	(9)	(10)
(1)	0.000	0.0407	−0.0013	−0.0042	−0.012	−0.0136	−0.0120	−0.0121	−0.0108	−0.0049
(2)	−0.0407	0.000	−0.0420	−0.0449	−0.0531	−0.0543	−0.0527	−0.0528	−0.0515	−0.0456
(3)	0.0013	0.0420	0.000	−0.0029	−0.0111	−0.0123	−0.0107	−0.0108	−0.0095	−0.0036
(4)	0.0042	0.0449	0.0029	0.000	−0.0081	−0.0093	−0.0077	−0.0078	−0.0065	−0.0006
(5)	0.0124	0.0531	0.0111	0.0081	0.000	−0.0012	0.0004	0.0003	0.0016	0.0075
(6)	0.0136	0.0543	0.0123	0.0093	0.0012	0.000	0.0016	0.0015	0.0028	0.0087
(7)	0.0120	0.0527	0.0107	0.0077	−0.0004	−0.0016	0.000	−0.0001	0.0012	0.0071
(8)	0.0121	0.0528	0.0108	0.0078	−0.0003	−0.0015	0.0001	0.000	0.0013	0.0072
(9)	0.0108	0.0515	0.0095	0.0065	−0.0016	−0.0028	−0.0012	−0.0013	0.000	−0.0022
(10)	0.0049	0.0456	0.0036	0.0006	−0.0075	−0.0087	−0.0071	−0.0072	0.0022	0.000
					For K=7					
	(1)	(2)	(3)	(4)	(5)	(6)	(7)	(8)	(9)	(10)
(1)	0.000	0.0838	0.0073	0.0014	−0.0229	−0.0244	−0.0286	−0.0237	−0.0196	−0.0135
(2)	−0.0838	0.000	−0.0765	−0.0823	−0.1067	−0.1082	−0.1124	−0.1075	−0.1034	−0.0973
(3)	−0.0073	0.0765	0.000	−0.0059	−0.0302	−0.0317	−0.0359	−0.0310	−0.0269	−0.0208
(4)	−0.0014	0.0823	0.0059	0.000	−0.0243	−0.0259	−0.0301	−0.0252	−0.0211	−0.0150
(5)	0.0229	0.1067	0.0302	0.0243	0.000	−0.0015	−0.0057	−0.0008	0.0032	0.0093
(6)	0.0244	0.1082	0.0317	0.0259	0.0015	0.000	−0.0042	0.0006	0.0047	0.0108
(7)	0.0286	0.1124	0.0359	0.0301	0.0057	0.0042	0.000	0.0048	0.0089	0.0150
(8)	0.0237	0.1075	0.0310	0.0252	0.0008	−0.0006	−0.0048	0.000	0.0041	0.0101
(9)	0.0196	0.1034	0.0269	0.0211	−0.0032	−0.0047	−0.0089	−0.0041	0.000	0.0016
(10)	0.0135	0.0973	0.0208	0.0150	−0.0093	−0.0108	−0.0150	−0.0101	−0.0016	−0.000
	(1)→ **QIGAMLTCI**		(2)→ **QISAMLTCI**		(3)→ **QIPSOMLTCI**		(4)→ **QIDEMLTCI**			
	(5)→ **GA**	(6)→ **SA**		(7)→ **PSO**		(8)→ **DE**		(9)→ **BSA**	(10)→ **CoDE**	

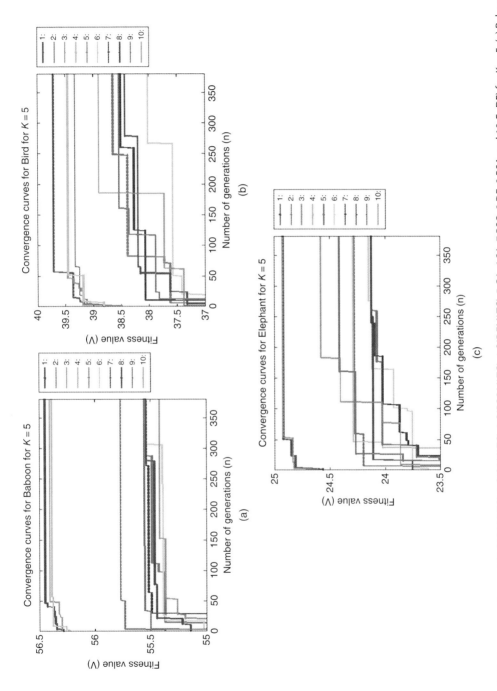

Figure 6.13 Convergence curves (1:QIGAMLTCI, 2:QISAMLTCI, 3:QIPSOMLTCI, 4:QIDEMLTCI, 5:GA, 6:SA, 7:PSO, 8:DE, 9:BSA and 10:CoDE) for $K = 5$, (a) Baboon, (b) Bird, (c) Elephant, (d) Seaport, and (e) Montreal for Kapur's method [140].

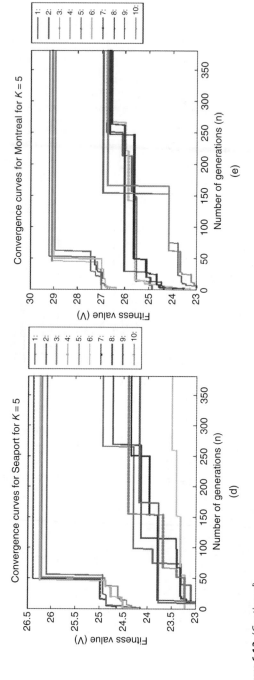

Figure 6.13 (*Continued*)

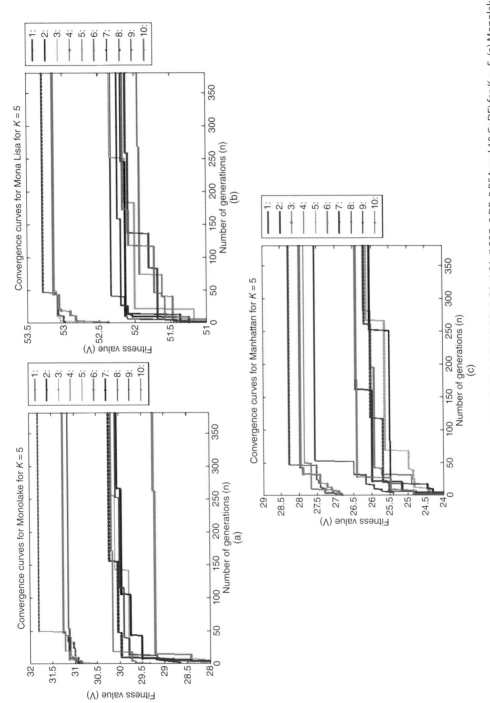

Figure 6.14 Convergence curves (1:QIGAMLTCI, 2:QISAMLTCI, 3:QIPSOMLTCI, 4:QIDEMLTCI, 5:GA, 6:SA, 7:PSO, 8:DE, 9:BSA and 10:CoDE) for $K = 5$, (a) Monolake, (b) Mona Lisa, (c) Manhattan, (d) Lighthouse, and (e) Daisies for Kapur's method [140].

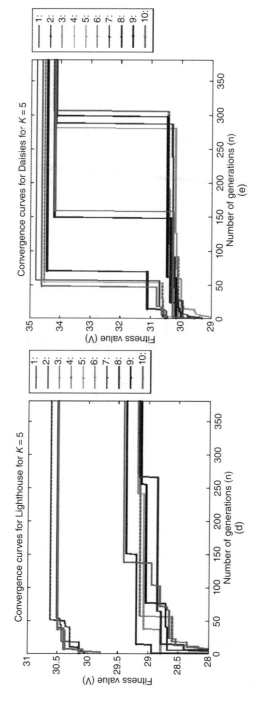

Figure 6.14 (*Continued*)

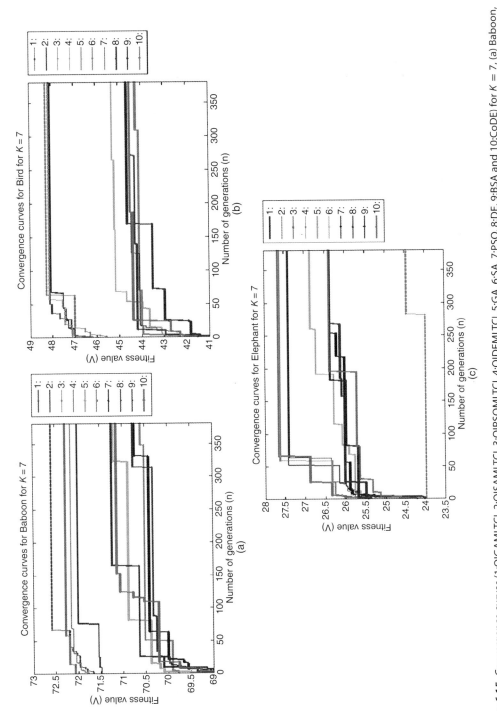

Figure 6.15 Convergence curves (1:QIGAMLTCI, 2:QISAMLTCI, 3:QIPSOMLTCI, 4:QIDEMLTCI, 5:GA, 6:SA, 7:PSO, 8:DE, 9:BSA and 10:CoDE) for $K = 7$, (a) Baboon, (b) Bird, (c) Elephant, (d) Seaport, and (e) Montreal for Kapur's method [140].

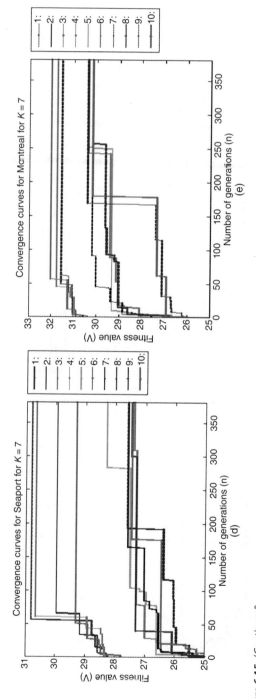

Figure 6.15 (*Continued*)

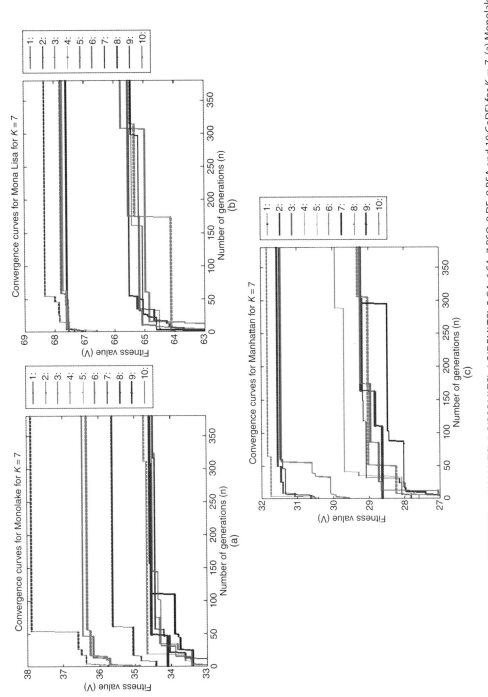

Figure 6.16 Convergence curves (1:QIGAMLTCI, 2:QISAMLTCI, 3:QIPSOMLTCI, 4:QIDEMLTCI, 5:GA, 6:SA, 7:PSO, 8:DE, 9:BSA and 10:CoDE) for $K = 7$, (a) Monolake, (b) Mona Lisa, (c) Manhattan, (d) Lighthouse, and (e) Daisies for Kapur's method [140].

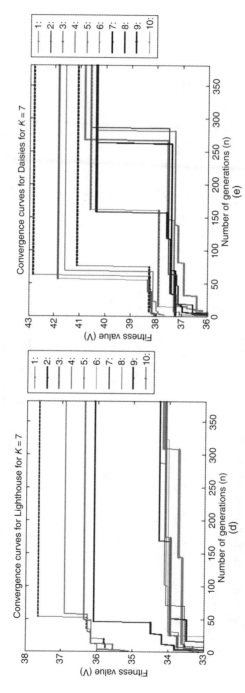

Figure 6.16 (*Continued*)

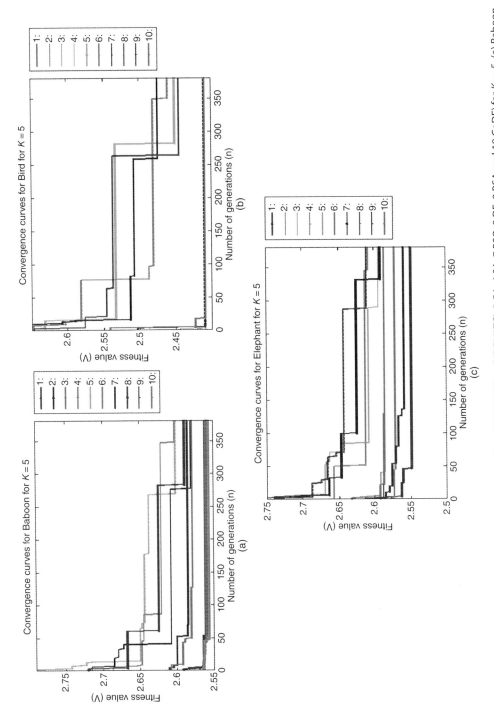

Figure 6.17 Convergence curves (1:QIGAMLTCI, 2:QISAMLTCI, 3:QIPSOMLTCI, 4:QIDEMLTCI, 5:GA, 6:SA, 7:PSO, 8:DE, 9:BSA and 10:CoDE) for $K = 5$, (a) Baboon, (b) Bird, (c) Elephant, (d) Seaport, and (e) Montreal for Huang's method [130].

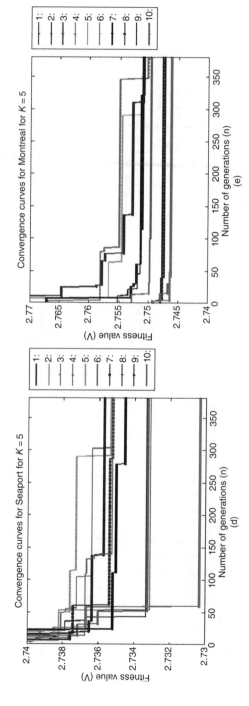

Figure 6.17 (*Continued*)

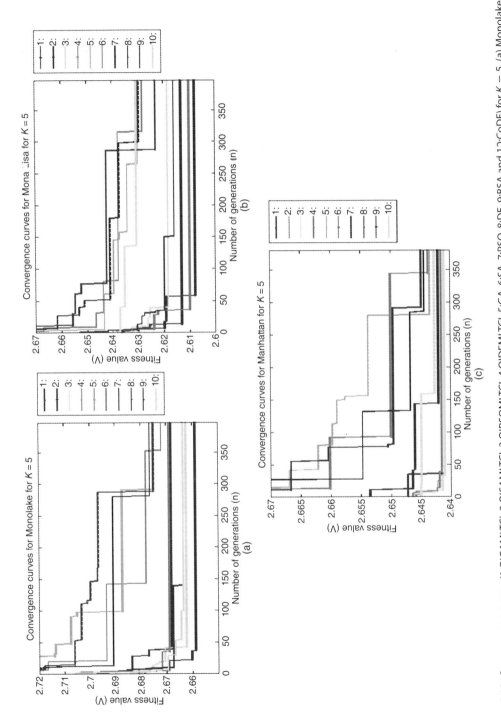

Figure 6.18 Convergence curves (1:QIGAMLTCI, 2:QISAMLTCI, 3:QIPSOMLTCI, 4:QIDEMLTCI, 5:GA, 6:SA, 7:PSO, 8:DE, 9:BSA and 10:CoDE) for $K = 5$, (a) Monolake, (b) Mona Lisa, (c) Manhattan, (d) Lighthouse, and (e) Daisies for Huang's method [130].

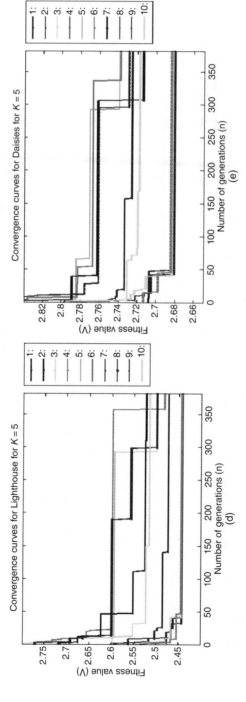

Figure 6.18 (*Continued*)

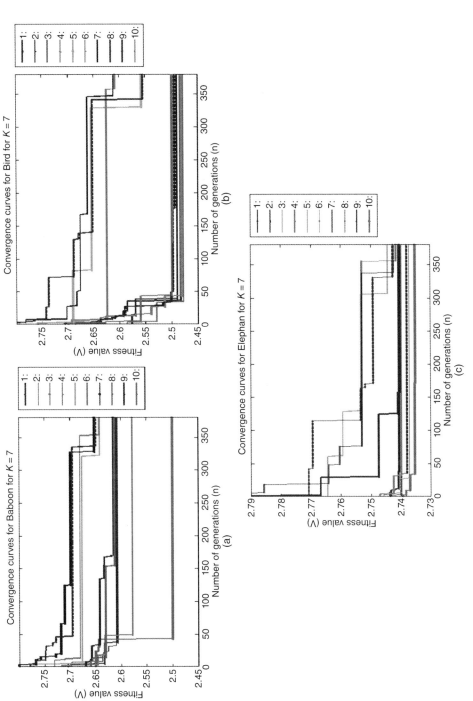

Figure 6.19 Convergence curves (1:QIGAMLTCI, 2:QISAMLTCI, 3:QIPSOMLTCI, 4:QIDEMLTCI, 5:GA, 6:SA, 7:PSO, 8:DE, 9:BSA and 10:CoDE) for K = 7, (a) Baboon, (b) Bird, (c) Elephant, (d) Seaport, and (e) Montreal for Huang's method [130].

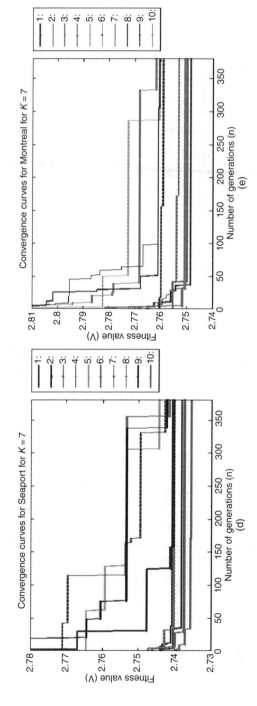

Figure 6.19 (*Continued*)

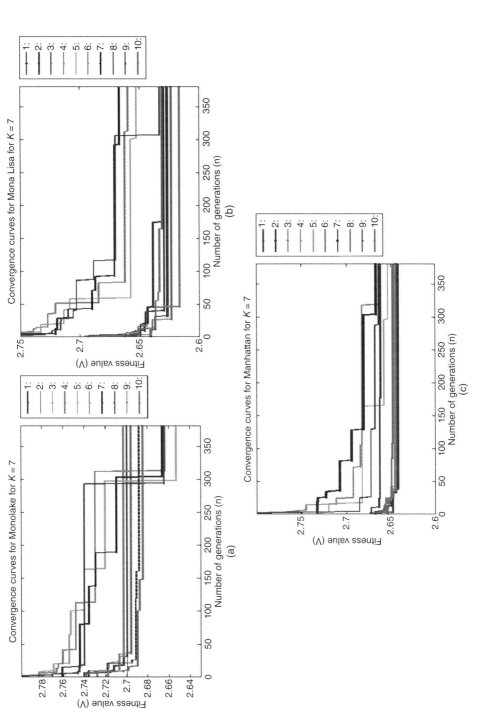

Figure 6.20 Convergence curves (1:QIGAMLTCI, 2:QISAMLTCI, 3:QIPSOMLTCI, 4:QIDEMLTCI, 5:GA, 6:SA, 7:PSO, 8:DE, 9:BSA and 10:CoDE) for $K = 7$, (a) Monolake, (b) Mona Lisa, (c) Manhattan, (d) Lighthouse, and (e) Daisies for Huang's method [130].

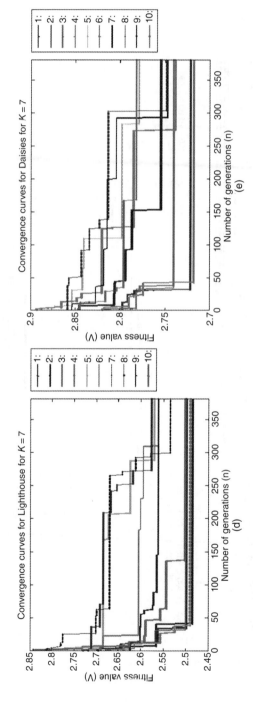

Figure 6.20 (*Continued*)

Table 6.53 Best results obtained for QITSMLTCI and TS for multi-level thresholding of color images.

	Tulips				
K	QITSMLTCI				
	θ_R	θ_G	θ_B	F_{best}	t
2	139	125	116	10909.76	01.14
4	56,124,195	69,120,175	61,116,181	12792.40	13.22
6	38,63,105,154,210	57,97,130,163,203	47,85,120,154,199	13119.14	16.41
	Tulips				
K	TS				
	θ_R	θ_G	θ_B	F_{best}	t
2	139	125	116	10909.76	01.14
4	57,122,199	70,121,172	62,113,176	12791.73	24.2
6	48,68,108,173,211	61,95,130,164,200	49,79,119,159,204	13116.47	24.2
	Barge				
K	QITSMLTCI				
	θ_R	θ_G	θ_B	F_{best}	t
2	106	99	92	5949.75	01.06
4	74,127,169	54,105,150	47,91,134	7258.03	13.49
6	64,107,142,164,187	46,81,122,151,179	46,78,104,133,152	7472.60	24.2
	Barge				
K	TS				
	θ_R	θ_G	θ_B	F_{best}	t
2	106	99	92	5949.75	01.06
4	74,127,169	55,110,150	44,92,133	7257.34	24.2
6	61,106,141,172,193	43,86,128,154,184	42,77,101,131,153	7470.81	24.2

Table 6.54 Mean fitness (F_{avg}) and standard deviation (σ) of QITSMLTCI and TS for multi-level thresholding of color images.

	Tulips			
K	QITSMLTCI		TS	
	F_{avg}	σ	F_{avg}	σ
2	10909.76	0	10909.76	0
4	12791.93	0.27	12790.84	1.33
6	13117.03	1.81	13112.86	2.78
	Barge			
K	QITSMLTCI		TS	
	F_{avg}	σ	F_{avg}	σ
2	5949.75	0	10909.76	0
4	7257.44	0.41	7255.72	1.01
6	7470.90	1.17	7468.61	1.85

confidence level, which means that the alternative hypothesis would be accepted as the test result if the desired value becomes less than 0.05. The test results are reported in Table 6.55.

Since both algorithms find identical threshold values for $K = 2$, there is no change in the mean value and standard deviation in this case. Hence, for lowest levels of thresholding, each algorithm sounds equally strong and accurate. From Table 6.54, it can easily be noted that, for $K = 4$ and 6, the proposed QITSMLTCI is more accurate and robust than its classical counterpart. From Table 6.55, the results of one-tailed t-test prove the QITSMLTCI outperforms TS.

Table 6.55 Test results of one-tailed t-test for Tulips and Barge at different levels of thresholding.

K	Tulips	Barge
2	Nan (Not-a-Number)	Nan (Not-a-Number)
4	0.038	0.016
6	0.005	0.023

6.10 Conclusion

The shortcomings of the processing capability of the quantum inspired classical algorithms [63, 67], in parallel, have been addressed by developing different mechanisms that can fix such flaws with a high efficacy. The computational capability of the techniques, introduced in Chapter 5, has been extended in parallel in the form of six novel quantum inspired meta-heuristics, presented in this chapter. The outline of this approach is the thresholding of color image information. Application of these quantum inspired versions is demonstrated with the thresholding of multi-level and color images. Image thresholding is carried out at the component levels by the proposed approaches with the help of different thresholding methods as objective functions.

The performance of the proposed approaches in the thresholding of true color images is evaluated by using several objective functions. The implementation results for the proposed algorithms have been reported in two phases in different aspects. In the first phase, the effectiveness of a set of proposed approaches is established quantitatively and visually with reference to the optimal threshold values along with the fitness values, the mean fitness measure, the standard deviation of the fitness values, the computational time and also the various convergence plots at different levels of thresholds. A statistical superiority test, called the Friedman test, has also been conducted to judge the superiority of the participating algorithms. The performance of the proposed algorithms is found to be superior to other comparable algorithms at the different level of thresholding, especially at higher levels of thresholding. It should be noted that the proposed QIACOMLTCI outperforms others quantitatively and visually. In the next phase, the effectiveness of another set of proposed approaches is established with respect to the optimal threshold values along with the fitness values, the computational time, and the convergence plots. Statistically their superiority is also proved by evaluating the Friedman test and a test based on the median-based estimation. In this phase, the proposed QIGAMLTCI is proved to be the best performing algorithm of them all. Hence, the results (PSNR values) reveal that the proposed algorithms outperform their conventional counterparts and two other comparable algorithms as far as the quality of segmentation is concerned.

6.11 Summary

- Six quantum inspired meta-heuristic techniques for color image thresholding have been introduced in this chapter.
- Parallel extensions to the quantum inspired classical algorithms, introduced in Chapter 5, are presented in this chapter.
- The introduction of this approach causes thresholding of color image information.

- Application of these versions of quantum inspired techniques is exhibited with thresholding of multi-level and color images.
- As a sequel to the proposed techniques, separate experiments have also been carried out with six others algorithms, to prove the effectiveness of the proposed techniques.
- The efficacy of the proposed techniques has been proved in a variety of ways, such as finding stability and accuracy, finding peak signal-to-noise ratio (PSNR) values at different levels, and performing a statistical test, called the Friedman test, etc.

Exercise Questions

Multiple Choice Questions

1 The term "PSNR" stands for
    (a) peak source-to-noise ratio
    (b) peak signal-to-noise ratio
    (c) primary signal-to-noise ratio
    (d) none of the above

2 Which are the primary color components, used to demonstrate pure color image information?
    (a) red, green and blue
    (b) red, grey and blue
    (c) red, green and black
    (d) none of the above

3 "CoDE" stands for
    (a) component DE
    (b) collaborative DE
    (c) composite DE
    (d) none of the above

4 In the proposed QIACOMLTCI, the dimension of a pheromone matrix (τ_{PH}) was taken as follows
    (a) same as the population P
    (b) greater than the population P
    (c) less than the population P
    (d) none of the above

5 In the proposed QISAMLTCI, the population P
    (a) comprises three configurations, one for each primary component of the input image
    (b) requires the number of configurations to be selected by the users
    (c) is the same as QIGAMLTCI
    (d) all of the above

6 In BSA, generation-wise, the population is guided by which of the following genetic operators?
    (a) crossover and selection

(b) selection and mutation
(c) crossover and mutation
(d) none of the above

7 What strategy is used to eliminate a string from the list in QITSMLTCI?
(a) LIFO
(b) FIFO
(c) both (a) and (b)
(d) neither (a) nor (b)

Short Answer Questions

1 What is the role of PSNR in the quality assessment of the proposed techniques?

2 How would you define a pure color image?

3 What is the role of the pheromone matrix in QIACOMLTCI?

4 Describe, in brief, the mutation and crossover operations in QIGAMLTCI.

5 How is the Boltzmann distribution used to find a better solution for QISAMLTCI?

6 Why is parameter tuning important in the meta-heuristics/quantum inspired meta-heuristics?

Long Answer Questions

1 Write short notes on CoDE and BSA algorithms.

2 How is median-based estimation used to prove the efficacy of the proposed techniques?

3 Compare the time complexities of QIACOMLTCI, QIPSOMLTCI, QIDEMLTCI, and QIGAMLTCI.

4 Discuss the algorithm of the proposed QISAMLTCI. Derive its time complexity.

5 Discuss the algorithm of the proposed QITSMLTCI. Derive its time complexity.

Coding examples

The functionality of the proposed techniques of Chapter 5 has been enhanced to the multi-level and color domains. Six novel quantum inspired meta-heuristics for color image thresholding have been presented in this chapter. The codes for the proposed QIPSO and QISA are presented in Figures 6.21–6.27. As in Chapter 5, the code for the same functions have been presented for one technique only.

```
Func¹

function [tr1,tr2,tr3,fbs]=QIPSOMLTCI(n,c1,c2,w,K)
%n-Number of particles, c1,c2-acceleration coefficients
%w-inertia weight
tic ; %Computational time calculation
t=0;
ntrials = 0 ;
L=256;
[Ha,Hb,Hc]=insert(L);%Histogram of the gray scale image is calculated
M=Generate(L,n);%Function used for generating initial population
s=M;
maxf=1000;
for k=1:3
for i=1:n
    for j=1:L
        v(i,j,k)=0;
        p(i,j,k)=0;
        p1(i,j,k)=0;
    end
end
end

for i=1:50
    if(i==1)
    [tr1,tr2,tr3,fr,fg,fb,fv]=Fitness(L,s,Ha,Hb,Hc,K,n); %Function
                              used for computing fitness of individuals

    p=s;
    pos=ones(1,3);
    fupr=fr;
    fupg=fg;
    fupb=fb;
    bs1=fr(1);
    bs2=fg(1);
    bs3=fb(1);
    for y=2:n
        if(bs1<fr(y))
            bs1=fr(y);
            pos(1)=y;
        end
        if(bs2<fg(y))
            bs2=fg(y);
            pos(2)=y;
        end
        if(bs3<fb(y))
            bs3=fb(y);
```

```
pos(3)=y;
    end
end
for k=1:3
for z=1:n
g(z,:,k)=s(pos(k),:,k);
end
end
maxf=fv(1);
for k=1:3
    for j=1:n
        v(j,:,k)=w*v(j,:,k)+c1*rand*(p(j,:,k)-
            s(j,:,k))+c1*rand*(g(j,:,k)-s(j,:,k));
        s(j,:,k)=s(j,:,k)+v(j,:,k);
    end
end
end
if(i>1)
[tr1,tr2,tr3,fr,fg,fb,fv]=Fitness(L,s,Ha,Hb,Hc,K,n);
if(maxf>=fv(1))
    maxf=fv(1);
    T1=tr1;
    T2=tr2;
    T3=tr3;
    t = toc;
end
[p1,g1]=pgbest(p,s,n,fupr,fupg,fupb,fr,fg,fb); %Function used
                        finding particle best and swarm best value
for k=1:3
    for j=1:n
        v(j,:,k)=w*v(j,:,k)+c1*rand*(p1(j,:,k)-
            s(j,:,k))+c2*rand*(g1(j,:,k)-s(j,:,k));
        s(j,:,k)=s(j,:,k)+v(j,:,k);
    end
end
p=p1;
fupr=fr;
fupg=fg;
fupb=fb;
end
tr1=T1;
tr2=T2;
tr3=T3;
fbs=maxf;
end
```

Figure 6.21 Sample code of Func¹ for QIPSOMLTCI.

```
Func²

function [Ha,Hb,Hc]=insert(L)
img=imread('D:\pic\tahoeC.png');
for k=1:3
is=img(:,:,k);
val=zeros(1,L);
c=0;
d=1;
while(d<=L*L)
    if(is(d)==0)
        c=c+1;
    end
for j=1:(L-1)
    if(is(d)==j)
        val(j)=val(j)+1;
    end
end
d=d+1;
end
val(L)=c;
for i=1:L
    if(k==1)
        ima(i)=val(i)/(L*L);
    end
    if(k==2)
        imb(i)=val(i)/(L*L);
    end
    if(k==3)
        imc(i)=val(i)/(L*L);
    end
end
end
Ha=ima;
Hb=imb;
Hc=imc;
end

Func³

function M=Generate(L,n)
for k=1:3
for i=1:n
    for j=1:L
        a(i,j,k)=rand;
    end
end
end
```

```
b=sqrt(1-a.^2);
M=b;
end


Func⁴

function [tr1,tr2,tr3,fr,fg,fb,fv]=Fitness(L,s,Ha,Hb,Hc,K,n)
[tr1,tr2,tr3,fr,fg,fb,fv]=thcal(L,s,Ha,Hb,Hc,n,K);
end

Func⁵

function [tr1,tr2,tr3,fr,fg,fb,fv]=thcal(L,s,Ha,Hb,Hc,n,K)
t1=0;
t2=0;
t3=0;
s1=0;
T=zeros(1,K);
for k=1:3
    mx=100000;
    d1=z(:,:,k);
for i=1:n
    d2=c1(i,:);
[tr,fv1]=partition(K,d2,L,k,n,Ha,Hb,Hc);
if(mx>fv1(1))
    mx=fv1(1);
    T=tr;
end
end
s1=s1+mx;
if(k==1)
    t1=tr;
    fr(i)=fv1;
elseif(k==2)
    t2=tr;
    fg(i)=fv1;
else
    t3=tr;
    fb(i)=fv1;
end
tr1=t1;
tr2=t2;
tr3=t3;
fv=s1;
end
```

```
Func⁶

function [tr,fv1]=partition(K,d2,L,k,n,Ha,Hb,Hc)
nx=100000;
st=zeros(1,255);
j=0;
t=[];
for i=1:255
        pm=rand;%probability factor
    if(pm>d2(i)*d2(i))
        j=j+1;
        st(j)=1;
    end
end
    an=randi([4,n],1,1);
    for k1=1:an
    for k2=1:(K-1)
    p(k2)=randi([st(1),st(j)],1,1);
    end
    P=sort(p);
    [tr,fv1]=thlic(L,Ha,Hb,Hc,P,k);
if(mx>fv1(1))
    mx=fv1(1);
    T=tr;
end
end
tr=T;
fv1=mx;
end
```

Figure 6.22 Sample code of Func², Func³, Func⁴, Func⁵, and Func⁶ for QIPSOMLTCI.

```
Func⁷

function [p1,g1]=pgbest(p,s,n,fupr,fupg,fupb,fr,fg,fb)    pos2r=1;                          end
for k=1:3                                                 pos2g=1;                      end
for i=1:n                                                 pos2b=1;                      end
    if(k==1)                                              br2=fupr(1);
    if(fupr(i)>fr(i))                                     bg2=fupg(1);                  p1=p;
        p(i,:,k)=p(i,:,k);                                bb2=fupb(1);                  g1=g;
    else                                                  for y=2:n                     end
        p(i,:,k)=s(i,:,k);                                    if(br2<fupr(y))
    end                                                           br2=fupr(y);
    end                                                           pos2r=y;
    if(k==2)                                                  end
    if(fupg(i)>fg(i))                                         if(bg2<fupg(y))
        p(i,:,k)=p(i,:,k);                                        bg2=fupg(y);
    else                                                          pos2g=y;
        p(i,:,k)=s(i,:,k);                                    end
    end                                                       if(bb2<fupb(y))
    end                                                           bb2=fupb(y);
    if(k==3)                                                      pos2b=y;
    if(fupb(i)>fb(i))                                         end
        p(i,:,k)=p(i,:,k);                                end
    else
        p(i,:,k)=s(i,:,k);                               if(fupr(pos2r)>fr(pos1r))
    end                                                      pos(1)=pos2r;
    end                                                  else
end                                                          pos(1)=pos1r;
end                                                      end
                                                         if(fupg(pos2g)>fr(pos1g))
pos=ones(1,3);                                               pos(2)=pos2g;
pos1r=1;                                                 else
pos1g=1;                                                     pos(2)=pos1g;
pos1b=1;                                                 end
br1=fr(1);                                               if(fupb(pos2b)>fr(pos1b))
bg1=fg(1);                                                   pos(3)=pos2b;
bb1=fb(1);                                               else
for y=2:n                                                    pos(3)=pos1b;
    if(br1<fr(y))                                        end
        br1=fr(y);
        pos1r=y;                                         for k=1:3
    end                                                  for z=1:n
    if(bg1<fg(y))                                          if(k==1)
        bg1=fg(y);                                         g(z,:,k)=s(pos(1),:,k);
        pos1g=y;                                          end
    end                                                   if(k==2)
    if(bb1<fb(y))                                          g(z,:,k)=s(pos(2),:,k);
        bb1=fb(y);                                        end
        pos1b=y;                                          if(k==3)
    end                                                   g(z,:,k)=s(pos(3),:,k);
end
```

Figure 6.23 Sample code of Func⁷ for QIPSOMLTCI.

```
Func⁸

function [tr,fv1]=thlic(L,Ha,Hb,Hc,P,k)     for j=1:Q
    if(k==1)                                    if(j==1)
        im=Ha;                                      for i=1:P(j)
    end                                                 if(mu(j)~=0 && (i/mu(j)>0))
    if(k==2)                                                si(j)=si(j)+im(i)*i*log(i/mu(j));
        im=Hb;                                          else
    end                                                     si(j)=1000;
    if(k==3)                                            end
        im=Hc;                                  end
    end                                     end
Q=length(P)+1;                              if(j==Q)
mun=zeros(1,Q);                                 for i=(P(j-1)+1):L
mud=zeros(1,Q);                                     if(mu(j)~=0 && (i/mu(j)>0))
si=zeros(1,Q);                                          si(j)=si(j)+im(i)*i*log(i/mu(j));
mu=zeros(1,Q);                                      else
sigb=0;                                                 si(j)=1000;
                                                    end
for j=1:Q                                       end
    if(j==1)                                end
        for i=1:(P(j)-1)                    if(j>1 && j<Q)
            mun(j)=mun(j)+im(i)*i;              for i=(P(j-1)+1):P(j)
            mud(j)=mud(j)+im(i);                   if(mu(j)~=0 && (i/mu(j)>0))
        end                                            si(j)=si(j)+im(i)*i*log(i/mu(j));
    end                                            else
    if(j==Q)                                           si(j)=1000;
        for i=(P(j-1)+1):(L-2)                      end
            mun(j)=mun(j)+im(i)*i;              end
            mud(j)=mud(j)+im(i);            end
        end                             end
    end
    if(j>1 && j<Q)                          s=sum(si);
        for i=(P(j-1)+1):(P(j)-1)            fv1=s;
            mun(j)=mun(j)+im(i)*i;           tr=P;
            mud(j)=mud(j)+im(i);        end
        end
    end
end

for j=1:Q
    mu(j)=mun(j)/mud(j);
end
```

Figure 6.24 Sample code of Func⁸ for QIPSOMLTCI.

```
Func¹

function [tr1,tr2,tr3,fbs]=QISAMLTCI (Tmax,Tmin,r,K)     Tmax=Tmax*r;
%n-Number of chromosomes, Tmax-Initial temperature, Tmin-Final    t=toc ;
temperature                                             end
%r-Reduction fraction                                   tr1=T1;
tic ; %Computational time calculation                   tr2=T2;
t = 0 ;                                                  tr3=T3;
ntrials = 0 ;                                            fbs=maxf;
o=1;                                                     end
L=512;
p=1;                                                    Func²
[Ha,Hb,Hc]=insert(L);%Histogram of the gray scale image is calculated
M=Generate(L); %Function used for generating initial population    function M=Generate(L)
b=M;                                                    for k=1:3
[tr1,tr2,tr3,fr,fg,fb,fv]=Fitness(L,b,Ha,Hb,Hc,K); %Function used for    for i=1:1
                        computing  fitness of individual        for j=1:L/2
fv2=fv(1);                                                       a(i,j,k)=rand;
maxf=fv2(1);                                                 end
while(Tmax>=Tmin)                                        end
  for i=1:1                                              end
    s=Perturbation(b,L); %Function used for modifying the    b=sqrt(1-a.^2);
                        characteristics of individual      M=b;
    b=s;                                                end
    [tr1,tr2,tr3,fr,fg,fb,fv]=Fitness(L,b,Ha,Hb,Hc,K);
    if(maxf<fv(1))                                      Func³
        maxf=fv(1);
        T1=tr1;                                         function [tr1,tr2,tr3,fr,fg,fb,fv]=Fitness(L,b,Ha,Hb,Hc,K)
        T2=tr2;                                         [tr1,tr2,tr3,fr,fg,fb,fv]=thcal(L,b,Ha,Hb,Hc,K);
        T3=tr3;                                         end
    end
    if(fv(1)-fv2(1)>0)
        b=s;
        fv2=fv(1);
        elseif(exp(-(fv(1)-fv2(1))/Tmax)>rand)
        b=s;
        fv2=fv(1);
        end
  end
```

Figure 6.25 Sample code of Func¹, Func², and Func³ for QISAMLTCI.

```
Func⁴

function [tr1,tr2,tr3,fr,fg,fb,fv,pr,pg,pb]=thcal(L,b,Ha,Hb,Hc,K)
t1=0;
t2=0;
t3=0;
ft1=0;
ft2=0;
ft3=0;
j=1;
p=1;
s1=0;
pos=zeros(1,3);
T=zeros(1,K);
for k=1:3
    j=1;
    mx=0;
    d1=b(:,:,k);
for i=1:1
    d2=d1(i,:);
[tr1,fv1]=partition(K,d2,L,k,Ha,Hb,Hc);
if(k==1)
    fr(i)=fv1;
elseif(k==2)
    fg(i)=fv1;
else
    fh(i)=fv1;
end

if(mx<fv1(1))
    mx=fv1(1);
    T=tr1;
    p=j;
end
j=j+1;
end
```

```
s1=s1+mx;
pos(k)=p;
if(k==1)
        t1=tr1;
elseif(k==2)
        t2=tr1;
else
        t3=tr1;
end
end
tr1=t1;
tr2=t2;
tr3=t3;
fv=s1;
fr=fr;
fg=fg;
fb=fb;
pr=pos(1);
pg=pos(2);
pb=pos(3);
end

Func⁵

function [tr1,fv1]=partition(K,d2,L,k,Ha,Hb,Hc)
mx=0;
st=zeros(1,255);
j=0;
for i=1:255
        pm=rand;%probability factor
    if(pm>d2(i)*d2(i))
        j=j+1;
        st(j)=i;
    end
end
```

```
for k1=1:1
    for k2=1:(K-1)
        p(k2)=randi([st(1),st(j)],1,1);
    end
    P=sort(p);
    [tr1,fv1]=thckK(L,Ha,Hb,Hc,K,P,k);
    if(mx<fv1(1))
        mx=fv1(1);
        T=tr1;
    end
end
tr1=T;
fv1=mx;
end
```

Figure 6.26 Sample code of Func⁴ and Func⁵ for QISAMLTCI.

```
Func⁶

function [tr1,fv1]=thckK(L,Ha,Hb,Hc,K,P,k)
    if(k==1)
        im=Ha;
    end
    if(k==2)
        im=Hb;
    end
    if(k==3)
        im=Hc;
    end
w=zeros(1,K);
y=zeros(1,K);
Y=0;
    for j=1:K
        if(j==1)
            for i=1:P(j)
                w(j)=w(j)+im(i);
            end
        end
        if(j==K)
            for i=(P(j-1)+1):255
                w(j)=w(j)+im(i);
            end
        end
        if(j>1 && j<K)
            for i=(P(j-1)+1):P(j)
                w(j)=w(j)+im(i);
            end
        end
    end
```

```
for j=1:K
    if(j==1)
        for i=1:P(j)
            if(im(i)~=0 && w(j)~=0)
                y(j)=y(j)+(im(i)/w(j))*log(im(i)/w(j));
            end
        end
    end
    if(j==K)
        for i=(P(j-1)+1):255
            if(im(i)~=0 && w(j)~=0)
                y(j)=y(j)+(im(i)/w(j))*log(im(i)/w(j));
            end
        end
    end
    if(j>1 && j<K)
        for i=(P(j-1)+1):P(j)
            if(im(i)~=0 && w(j)~=0)
                y(j)=y(j)+(im(i)/w(j))*log(im(i)/w(j));
            end
        end
    end
end

for i=1:K
Y=Y+y(i);
end
Y=-Y;
fv1=Y;
tr1=P;
end
```

```
Func⁷

function s=Perturbation(b,L)
for k=1:3
b1=b(1,:,k);
r1=randi(L);
r2=randi(L);
if(r1>r2)
    for i=r2:r1
        b1(1,i,k)=rand;
        b1(1,i,k)=sqrt(1-b1(1,i,k).^2);
    end
end
if(r2>r1)
    for i=r1:r2
        b1(1,i,k)=rand;
        b1(1,i,k)=sqrt(1-b1(1,i,k).^2);
    end
end
s=b1;
end
end
```

Figure 6.27 Sample code of Func⁶ and Func⁷ for QISAMLTCI.

7

Quantum Inspired Multi-objective Algorithms for Multi-level Image Thresholding

7.1 Introduction

Chapter 4 focuses on introducing two quantum inspired evolutionary techniques for bi-level image thresholding [28, 61, 64, 68]. Application of these techniques was extended and expanded to the multi-level and gray-scale spectrum [63, 67] in Chapter 5. Thereafter, the applicability of these techniques has again been extended to the multi-level and color domain [62, 65, 70, 72] in Chapter 6.

This chapter introduces several quantum inspired multi-objective algorithms using different approaches [60, 69, 74]. First, a NSGA-II-based quantum inspired algorithm is introduced in a multi-objective framework. Later, several quantum inspired classical algorithms are presented in a multi-objective flavor for bi-level, multi-level, and gray-scale image thresholding.

There exist certain problems in real-life situations, where a number of incommensurable constraints of the problems are simultaneously optimized in order to obtain apposite solutions of them. In such cases, rather focusing on a single objective, a trade-off between multiple objectives is taken into account. As a result of this, these particular problems have several alternative solutions instead of having a single optimal solution. Considering all constraints, no other solution is found to be superior to the existing solutions, and hence, this solution set is termed the optimal set of solutions. The most challenging task in reference to multi-objective optimization is probably defining optimality for a certain problem. No accepted definition with regard to optimal solutions can be acknowledged in this case. So, it is not easy to compare one solution to another when selecting the optimal set of solutions, which are basically called Pareto-optimal solutions.

Image thresholding is known to be an important tool to solve different real-life problems these days. There exist a number of segmentation algorithms, which can be successfully applied for optimization purposes using single or multiple evaluation functions. With regard to multiple thresholding, it is obligatory to optimize several single objectives simultaneously and handle various features of segmentation. This may lead to good segmented images if the constraints of the problem are carefully considered. In MOO [54] problems, several objective functions, generally conflicting, are invoked to search for solutions to certain problems.

This chapter is aimed at the development of different quantum inspired multi-objective algorithms to the gray-scale domain in order to facilitate it for bi-level/multi-level image thresholding of Berkeley images [1]/real-life gray-scale

Quantum Inspired Meta-heuristics for Image Analysis, First Edition.
Sandip Dey, Siddhartha Bhattacharyya, and Ujjwal Maulik.
© 2019 John Wiley & Sons Ltd. Published 2019 by John Wiley & Sons Ltd.

images [66, 71, 73]. This chapter is focused on adopting different approaches to develop several quantum inspired algorithms on the multi-objective framework. Several quantum inspired algorithms are introduced in different parts of the proposed methodology section.

As a sequel to the proposed quantum inspired NSGA-II (QINSGA-II), experiments have been conducted with the NSGA-II [56] and SMS-EMOA [84]. For other sets of proposed versions, a comparison has been made between the proposed approaches and their respective classical counterparts. The results of the proposed algorithms show better performances over other comparable algorithms quantitatively and visually. The set of proposed algorithms is illustrated with number of bi-level/multi-level Berkeley images [1]/real-life gray-scale images.

The chapter is organized as follows. A brief overview of the multi-objective optimization problem is presented in Section 7.2. The experimental methodology of different versions of multi-objective algorithms, is illustrated in Section 7.3. The results of the application of the proposed algorithms using Berkeley images [1]/real-life gray-scale images is illustrated in Section 7.4. The chapter ends with some relevant conclusions, which is presented in Section 7.5. In Section 7.6, the chapter summary is presented and a set of exercise questions with regards to the theme of the chapter. At the end of this chapter, a few sample coding examples are presented.

7.2 Multi-objective Optimization

Unlike a single-objective optimization problem, a group of objective functions are considered simultaneously for optimization in multi-objective optimization problem. Initially, the decision space is explored to find the vector of decision variables and henceforth the objective functions are simultaneously optimized. In addition, there may be a few constraints, which must be satisfied by any solution in a multi-objective optimization problem [21, 55, 56, 290]. The formal definition of a multi-objective optimization problem can be presented as follows [21, 47, 55, 56, 218, 290].

$$\mathcal{M} \ \mathbf{y} = f(\mathbf{x}) = [f_1(\mathbf{x}), f_2(\mathbf{x}), \dots, f_p(\mathbf{x})]^T$$
$$\text{subject to } \phi_i(\mathbf{x}) \geq 0, \qquad i = 1, 2 \dots, q$$
$$\psi_j(\mathbf{x}) = 0, \qquad j = 1, 2 \dots, r \tag{7.1}$$

According to the nature of optimization, $\mathcal{M}$ is considered as either a minimization or a maximization problem. Here, $\mathbf{x} = (x_1, x_2, \dots, x_s) \in X$ denotes a vector, or sometimes called a decision vector, comprising s number of decision variables, whereas $\mathbf{y} = (y_1, y_2, \dots, y_t) \in Y$ is the called the objective vector. X and Y denote the parameter space and the objective space, respectively. As mentioned in Eq. (7.1), the decision vector must satisfy q number of inequality constraints and r number of equality constraints for multi-objective optimization.

The thought of dominance is exceedingly significant in MOO. For the sake of clarity, in a minimization framework, let us assume that $\mathcal{Z} = \{z_1, z_2, \dots, z_u\}$ is the set of solutions for a MOO problem comprising $\mathcal{H}$ number of objective functions.

The solution $z_i \in \mathcal{Z}$ dominates the solution $z_j \in \mathcal{Z}$ if both of the following criteria are satisfied.

1. $f_l(z_i) \leq f_l(z_j)$, $\forall \ell \in 1, 2, \dots, \mathcal{H}$.
2. $f_l(z_i) < f_l(z_j)$, if $\exists$ at least one $\ell \in 1, 2, \dots, \mathcal{H}$.

As a general rule, a MOO algorithm acknowledges a set of non-dominated solutions.

7.3 Experimental Methodology for Gray-Scale Multi-Level Image Thresholding

In this chapter, four different quantum inspired multi-objective algorithms are presented and discussed in different phases. In the first phase, an NSGA-II-based quantum inspired multi-objective algorithm (QINSGA-II) is developed. This algorithm can efficiently be applied for multi-level image thresholding in multi-objective framework. Thereafter, an alternative version of simulated annealing-based quantum behaved multi-objective algorithm (QISAMO) is introduced for bi-level thresholding. Lastly, another two meta-heuristics-based quantum inspired multi-objective algorithms, called Quantum Inspired Multi-objective Particle Swarm Optimization (QIMOPSO) and Quantum Inspired Multi-objective Ant Colony Optimization (QIMOACO), are also presented in this chapter. These two algorithms are basically designed for multi-level image thresholding. These algorithms are applied on a set of gray-scale, real-life images assisted by some standard objective functions for the determination of optimum threshold values.

7.3.1 Quantum Inspired Non-dominated Sorting-Based Multi-objective Genetic Algorithm

In this subsection, a new variation of quantum inspired nondominated sorting-based multi-objective genetic algorithm is introduced. This algorithm can efficiently be applied for multi-level image thresholding. The details of QINSGA-II are explained in Algorithm 14. At the beginning, the population of chromosome (POP) is filled with $\mathcal{A}$ number of chromosomes. The length of each chromosome in POP is taken as $\mathcal{L} = \sqrt{L}$, where L represents the maximum intensity value of the image. A real encoding scheme using the basic features of QC is employed to encode real numbers between $(0,1)$ for each image pixel in POP. This results in an encoded population matrix, called POP'. At each generation, POP' passes through a quantum rotation gate to ensure faster convergence. Successively, each element in POP' must ensure the fundamental property of QC, known as quantum orthogonality, which produces POP''. Three basic genetic operators (selection, crossover, and mutation) are successively applied in POP', which results in an offspring population ($OPOP'$) of size $\mathcal{A}$. Thereafter, the parent population is combined with its offspring population to have a combined population matrix (FP) of size $2 \times \mathcal{A}$. This combined population is then sorted according to its non-domination. A fast non-dominated sorting strategy (as used in NSGA-II [56]) is applied for this purpose. This sorting approach finds the following measures:

- X_k: represents a set of all individuals which are dominated by k.
- c_k: stands for the number of individuals which dominate k.
- F_j: It signifies the set individuals in jth front.

Algorithm 14: Steps of QINSGA-II for multi-objective multi-level thresholding

---

Input: Number of generation: $\mathcal{G}$
Size of the population: $\mathcal{A}$
No. of thresholds: K
Output: Optimal threshold values: θ

1: First, pixel intensity values from the input image are randomly selected to generate $\mathcal{A}$ number of chromosomes (POP). Here, each chromosome in POP is of length $\mathcal{L} = \sqrt{L}$, where L stands for maximum intensity value of the input image.

2: Using the fundamental concept of quantum computing, each pixel in POP is encoded with a positive real number less than unity. Let us assume that POP becomes POP' after this real encoding.

3: Set the number of generation, $g = 1$.

4: The quantum rotation gate is employed to update POP' as explained in Eq. (1.10).

5: Each chromosomal ingredient in POP' endures *quantum orthogonality* to form POP'' the.

6: Using a probability criterion, K number of thresholds as pixel intensity are found from population. Let it create POP^*.

7: Evaluate the fitness of each chromosome in POP^* using Eq. (2.44).

8: Three basic operators, namely, selection, crossover and mutation are successively applied in POP'' to generate its offspring population of size $\mathcal{A}$. Let it be called $OPOP'$.

9: Repeat steps 4–5 to generate $OPOP''$.

10: $FP = POP''g \cup OPOP''g$.

11: **repeat**

12:     Use fast-non-dominated-sort mechanism to create F_j (say, j number of front).

13:     Set $POP_{g+1} = \emptyset$.

14:     Use crowding-distance-computation mechanism in $F_d, d \in [1, j]$ to populate POP_{g+1}.

15:     Use crowded-comparison operator $(\prec_n)$ to sort F_d in descending order (assuming maximization problem).

16:     $POP_{g+1} = POP_{g+1} \cup F_d[1 : (\mathcal{A} - |POP_{g+1}|)]$.

17:     Use Eq. (2.44) to evaluate POP_{g+1}. The threshold value of the input image is recorded in T_C.

18:     The thresholds possessing best fitness value are recorded in T_B.

19:     Repeat step 6: to generate a new population $OPOP_{g+1}$.

20:     $FP = POP_{g+1} \cup OPOP_{g+1}$.

21:     Repeat steps 4–5 for the chromosomes in FP.

22:     $g = g + 1$.

23: **until** $g < \mathcal{G}$

24: The optimal threshold values for MOO algorithm is reported in $\theta = T_B$.

The aim of this sorting strategy is to assign a rank for each solution according to the non-domination level. For example, rank 1 is assigned to the solutions at the best level, likewise, rank 2 to the second-best level, and so on. In the next phase, crowding distances for every fronts are computed. Since the individuals are selected based on rank and crowding distance, all the individuals in the population are assigned a crowding distance value.

7.3.2 Complexity Analysis

In this section, the time complexity (worst case) of QINSGA-II is presented in detail. The stepwise complexity analysis is given below.

1. Suppose the population contains $\mathcal{V}$ number of chromosomes. The time complexity to generate the initial population becomes $O(\mathcal{V} \times \mathcal{L})$. Note that, $\mathcal{L}$ represents the maximum pixel intensity value of the image.
2. Using the basic principle of qubit, each pixel in the gray-scale image is encoded with a real number between (0,1), which introduces the time complexity as $O(\mathcal{V} \times \mathcal{L})$.
3. To implement the quantum rotation gate and apply the property called quantum orthogonality in each member of the population, the time complexity for each operation requires $O(\mathcal{V} \times \mathcal{L})$.
4. The method finds a predefined number of threshold values from POP'' and produces POP^*. The time complexity for this process turns into $O(\mathcal{V} \times \mathcal{L})$.
5. Again, to compute the fitness values using POP^*, the time complexity becomes $O(\mathcal{V} \times C)$. Note that C is the required number of classes.
6. Three genetic operators, namely, selection, crossover, and mutation are applied in sequence in POP'' to generate $OPOP'$. The time complexity for performing each operation turns out to be $O(\mathcal{V} \times \mathcal{L})$.
7. In the next step, steps 4–5 are repeated to generate $OPOP''$. The time complexity for each operation is $O(\mathcal{V} \times \mathcal{L})$.
8. POP'' and $OPOP''$ are combined together to form FP. The time complexity to produce FP becomes $O(2 \times \mathcal{V}) = O(\mathcal{V})$.
9. For fast, non-dominated sort, one individual can only be a member of a single front. Suppose the maximum ($\mathcal{Y}$) number of comparisons is required for each domination checking in the population. Hence, to perform the fast non-dominated sort, the time complexity becomes $O(\mathcal{V} \times \mathcal{Y}^2)$.
10. The crowding distance computation in any single front may require at most $\mathcal{Y}$ number of sorting for $\frac{\mathcal{V}}{2}$ solutions. Therefore, the time complexity becomes $O(\mathcal{Y} \times \mathcal{V} \times \log(\mathcal{V}))$.
11. The crowded comparison operator ($<_n$) is used for sorting, which requires $O(\mathcal{V} \times \log(\mathcal{V}))$ computational complexity.
12. The algorithm runs for a predefined number of generations ($\mathcal{G}$). Hence, the overall time complexity to run the proposed algorithm turns into $O(\mathcal{V} \times \mathcal{L} \times \mathcal{Y}^2 \times \mathcal{G})$.

Therefore, summarizing, the worst case time complexity of QINSGA-II happens to be $O(\mathcal{V} \times \mathcal{L} \times \mathcal{Y}^2 \times \mathcal{G})$.

7.3.3 Quantum Inspired Simulated Annealing for Multi-objective Algorithms

The proposed QISAMO is presented in this subsection. The outline of this proposed approach is described as follows. At the outset, the symbols used in QISAMO are addressed below.

- Starting temperature: τ_s
- Closing temperature: τ_c
- Number of iterations: ι
- Reduction fraction: ϵ

- Objective functions required to find non-dominated set of solutions: φ_r, r represents number of such functions.
- The objective function used on the set of Pareto-optimal solutions to find the optimal threshold value: ϕ
- Non-dominated set of solutions: S_r
- Optimal threshold value: θ

Of these eight symbols, the first seven are the input symbols and the last one is referred to as the output symbol. The proposed QISAMO is briefly described in this subsection. In QISAMO, first, an initial population comprising one configuration ($\mathcal{P}$) is generated by randomly choosing pixel intensity values from the image. The length of the configuration is selected as $\mathcal{L} = \lceil \max(\sqrt{L}) \rceil$, L is the maximum pixel intensity value. Then, each pixel of this configuration is encoded with a real value between $(0,1)$ randomly using the basic theory of QC, this creates $\mathcal{P}'$. A basic quantum property, called the *quantum orthogonality*, is applied to $\mathcal{P}'$ to create $\mathcal{P}''$. Afterwards, a quantum *rotation gate* is applied to $\mathcal{P}''$ to achieve quick convergence. Then, based on probability measures, all positions in the configuration are selected to find solutions. Let it produce $\mathcal{P}^+$. Thereafter, a Pareto-optimal solution set (S_r) is determined using a number of (say, r) fitness functions (φ_r). Initially, the proposed algorithm starts exploring its search space at a high temperature (τ_s). At each temperature, this algorithm is executed for successive ι number of iterations. Thereafter, the temperature is reduced using a reduction fraction (ϵ) as given in step 24. Better movement towards the optimal solution is always accepted. If ($\mathcal{F}(Q^+) > \mathcal{F}(S_r^+)$) (as shown in step 21), the configurations Q^+ and Q are admitted. Otherwise, these newly created configurations can also be accepted as non-improving movement with a probability $\exp(-(\mathcal{F}(S_r^+) - \mathcal{F}(Q^+)))/\mathcal{T}$. It stops its execution when the temperature crosses a predefined low temperature value (τ_c). The working steps of the proposed QISAMO are illustrated below.

1. In the first step, a population of single configuration ($\mathcal{P}$) is initially created by the random selection of a pixel intensity value from the input image. The length of this configuration is taken as $\mathcal{L} = \lceil \max(L) \rceil$, L is the maximum pixel intensity value of the image.
2. For pixel encoding, each pixel in $\mathcal{P}$ is encoded with a random real number between $(0,1)$ by using the theory of qubit in QC. Let this pixel encoding scheme create a new configuration, called $\mathcal{P}'$.
3. Afterwards, the property of QC, called *quantum orthogonality*, is maintained for each location of the configuration in $\mathcal{P}$, which creates $\mathcal{P}''$.
4. Then, for faster convergence, each position of the configuration in $\mathcal{P}''$ is updated by applying the quantum rotation gate in QC.
5. Then, each position of the configuration in $\mathcal{P}''$ may be represented as a possible solution. A particular location which contains more value than a randomly generated number between $(0, 1)$, may be considered a possible solution. Let the set of all possible solutions in $\mathcal{P}''$ create $\mathcal{P}^+$.
6. Count the number of such positions and save this number in ℓ.
7. Chose a number, j between $[1, \ell]$ at random.
8. Using $\mathcal{P}^+$, j threshold values in the form of pixel intensity are computed as solutions. Let it create $\mathcal{P}^*$.
9. For $g = 1$ to r, perform the following two steps.

10. (a) The threshold values in $\mathcal{P}^+$ are used r number of times to find the fitness values using r number of fitness functions, φ_r.
    (b) The best solution for each fitness function is found and recorded in S_g.
11. The Pareto-optimal solution set is determined using S_g and then these solutions are recorded in S_r.
12. The fitness value of each configuration in S_r is computed using ϕ. Let it be denoted by $\mathcal{F}(S_r)$.
13. Store the best configuration, $b_s \in \mathcal{P}''$, its threshold value in $T_s \in S_r$ and the corresponding fitness value in F_s, respectively.
14. Repeat steps 2–5 to create S_r^+.
15. Set, $\mathcal{T} = \tau_s$.
16. Run the following steps (steps 17–24) until $\mathcal{T} \geq \tau_c$
17. For $h = 1$ to ι, perform the following steps (steps 18–23).
18. Perturb S_r. Let it make Q.
19. Repeat steps 2–5 to create Q^+.
20. Repeat steps 6–12 and use Eq. (2.56) to calculate the fitness value $\mathcal{F}(Q^+, T)$.
21. If $(\mathcal{F}(Q^+) - \mathcal{F}(S_r^+) > 0)$ holds
22. Set $S_r^+ = Q^+$, $S_r = Q$ and $\mathcal{F}(S_r^+) = \mathcal{F}(Q^+)$.
23. Otherwise, set $S_r^+ = Q^+$, $S_r = Q$ and $\mathcal{F}(S_r^+) = \mathcal{F}(Q^+)$ with probability $\exp(-(\mathcal{F}(S_r^+) - \mathcal{F}(Q^+)))/\mathcal{T}$.
24. $\mathcal{T} = \mathcal{T} \times \epsilon$.
25. Report the optimal threshold value, $\theta = S_r^+$.

7.3.3.1 Complexity Analysis

Step-wise analysis of time complexity (worst case) of the proposed QISAMO is given below.

- Initially (step 1), the population ($\mathcal{P}$) of QISAMO is formed with a single configuration of length $\mathcal{L} = \lceil \max(\sqrt{L}) \rceil$, where L represents the pixel intensity of a gray-scale image. For this step, the time complexity turns into $O(\mathcal{L})$.
- For the pixel encoding part (as described in step 2), the time complexity becomes $O(\mathcal{L})$.
- The time complexity for performing each of the next three steps (3–5) is $O(\mathcal{L})$.
- Similarly, the time complexity to perform step 8 turns out to be $O(\mathcal{L})$.
- The non-dominated set of solutions are determined through steps 10 and 11. The overall time complexity for performing these steps turns into $O(\mathcal{L})$.
- Similarly, the time complexity to perform the next step (step 12) is $O(\mathcal{L})$.
- In step 14, steps 2–5 are repeated. To execute each of these steps, the time complexity is $O(\mathcal{L})$.
- Steps 18–20 are executed for ι number of iterations. For each iteration, the time complexity to execute each step turns into $O(\mathcal{L})$.
- Let the outer loop (step 15) and the inner loop (step 16) of this algorithm be executed ξ and ι number of times, respectively. Hence, the time complexity to execute these steps of the proposed algorithm is $\xi \times \iota$. Therefore, aggregating (summarizing) the steps stated above, the overall time complexity (worst case) of QISAMO happens to be $\mathcal{L} \times \xi \times \iota$.

7.3.4   Quantum Inspired Multi-objective Particle Swarm Optimization

In this subsection, the proposed QIMOPSO is discussed. The list of symbols used in this proposed approach is given below.

- Number of generation: $\mathcal{G}$.
- Population size: $\mathcal{V}$.
- Number of classes: C.
- Acceleration coefficients: ξ_1 and ξ_2.
- Inertia weight: ω.
- Set of intermediary objective functions: $\phi_f()$, f is the user-defined number of objective functions used to determine Pareto-optimal solutions.
- The objective function used to find the optimal threshold values for Pareto-optimal solutions: φ.
- Set of Pareto-optimal solutions: S_f.
- Optimal threshold values: θ.

The brief summary of the proposed algorithm is described here. First, pixel intensity values are selected at random to generate a population *POP* of initial particles. *POP* comprises $\mathcal{V}$ number of particles, each of length $\mathcal{L} = \sqrt{L_w \times L_h}$, where L_w and L_h are the width and height of the input image. Using the fundamental concept of QC, a real encoding scheme is initiated by assigning a random value between $(0,1)$ for each pixel in *POP*, which produces *POP′*. Afterwards, *POP′* undergoes a basic feature of QC called, *quantum orthogonality*, which creates *POP″*. The elements of *POP″* are updated by applying the quantum rotation gate to accelerate its convergence. Using a probability criteria, a user-defined number of threshold values as pixel intensity of the image are determined from *POP″*, which produces *POP⁺*. A random number between $[\frac{\mathcal{V}}{2}, \mathcal{V}]$ is generated to create a set of solutions using different objective functions. Thereafter, the non-dominated set of solutions is produced and recorded in S_{nd}. Up to $\mathcal{V}$ number of runs, for each particle $j \in POP''$, the best position of the particle visited so far and the swarm best position are recorded in p_j and p_g, respectively. Then the particles in *POP″* are updated at each generation. The working procedure of QIMOPSO is summed up below.

1. The pixel intensity values are randomly selected from the gray-scale image to generate $\mathcal{V}$ number of preliminary particles, *POP*. The particle length is selected as $\mathcal{L} = \sqrt{L_w \times L_h}$, where, L_w is the image width where L_h is the height of the image.
2. For pixel encoding in *POP*, the fundamental unit of QC (qubit) is used to allocate positive a real number between $(0,1)$ to each pixel, which produces *POP′*.
3. Each element in *POP′* experiences a *quantum orthogonality* to create *POP″*.
4. Update *POP″* by using the quantum rotation gate.
5. Any position in *POP″* satisfying $POP'' > rand(0, 1)$, may lead to a possible solution. Find all possible solutions from *POP″*. Let it create *POP⁺*.
6. Generate a random integer, g between $[\frac{\mathcal{V}}{2}, \mathcal{V}]$.
7. Using *POP⁺*, for each particle in *POP″*, a set of g number of solutions is determined. Each solution comprises $(C - 1)$ number of threshold values as pixel intensities. Let it produce *POP⁺⁺*.

8. For $e = 1$ to f

    (a) Use POP^{++}, to evaluate the fitness of each solution using an objective function, $\phi_e()$.

    (b) Record the best solution for each particle in S_e.

9. Use S_f, to find the set of all non-dominated solutions. These solutions are recorded in S_{nd}.

10. Use φ, to determine the fitness value for each element in S_{nd}.

11. Save the best string $b_s \in POP''$, the respective threshold values in $T_s \in S_{nd}$ and its fitness value in F_s, respectively.

12. For $t = 1$ to $\mathcal{G}$

    (a) For all $j \in POP''$

    (i) Find the best position of the particle j visited so far and is recorded in p_j. Record the swarm best position at p_g.

    (ii) Update the swarm using the following equations:

    $v_j = \omega * v_j + \xi_1 * rand(0, 1) * (p_j - y_j) + \xi_2 * rand(0, 1) * (p_g - y_j). \, y_j = y_j + v_j.$

    (b) Repeat steps 3 to 10.

    (c) Store the best string $b_{sb} \in POP''$, the respective threshold values in $T_{sb} \in S_{nd}$ and, finally, its fitness value in F_{sb}, respectively.

    (d) Compare the fitness values of b_s and b_{sb}. The better fitness value is recorded in b_s. The corresponding threshold values is also recorded in T_s.

13. Report the optimal threshold values in $\theta = T_s$.

7.3.4.1 Complexity Analysis

The worst case time complexity of QIMOPSO is described below.

- For the step 1, the time complexity becomes $O(\mathcal{U} \times \mathcal{L})$, where $\mathcal{U}$ stands for the population size and length of each particle is $\mathcal{L} = \sqrt{L_w \times L_h}$. Note that L_w and L_h represent the width and height of the image, respectively.
- For the step 2, the time complexity turns into $O(\mathcal{U} \times \mathcal{L})$.
- The time complexity to perform each step from 3–5 becomes $O(\mathcal{U} \times \mathcal{L})$.
- For step 7, the time complexity is $O(\mathcal{U} \times \mathcal{L})$.
- To evaluate a fitness function, the time complexity turns into $O(\mathcal{U} \times (C - 1))$. The set of Pareto-optimal solutions are evaluated through steps 8 and 9. To compute each of them, the time complexity turns into $O(\mathcal{U} \times (C - 1) \times f) = O(\mathcal{U} \times (C - 1))$.
- Similarly, for step 1, the time complexity turns into $O(\mathcal{U} \times (C - 1))$.
- In step 12, the outer loop is executed $\mathcal{G}$ times. For step 12, the time complexity turns out to be $O(\mathcal{U} \times \mathcal{L} \times \mathcal{G})$, where, $\mathcal{G}$ is the number of generations.

Therefore, summarizing the above discussion, the worst case time complexity for QIMOPSO happens to be $O(\mathcal{U} \times \mathcal{L} \times \mathcal{G})$.

7.3.5 Quantum Inspired Multi-objective Ant Colony Optimization

Similar to the above algorithm, the symbols used in QIMOACO, are listed below.

- Number of generation: $\mathcal{G}$.
- Population size: $\mathcal{U}$.
- Number of classes: C.

- Priori defined number: ϖ.
- Persistence of trials: ϱ.
- Set of intermediary objective functions: $\phi_f()$, f is the user-defined number of objective functions used to determine Pareto-optimal solutions.
- The objective function used for the optimal threshold values for Pareto-optimal solutions: φ.
- Set of Pareto-optimal solutions: S_f.
- Optimal threshold values: θ.

The short summary of the proposed algorithm is illustrated in this subsection. The first eleven steps for QIMOACO and QIMOPSO are identical. The discussion here is only confined to the remaining parts of the proposed algorithm. A pheromone matrix, τ is randomly produced by using random real numbers between $(0, 1)$. Note that the dimension of τ and participating string of the population are identical. Moreover, POP'' is updated at each generation using the value of τ. At the same time, the best string, its fitness and the threshold values are also recorded. τ is also updated at each generation. The details of QIMOACO are discussed through the following steps.

1. The first eleven steps of QIMOACO are identical to QIMOPSO.
2. Produce a pheromone matrix, τ.
3. For $i = 1$ to $\mathcal{G}$
   (a) For all $j \in POP''$
   (i) For each kth place in j.
   (ii) If $(random(0, 1) > \varpi)$
   (iii) $POP'' = \arg \max \tau_{jk}$.
   (iv) Else
   (v) $POP'' = random(0, 1)$.
   (b) Repeat steps 5 to 10.
   (c) Store the best string $b_{sb} \in POP''$, the respective threshold values in $T_{sb} \in S_{nd}$ and finally, its fitness value in F_{sb}, respectively.
   (d) Compare the fitness values of b_s and b_{sb}. Record the best among them in b_s and the corresponding threshold values in T_s.
   (e) For all $j \in POP''$
   (i) For each kth place in j.
   (ii) $\tau_{jk} = \varrho \tau_{jk} + (1 - \varrho) \tau_{jk}$.
4. The optimal threshold values are reported in $\theta = T_s$.

7.3.5.1 Complexity Analysis

The worst case time complexity of QIMOACO is discussed here. The first eleven steps of this algorithm have already been discussed in the respective section of proposed QIMOPSO. The remaining time complexity analysis is presented below.

- Pheromone matrix is produced at step 2. The time complexity for performing this step is $O(\mathcal{U} \times \mathcal{L})$.
- In step 3, the outer loop is executed $\mathcal{G}$ times. So to conduct this step, the time complexity turns out to be $O(\mathcal{U} \times \mathcal{L} \times \mathcal{G})$, where, $\mathcal{G}$ is the number of generations.

Therefore, the overall worst case time complexity for QIMOACO happens to be $O(\mathcal{U} \times \mathcal{L} \times \mathcal{G})$.

7.4   Implementation Results

Application of the proposed algorithms is demonstrated with thresholding of bi-level, multi-level, and Berkeley images (Benchmark dataset)/real-life gray-scale images. To establish the effectiveness of the proposed approaches, a set of real-life gray-scale images has been used as the test images. The original real-life images, namely, Anhinga, Couple, Desert, Greenpeace, Monolake, Oldmill, Stonehouse are presented in Figures 7.1 (a)–(g), the images, namely, Lena, Peppers, Baboon, and Barbara, are shown in Figures 4.7 (a), (b), (d) and (f) and the images, namely, Boat, Cameraman, Jetplane are depicted in Figures 5.5 (b)–(d). The Berkeley images [1] are thereafter presented in Figures 7.1 (h)–(j). The dimensions of real-life images and Berkeley images are selected as 256×256 and 120×80, respectively. The proposed quantum inspired multi-objective algorithms are described in detail in Section 7.3.

7.4.1   Experimental Results

In the first part, the performance of QINSGA-II has been assessed with reference to optimal threshold values of three real-life gray-scale images and three Berkeley

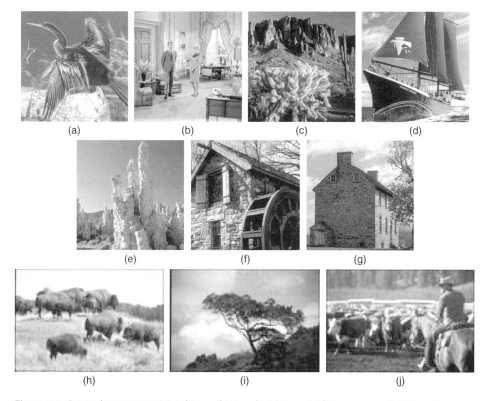

Figure 7.1 Original test images (a) Anhinga, (b) Couple, (c) Desert, (d) Greenpeace, (e) Monolake, (f) Oldmill, (g) Stonehouse, (h) #38092, (i) #147091, and (j) #220075.

Table 7.1 Parameters specification for QINSGA-II, NSGA-II, and SMS-EMOA.

QINSGA2	NSGA2	SMS-EMOA
Number of generations: $G = 100$	Number of generations: $G = 500$	Number of generations: $G = 1000$
Population size: $S = 20$	Population size: $S = 20$	Population size: $S = 20$
Crossover probability: $\varsigma = 0.95$	Crossover probability: $\varsigma = 0.95$	Crossover probability: $\varsigma = 0.95$
Mutation probability: $\varrho = 0.05$	Mutation probability: $\varrho = 0.05$	Mutation probability: $\varrho = 0.05$
No. of thresholds: $K = 3, 5, 7$	No. of thresholds: $K = 3, 5, 7$	No. of thresholds: $K = 3, 5, 7$

images [1]. The proposed algorithm has been constructed using NSGA-II [56] as a reference algorithm. The parameter specification (best combinations) for all the algorithms is listed in Table 7.1. The experiment has been conducted on six test images; (a) Lena, (b) Peppers, (c) Cameraman, and (d) ♯38092 [1], (e) ♯147091 [1], and (f) ♯220075 [1]. In this chapter, the experiments have been conducted for the proposed QINSGA-II, NSGA-II [56] proposed by Deb et al. and SMS-EMOA [84] proposed by Emmerich et al. Each algorithm has been evaluated for 30 different runs. The optimum threshold values for different gray-scale images have been documented at various levels using multi-objective flavor. At the outset, two different conflicting type objective functions were evaluated to determine the non-dominates set of solutions. Thereafter, this set of non-dominated solutions is used to find the optimal threshold values assisted by a standard objective function, called Kittler's method [148]. In this subsection, the overall results are presented on the basis of the following aspects.

1. The results of multilevel thresholding using Kittler's method [148].
2. The stability of each of them.
3. Their performance evaluation.

7.4.1.1 The Results of Multi-Level Thresholding for QINSGA-II, NSGA-II, and SMS-EMOA

The best results for different images are reported in reference to optimum thresholds (θ) and objective (fitness) value ($\mathcal{F}_{best}$) in Table 7.2. The time for computations (t) (in seconds) for each of them in presented in Table 7.3. The experimental results have been reported for four-level, six-level and eight-level of computation (K). The thresholded images for QINSGA-II are presented in Figures 7.2 (a)–(r).

7.4.1.2 The Stability of the Comparable Methods

The stability and accuracy of each algorithm are examined in terms of mean fitness ($\mathcal{F}_{avg}$) value and standard deviation (σ) over different runs. These results are reported in Table 7.4. As the level increases, the values of ($\mathcal{F}_{avg}$) and (σ) vary significantly. The proposed algorithm possesses the lowest (σ) value for all levels of thresholding, which proves that QINSGA-II is the most stable algorithm of them all. Moreover, from Tables 7.2 and 7.4, it can be found that QINSGA-II has very little deviation in $\mathcal{F}_{avg}$ values for each case. Hence, the proposed algorithm possesses the highest level of accuracy compared to the others.

Table 7.2 Best results obtained for QINSGA-II, NSGA-II, and SMS-EMOA for multi-level thresholding of Lena, Peppers, Cameraman, #38092, #147091, and #220075.

Lena

K	QINSGA-II θ	F_{best}	NSGA-II θ	F_{best}	SMS-EMOA θ	F_{best}
4	68,94,114	9.560	64,93,115	9.645	66,87,109	9.614
6	71,93,111,127,144	10.566	58,79,111,130,161	11.069	84,118,139,151,161	10.956
8	67,89,108,127,139,156,170	11.747	75,100,104,137,145,158,183	12.549	46,77,97,111,126,150,165	12.307

Peppers

K	QINSGA-II θ	F_{best}	NSGA-II θ	F_{best}	SMS-EMOA θ	F_{best}
4	64,97,116	9.770	56,95,122	9.859	64,98,117	9.775
6	62,93,110,131,152	10.806	27,63,93,119,151	11.204	49,77,102,127,148	11.042
8	62,84,104,115,142,160,178	12.038	56,92,116,145,167,173,189	12.466	56,86,102,128,150,165,186	12.271

Cameraman

K	QINSGA-II θ	F_{best}	NSGA-II θ	F_{best}	SMS-EMOA θ	F_{best}
4	22,93,126	9.331	26,91,126	9.349	30,95,127	9.388
6	25,81,115,135,152	10.385	28,77,115,130,152	10.491	27,87,128,152,169	10.604
8	19,71,117,127,144,157,169	11.681	11,22,69,106,121,139,149	12.063	24,44,61,115,129,148,155	12.228

Table 7.2 *Continued*

#38092

K	QINSGA-II		NSGA-II		SMS-EMOA	
	θ	$\mathcal{F}_{best}$	θ	$\mathcal{F}_{best}$	θ	$\mathcal{F}_{best}$
4	84,134,177	4.757	104,145,188	4.769	79,133,167	4.765
6	67,113,150,194,229	6.587	70,115,144,173,208	6.600	27,64,102,147,184	6.587
8	53,88,121,145,171,210,241	8.501	49,100,133,158,182,207,234	8.524	60,87,103,139,166,196,228	8.520

#147091

K	QINSGA-II		NSGA-II		SMS-EMOA	
	θ	$\mathcal{F}_{best}$	θ	$\mathcal{F}_{best}$	θ	$\mathcal{F}_{best}$
4	50,112,154	4.734	43,93,143	4.748	51,114,160	4.739
6	34,70,120,151,193	6.554	27,71,102,138,170	6.602	38,61,112,147,183	6.572
8	32,54,96,127,153,180,220	8.494	39,63,106,133,162,211,232	8.524	33,73,104,129,159,201,252	8.510

#220075

K	QINSGA-II		NSGA-II		SMS-EMOA	
	θ	$\mathcal{F}_{best}$	θ	$\mathcal{F}_{best}$	θ	$\mathcal{F}_{best}$
4	49,86,124	4.706	50,94,139	4.714	50,90,129	4.707
6	41,67,97,126,165	6.553	40,67,103,145,209	6.566	38,70,101,119,147	6.573
8	27,48,74,94,122,153,212	8.490	25,55,75,94,112,144,207	8.501	30,44,58,88,115,159,224	8.505

Table 7.3 Execution time (t) of QINSGA-II, NSGA-II, and SMS-EMOA for multi-level thresholding of Lena, Peppers, Cameraman, ♯38092, ♯147091, and ♯220075.

	Lena			Peppers		
K	(a)	(b)	(c)	(a)	(b)	(c)
4	04.59	06.36	06.17	04.09	06.14	05.20
6	07.39	09.10	08.09	06.24	08.12	09.01
8	08.46	12.27	11.03	08.12	11.01	10.30
	Cameraman			♯38092		
K	(a)	(b)	(c)	(a)	(b)	(c)
4	04.27	05.56	05.02	02.49	03.10	03.13
6	06.19	08.16	07.25	05.50	06.35	06.24
8	08.26	10.24	10.49	06.25	10.03	09.11
	♯147091			♯220075		
K	(a)	(b)	(c)	(a)	(b)	(c)
4	02.11	04.04	03.23	02.22	03.16	03.27
6	04.56	05.38	05.55	05.14	06.12	06.01
8	06.03	09.25	08.04	06.23	09.21	08.12
(a):→ QINSGA-II (b):→ NSGA-II						
(c):→ SMS-EMOA						

7.4.1.3 Performance Evaluation

Here, the performance of the proposed method has been evaluated in two directions. First, two convergence-diversity based metrics, known as Inverted Generational Distance (IGD) [265, 292] and Hypervolume (HV) [290, 291] are introduced to establish the superiority of the proposed method from a multi-objective viewpoint. In addition, in connection with multi-objective optimization, a statistical comparison methodology has also been used for that perspective. IGD and HV are basically used to judge the quality of the optimal set of solutions with regard to convergence and diversity. This statistical comparison test finds the number of participating methods significantly dominating the selected method. Later, the effectiveness of the proposed method has also been statistically proved by using a statistical superiority test, called the Friedman test [94, 95]. Generally, this test is conducted to compare the performance of several methods using a group of data set. This test determines the average rank of each method as output. The results of this test are presented in Table 7.7. This test finds the lowest average rank value for the proposed method in each case as compared to NSGA-II [56] and SMS-EMOA [84]. A brief overview of IGD, HV, and the statistical comparison test in connection with multi-objective optimization is discussed below.

1. Inverted Generational Distance: IGD [265, 292] is a superior version of Generational Distance [56, 265], which considers solutions with reference to diversity and convergence. IGD considers the obtained solution set and sample points from the Pareto front, and thereafter computes the average nearest distance between them. Formally,

Figure 7.2 For $K = 4, 6$ and 8, images (a)–(c), for Lena, (d)–(f), for Peppers, (g)–(i), for Cameraman, (j)–(l), for ♯38092, (m)–(o), for ♯147091, and (p)–(r), for ♯220075, after using QINSGA-II, for multi-level thresholding.

IGD can be defined by

$$IGD(P, S) = \frac{\left[\sum_{j=1}^{|P|} d_j^q \right]^{\frac{1}{q}}}{|P|} \tag{7.2}$$

Table 7.4 Average fitness (F_{avg}) and standard deviation (σ) of QINSGA-II, NSGA-II, and SMS-EMOA for multi-level thresholding of Lena, Peppers, Cameraman, ♯38092, ♯147091, and ♯220075.

Lena						
K	QINSGA-II		NSGA-II		SMS-EMOA	
	F_{avg}	σ	F_{avg}	σ	F_{avg}	σ
4	9.579	0.013	10.389	0.734	10.209	0.474
6	10.635	0.089	12.147	0.770	11.751	0.559
8	11.937	0.094	13.483	0.650	13.291	0.501
Peppers						
K	QINSGA-II		NSGA-II		SMS-EMOA	
	F_{avg}	σ	F_{avg}	σ	F_{avg}	σ
4	9.782	0.010	10.521	0.366	10.348	0.454
6	10.876	0.035	12.055	0.698	11.732	0.389
8	12.187	0.082	13.332	0.614	13.049	0.335
Cameraman						
K	QINSGA-II		NSGA-II		SMS-EMOA	
	F_{avg}	σ	F_{avg}	σ	F_{avg}	σ
4	9.337	0.003	9.744	0.311	9.671	0.227
6	10.444	0.035	11.205	0.617	11.225	0.485
8	11.816	0.082	12.812	0.453	12.868	0.463
♯38092						
K	QINSGA-II		NSGA-II		SMS-EMOA	
	F_{avg}	σ	F_{avg}	σ	F_{avg}	σ
4	4.757	0.001	4.796	0.033	4.791	0.023
6	6.592	0.002	6.662	0.038	6.651	0.034
8	8.512	0.003	8.572	0.028	8.569	0.023
♯147091						
K	QINSGA-II		NSGA-II		SMS-EMOA	
	F_{avg}	σ	F_{avg}	σ	F_{avg}	σ
4	4.735	0.001	4.766	0.021	4.766	0.020
6	6.561	0.004	6.627	0.026	6.619	0.041
8	8.503	0.004	8.541	0.020	8.550	0.025
♯220075						
K	QINSGA-II		NSGA-II		SMS-EMOA	
	F_{avg}	σ	F_{avg}	σ	F_{avg}	σ
4	4.706	0.001	4.758	0.042	4.734	0.019
6	6.555	0.004	6.610	0.030	6.597	0.022
8	8.494	0.003	8.527	0.019	8.540	0.028

where S represents the optimal set of solutions and P is the finite number of Pareto-optimal set of solutions that approximates the true Pareto front. d_j is the smallest Euclidean distance of a point in P to the nearest solutions in S.

2. Hypervolume: Zitzler et al. [290, 291] introduced the most popular performance metric, called Hypervolume (HV). In this metric, first, a set of hypercubes (HC_j) is formed by using a nadir point and a solution j from its non-dominated solution

set. To determine the HV value, the volume of each member of HC_j is accounted separately and added together. Formally, HV can be defined as follows:

$$HV = volume \left[\bigcup_{j=1}^{|Q|} HC_j \right] \tag{7.3}$$

where Q denotes the solution set. A higher HV value designates a better solution as regards convergence and diversity. Note that a nadir point is created by considering the worst objective values from the non-dominated solution set.

3. Statistical Comparison Methodology: The hypervolume indicator can be used to judge the quality of the Pareto set of solutions. In the case of multi-objective problems dealing with less than six objectives, the values of this hypervolume indicator are computed exactly according to the HV values. Otherwise, the concept of Monte Carlo sampling [15] is used for this approximation. Let B_k denote the set of participating methods to be compared, where $1 \leq k \leq m$ and m is the number of methods. Each method B_k has been executed for 30 independent runs for 500 generations. Investigation is carried out on the statistical superiority of the participating methods. The null hypothesis (H_0) for this statistical test establishes equality in the approximation of the Pareto-optimal set of solutions. To reject this hypothesis, an alternative hypothesis (H_1) is introduced, which establishes the unequal behavior in approximation of the Pareto-optimal solution set.

Thereafter, the median of the HV values for each pair of methods is compared. The significance of this difference is tested by using the Conover-Inman method [48]. To judge this significance, a performance index $P(B_k)$ is introduced. This index is formally defined by

$$P(B_k) = \sum_{k=1, q \notin p}^{m} \delta_{p,q} \tag{7.4}$$

Let $B_i, B_j \in B_k$. If B_i is significantly better than B_j, $\delta_{p,q}$ is 1, otherwise, $\delta_{p,q}$ is 0. This value finds the number of better performing methods than the corresponding method for a particular test case. Note that, the smaller the performance index value, the better the method. A zero index indicates a significantly best approximation for Pareto-set with reference to hypervolume indicator.

The values for IGD and HV for different algorithms have been reported in Table 7.5. Compared to other algorithms, the proposed algorithm possesses better performance metric values in all respects except for a very few occasions. For a statistical comparison among the participants, a performance index is introduced indicating the number of other participating algorithms significantly dominating the selected algorithm. These index values are reported in Table 7.6. The proposed algorithm significantly dominates other algorithms in all cases. For $K = 6$, neither of the proposed QINSGA-II and SMS-EMOA dominates the other for ♯38092. For most of the cases, SMS-EMOA dominates NSGA-II except for a few occasions. The computational times for each algorithm are reported in Table 7.3. It is evident from Table 7.3 that the proposed algorithm takes the least amount of time to compute compared to the others for each level of thresholds. Hence, the superiority of the proposed algorithm is established in reference to computational time. Moreover, the convergence curves for different algorithms are presented

Table 7.5 IGD and HV values of QINSGA-II, NSGA-II, and SMS-EMOA for multi-level thresholding of Lena, Peppers, Cameraman, ♯38092, ♯147091, and ♯220075.

Lena						
K	QINSGA-II		NSGA-II		SMS-EMOA	
	IGD	HV	IGD	HV	IGD	HV
4	0.06200	21.476	0.00480	12.089	0.00210	2.568
6	0.00004	21.045	0.00032	17.088	0.00015	3.208
8	0.00009	39.838	0.00024	2.944	0.00013	2.236

Peppers						
K	QINSGA-II		NSGA-II		SMS-EMOA	
	IGD	HV	IGD	HV	IGD	HV
4	0.00598	4.578	0.00747	1.170	0.00133	1.538
6	0.00010	26.029	0.00019	1.116	0.00031	1.428
8	0.00015	3.500	0.00018	1.634	1.358	0.00045

Cameraman						
K	QINSGA-II		NSGA-II		SMS-EMOA	
	IGD	HV	IGD	HV	IGD	HV
4	0.00273	5.579	0.00562	3.031	0.00284	1.894
6	0.00054	1.849	0.00092	1.071	0.00093	1.653
8	0.00025	2.588	0.00028	1.924	0.00034	2.465

♯38092						
K	QINSGA-II		NSGA-II		SMS-EMOA	
	IGD	HV	IGD	HV	IGD	HV
4	0.00067	7.626	0.00086	3.285	0.00014	4.974
6	0.00009	6.691	0.00028	4.784	0.00031	3.984
8	0.00240	5.222	0.00017	1.925	0.00039	3.129

♯147091						
K	QINSGA-II		NSGA-II		SMS-EMOA	
	IGD	HV	IGD	HV	IGD	HV
4	0.00129	5.293	0.00057	1.628	0.00055	5.014
6	0.01073	6.254	0.00280	1.377	0.00168	4.437
8	0.00331	8.066	0.00025	1.675	0.00023	3.376

♯220075						
K	QINSGA-II		NSGA-II		SMS-EMOA	
	IGD	HV	IGD	HV	IGD	HV
4	0.01530	20.872	0.02441	1.939	0.08618	13.531
6	0.00482	20.370	0.00927	4.114	0.00070	15.013
8	0.00008	24.722	0.00005	1.719	0.00001	12.965

in Figures 7.3 and 7.4, respectively at different levels of thresholds. The convergence curves are illustrated using three different colors for three different algorithms. Among the three different colors, the blue represents the convergence of QINSGA-II. For each level of threshold and each test image, it is clearly visible that the proposed QINSGA-II converges in the least number of generations. Hence, the superiority of the proposed algorithm is visually established with regards to the convergence among others. In an

Table 7.6 Statistical comparison of QINSGA-II to NSGA-II and SMS-EMOA with reference to the hypervolume indicator of Lena, Peppers, Cameraman, #38092, #147091, and #220075. The number reveals the performance score $\mathcal{P}$.

	Lena			Peppers			Cameraman		
K	(a)	(b)	(c)	(a)	(b)	(c)	(a)	(b)	(c)
4	0	2	1	0	2	1	0	1	2
6	0	2	1	0	2	1	0	1	2
8	0	2	1	0	2	1	0	1	2
	#38092			#147091			#220075		
K	(a)	(b)	(c)	(a)	(b)	(c)	(a)	(b)	(c)
4	0	2	1	0	2	1	0	2	1
6	0	1	0	0	2	1	0	1	2
8	0	2	1	0	2	1	0	1	2
(a):→ **QINSGA-II (b):→ NSGA-II**									
(c):→ **SMS-EMOA**									

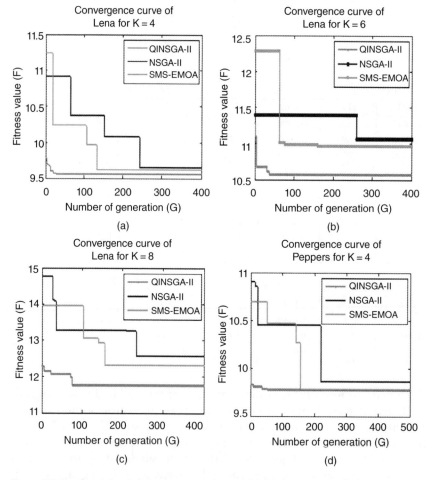

Figure 7.3 For $K = 4, 6,$ and 8, convergence curves (a)–(c), for Lena, (d)–(f), for Peppers, (g)–(i), for Cameraman, for QINSGA-II, NSGA-II, and SMS-EMOA.

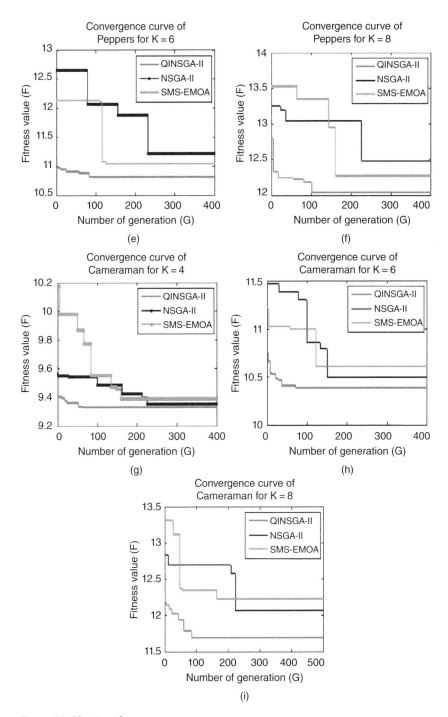

Figure 7.3 (*Continued*)

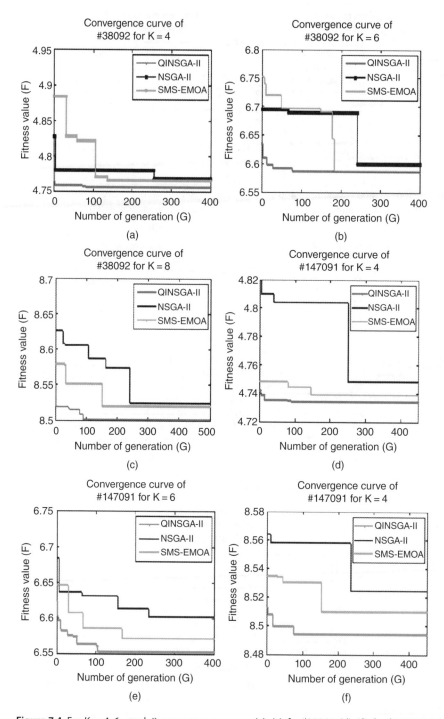

Figure 7.4 For $K = 4, 6,$ and $8,$ convergence curves (a)–(c), for ♯38092, (d)–(f), for ♯147091, (g)–(i), for ♯220075, for QINSGA-II, NSGA-II, and SMS-EMOA.

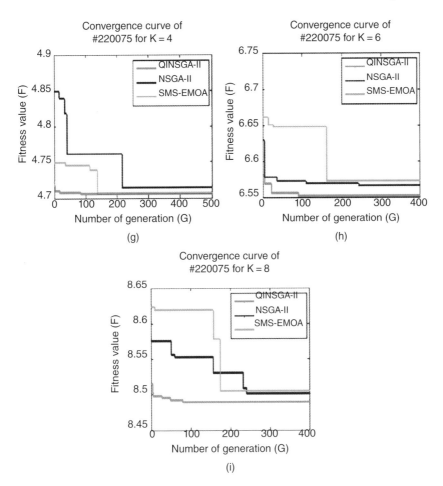

Figure 7.4 *(Continued)*

another approach, the application of QISAMO is exhibited with thresholding of bi-level and real-life gray-scale images. Twelve real images, namely, (a) Jetplane, (b) Anhinga, (c) Baboon, (d) Barbara, (e) Boat, (f) Cameraman, (g) Couple, (h) Desert, (i) Greenpeace, (j) Monolake, (k) Oldmill, and (l) Stonehouse used for this purpose. As a part of the comparative research, QISAMO has been compared with its classical counterpart. Each algorithm has been executed for 40 number of different runs. The optimal threshold values of the test images are reported in Table 7.8. Mean $(\mathcal{V}_{av})$ and standard deviation (σ) of fitness value are presented in Table 7.10. It can be shown from Table 7.10 that QISAMO has better $(\mathcal{V}_{av})$ and (σ) values in all cases. Hence, the accuracy and stability of the proposed QISAMO are proved. Moreover, PSNR and the computational time of each algorithm are presented in Table 7.9. The proposed algorithm has the higher PSNR values in all cases. The computational time is reported least compared to its counterpart. Therefore, the superiority of the proposed QISAMO is established in terms of computational time as well as segmentation point of view.

Table 7.7 Data sets used in QINSGA-II, NSGA-II, and SMS-EMOA for the Friedman test for K = 4, 6 and 8, respectively. The value in parentheses signifies the rank of the respective algorithm.

SN	Image	QINSGA-II	NSGA-II	SMS-EMOA
		For K=4		
1	Lena	9.560 (1)	9.645 (3)	9.614 (2)
2	Peppers	9.770 (1)	11.204 (3)	11.042 (2)
3	Cameraman	9.331 (1)	9.349 (2)	9.388 (3)
4	#38092	4.757 (1)	4.769 (3)	4.765 (2)
5	#147091	4.734 (1)	4.748 (3)	4.739 (2)
6	#220075	4.706 (1)	4.714 (3)	4.707 (2)
	Average rank	1.00	2.83	2.16
		For K=6		
1	Lena	10.566 (1)	11.069 (3)	10.956 (2)
2	Peppers	10.806 (1)	11.204 (3)	11.042 (2)
3	Cameraman	10.385 (1)	10.491 (2)	10.604 (3)
4	#38092	6.587 (1.5)	6.600 (3)	6.587 (1.5)
5	#147091	6.554 (1)	6.602 (3)	6.572 (2)
6	#220075	6.553 (1)	6.566 (2)	6.573 (3)
	Average rank	1.08	2.66	2.25
		For K=8		
1	Lena	11.747 (1)	12.549 (3)	12.307 (2)
2	Peppers	12.038 (1)	12.466 (3)	12.271 (2)
3	Cameraman	11.681 (1)	12.063 (2)	12.228 (3)
4	#38092	8.501 (1)	8.524 (3)	8.520 (2)
5	#147091	8.494 (1)	8.524 (3)	8.510 (2)
6	#220075	8.490 (1)	8.501 (2)	8.505 (3)
	Average rank	1.00	2.66	2.33

Table 7.8 Best results for QISAMO and SAMO for bi-level thresholding of Airplane, Anhinga, Baboon, Barbara, Boat, Cameraman, Couple, Desert, Greenpeace, Monolake, Oldmill, and Stonehouse.

SNo.	Image	QISAMO		SAMO	
		θ	ν_b	θ	ν_b
1	Airplane	136	0.176456	133	0.176661
2	Anhinga	139	0.165477	129	0.166681
3	Baboon	130	0.194270	132	0.194300
4	Barbara	128	0.171788	125	0.171969
5	Boat	110	0.176947	112	0.176956
6	Cameraman	128	0.176734	135	0.177782
7	Couple	108	0.190757	109	0.190770
8	Desert	135	0.169047	138	0.169265
9	Greenpeace	143	0.166374	149	0.166985
10	Monolake	158	0.175334	154	0.176172
11	Oldmill	125	0.164389	124	0.164437
12	Stonehouse	143	0.167608	146	0.167961

Table 7.9 PSNR and computational time of best results of QISAMO and SAMO of Airplane, Anhinga, Baboon, Barbara, Boat, Cameraman, Couple, Desert, Greenpeace, Monolake, Oldmill, and Stonehouse.

SNo.	Image	QISAMO		SAMO	
		PSNR	t	PSNR	t
1	Airplane	20.4128	2.01	20.310	8.15
2	Anhinga	18.6295	2.29	18.2940	7.48
3	Baboon	16.4678	1.58	16.4674	7.12
4	Barbara	17.4023	1.49	17.4015	8.12
5	Boat	17.6201	2.37	17.5175	7.52
6	Cameraman	16.0488	2.12	15.6279	8.00
7	Couple	15.7973	2.02	15.7467	8.38
8	Desert	17.0052	2.21	16.9312	8.07
9	Greenpeace	16.5847	1.51	16.4109	7.25
10	Monolake	17.3693	1.52	17.2862	7.45
11	Oldmill	11.4236	2.00	11.4179	8.30
12	Stonehouse	16.9577	1.49	16.8155	8.19

Table 7.10 Mean (V_{av}) and standard deviation (σ) of QISAMO and SAMO of Airplane, Anhinga, Baboon, Barbara, Boat, Cameraman, Couple, Desert, Greenpeace, Monolake, Oldmill, and Stonehouse.

SNo.	Image	QISAMO		SAMO	
		V_{av}	σ	V_{av}	σ
1	Airplane	0.176493	0.000063	0.179513	0.005877
2	Anhinga	0.165505	0.000069	0.168902	0.007204
3	Baboon	0.194296	0.000063	0.194374	0.006089
4	Barbara	0.171809	0.000051	0.174579	0.006741
5	Boat	0.176959	0.000024	0.178913	0.002810
6	Cameraman	0.171809	0.000051	0.182707	0.008145
7	Couple	0.190768	0.000023	0.192895	0.004393
8	Desert	0.169143	0.000130	0.178247	0.018049
9	Greenpeace	0.166405	0.000060	0.167743	0.001246
10	Monolake	0.175467	0.000220	0.187234	0.009932
11	Oldmill	0.164389	0.000000	0.16440	0.000008
12	Stonehouse	0.167710	0.000125	0.168541	0.000952

In other approaches, the application of the proposed QIMOPSO and QIMOACO has been exhibited on two real-life gray-scale images, namely, Lena and Peppers. The proposed algorithms have been compared with their respective conventional algorithms to judge their effectiveness. Each algorithm has been executed for 20 different runs. The best results of all the participating algorithms are reported in terms of threshold selection values (optimal)(θ), fitness value (V_{best}), and computation time (t) (in seconds) for three-level, five-level, and seven-level field of thresholds (K) in Table 7.11. The average fitness (V_{avg}) and standard deviation (σ) are reported for each algorithm in Table 7.12. In addition, the performance of each algorithm has been evaluated using the PSNR values. The results for PSNR values are reported in Table 7.13.

Table 7.11 Best results of QIMOPSO, PSO, QIMOACO, and ACO.

	Lena					
K	QIMOPSO			PSO		
	θ	V_{best}	t	θ	V_{best}	t
3	85,136	1.516	2.61	109,138	2.114	7.11
5	72,88,118,166	0.809	4.48	70,120,135,149	1.044	18.46
7	39,67,87,121,130,168	0.694	7.44	66,78,82,121,164,244	0.815	32.18

	Lena					
K	QIMOACO			ACO		
	θ	V_{best}	t	θ	V_{best}	t
3	84,135	1.526	3.11	95,100	2.778	8.25
5	71,99,111,161	0.881	5.13	39,75,97,164	1.249	20.20
7	67,82,99,119,162,199	0.726	8.01	71,76,81,108,113,178	0.939	31.08

	Peppers					
K	QIMOPSO			PSO		
	θ	F_{best}	t	θ	F_{best}	t
3	64,135	1.653	3.11	80,116	2.246	5.29
5	49,84,120,152	0.917	4.16	45,92,130,210	1.286	15.31
7	37,58,71,93,136,185	0.809	6.19	24,41,63,73,97,138	1.026	35.14

	Peppers					
K	QIMOACO			ACO		
	θ	F_{best}	t	θ	F_{best}	t
3	64,133	1.659	3.19	105,143	2.944	7.16
5	50,89,142,181	1.036	5.05	29,51,76,146	1.412	14.22
7	46,54,94,98,133,175	0.833	7.24	6,43,60,89,101,134	1.074	37.23

Table 7.12 Mean fitness (V_{avg}) and standard deviation (σ) of QIMOPSO, PSO, QIMOACO, and ACO.

	Lena							
K	QIMOPSO		PSO		QIMOACO		ACO	
	V_{avg}	σ	V_{avg}	σ	V_{avg}	σ	V_{avg}	σ
3	1.575	0.051	3.226	0.524	1.582	0.059	3.403	0.488
5	0.944	0.067	1.621	0.560	0.986	0.075	1.808	0.539
7	0.840	0.087	1.211	0.407	0.866	0.080	1.148	0.419

	Peppers							
K	QIMOPSO		PSO		QIMOACO		ACO	
	V_{avg}	σ	V_{avg}	σ	V_{avg}	σ	V_{avg}	σ
3	1.552	0.024	3.812	0.369	0.402	0.046	3.999	0.353
5	1.149	0.089	1.872	0.446	1.156	0.090	1.903	0.456
7	0.946	0.096	1.224	0.319	0.972	0.096	1.283	0.349

Table 7.12 shows that the proposed algorithm has the better V_{avg} and σ for each level compared to the others. Hence, QIMOPSO and QIMOACO are proved to be more accurate and more stable algorithms than their counterparts. Moreover, the proposed approaches possess higher PSNR values than the others in all cases. Hence, the quality of thresholding is proved to be better for the proposed approaches.

Table 7.13 PSNR values of QIMOPSO, PSO, QIMOACO, and ACO.

K	Lena				Peppers			
	A	B	C	D	A	B	C	D
3	19.142	18.512	18.997	16.667	20.604	19.732	20.598	19.238
5	23.631	21.110	22.985	22.874	24.602	23.975	24.351	21.196
7	25.796	24.179	25.669	23.609	25.184	21.663	24.399	21.638
A:→ QIMOPSO B:→ PSO C:→ QIMOACO D:→ ACO								

7.5 Conclusion

Four different versions of quantum inspired multi-objective algorithms for thresholding of gray-scale images are discussed in this chapter. In addition, the proposed QINSGA-II has been successfully applied for the thresholding of Berkeley images/gray-scale images. In the first section of this chapter, an NSGA-II-based quantum inspired multi-objective algorithm is introduced. The proposed approach is designed for multi-level image thresholding. Later, a simulated annealing-based multi-objective algorithm is developed for bi-level image thresholding. Two alternate versions of quantum inspired multi-objective algorithms based on particle swarm optimization and ant colony optimization, are proposed in the last section of this chapter. A distinctive approach has been adopted to develop each of the algorithms. The proposed QINSGA-II has been compared with two popular multi-objective optimization algorithms, referred to as NSGA-II [56] and SMS-EMOA [84] to evaluate its performance. The superiority of QINSGA-II is established with regards to fitness measure, standard deviation of fitness measures, average fitness value, and computational time. Thereafter, two different performance metrics and a statistical comparison methodology were introduced to evaluate their performances from a multi-objective point of view. A statistical superiority test, called the Friedman test, has also been used to judge their effectiveness. Later, the effectiveness of QINSGA-II has been proved visually by presenting convergence curves for all algorithms. It can be noted that the proposed QINSGA-II outperforms others in all respects. For other sets of proposed algorithms, a comparative study has been carried out between the proposed approach and its respective classical counterpart. The superiority of each of them is established quantitatively in different aspects.

7.6 Summary

- Four quantum inspired multi-objective algorithms for gray-scale image thresholding have been introduced in this chapter.
- In the first approach, a NSGA-II-based quantum inspired algorithm is introduced in a multi-objective framework.
- Later, several quantum inspired classical algorithms are presented in multi-objective flavor for bi-level, multi-level and gray-scale image thresholding.
- Application of these algorithms is demonstrated on several Berkeley images and real-life gray-scale images.

- As a sequel to the proposed algorithm (QINSGA-II), experiments have been conducted with the NSGA-II [56] and SMS-EMOA [84].
- Results of proposed algorithms show better performances over other comparable algorithms quantitatively and visually.

Exercise Questions

Multiple Choice Questions

1 The concept of "dominance" is significant in
   (a) single-objective optimization
   (b) multi-objective optimization
   (c) both (a) and (b)
   (d) none of the above

2 The term "NSGA" stands for
   (a) Non-determined Sorting Genetic Algorithm
   (b) Non-dominated Sorting Genetic Algorithm
   (c) Non-dominance Sorting Genetic Algorithm
   (d) none of the above

3 "SPEA" is a popular multi-objective optimization algorithm. It was introduced by
   (a) Zitzler and Thiele
   (b) Zitzler and Deb
   (c) Mukhopadhyay and Maulik
   (d) none of the above

4 Which of the following convergence-diversity based metrics are used to judge the quality of the optimal set of solutions with regard to convergence and diversity?
   (a) IGD
   (b) HV
   (c) both (a) and (b)
   (d) none of the above

5 The unit of "computational time" measured in the experiments of MOO problems is
   (a) picoseconds
   (b) seconds
   (c) nanoseconds
   (d) none of the above

6 The concept of the crowding distance computation mechanism, found in unit of "computational time", is measured in the experiments of MOO problems. It is suited for
   (a) QINSGA-II
   (b) QISAMO

(c) QIMOPSO

(d) QIMOACO

Short Answer Questions

1   What do you mean by a non-dominated set of solutions in multi-objective optimization problem?

2   How is the effectiveness of the proposed quantum inspired multi-objective algorithms quantitatively and visually established?

3   What do you mean by single-objective and multi-objective optimization problems? Illustrate with suitable examples.

4   What is the significance of using convergence curves in the performance evaluation of a problem?

5   How is the concept of "rank" associated in the proposed QINSGA-II?

6   What is the role of perturbation in QISAMO?

Long Answer Questions

1   What is the role of convergence-diversity based metrics in the performance evaluation of a MOO problem?

2   Why are the value of PSNR and computational time measured in the proposed algorithms?

3   Discuss the algorithm of the proposed QINSGA-II. Derive its time complexity.

4   Discuss the algorithm of the proposed QIMOPSO. Derive its time complexity.

Coding Examples

In this chapter, several quantum inspired multi-objective algorithms were introduced for multi-level image thresholding. Of these algorithms, the code for QINSGA-II is presented in this section. This algorithm uses two different conflicting type image thresholding methods to find non-dominated set of solutions. These methods have already been presented in the previous chapters. Figures 7.5–7.9 depict the code for QINSGA-II. Note that the common functions used in different techniques are presented only once.

```
Func¹

function [tr,fv]=QNSGA2(n,mu,cr,K)
%n-Number of chromosomes, Mu-Mutation probability, Cr-Crossover
probability
tic ; %Computational time calculation
t = 0;
L=256;
His=insert(L);%Histogram of the gray scale image is calculated
M=Generate(n,K); %Function used for generating initial population
s1=M;
maxf=1000;
n1=floor(n/2);
nc=0;

for v=1:300
[fo,fk,fa]=Fitness(L,s1,His,K,n); %Function used for computing
                              fitness of individual
[f1,f2,ff,cpp]=fndsI(s1,n,K,fo,fk); 
[nd,fu1,fu2]=nondom(nc,fo,fk,v,n); %These two functions used for
                          finding non-dominated set of solution
nc=nd;
s=ff;
S=selection(s1,n1,fa,cpp); %Function used for selecting best
individual
C=cross(S,cr,K,n1); %Function used for crossover operation
MU=mute(C,mu,n1,K); %Function used for mutation operation
s2=MU;
ss=totpop(s,s2,n1);
[ta,fb,fv,ff1,ff2]=fdv(n,ss,L,His,K);
if(maxf>fb)
    maxf=fb;
    T=ta;
end
t = toc ;
s1=ss;
end

tr=T;
fv=maxf;
end
```

```
Func²

function [fo,fk,fa]=Fitness(L,s1,His,K,n)
[fo,fk,fa]=thcal(L,s1,His,n,K);
end

Func³

function [fo,fk,fa]=thcal(L,s1,His,n,K)
fo=zeros(1,n);
fk=zeros(1,n);
mx=0;
for i=1:n
d=s1(i,:);
P=partition(K,d);
[tro,fvo]=thck(L,His,K,P);
[trk,fvk]=thli(L,His,P);
[tr,fv1]=thkit(L,His,P);
fo(i)=fvo;
fk(i)=fvk;
fa(i)=fv1;
end

k=1;
for i=2:n
    mx=fk(1);
    if(mx>fk(i))
        mx=fk(i);
        k=i;
    end
end
for i=1:n
    if(fk(i)>16)
        fo(i)=fo(k);
        fk(i)=fk(k);
        fa(i)=fa(k);
    end
end
pp=k;
end
```

Figure 7.5 Sample code of Func¹, Func², and Func³ for QINSGA-II.

```
Func⁴

function P=partition(K,d)
r=zeros(1,length(d));
w=zeros(1,K-1);
for j=1:(K-1)
    r=d(j*8-7:j*8);
    r=fliplr(r);
    r=bi2de(r);
    w(j)=r;
end
P=sort(w);
end

Func⁵
function [tr,fv1]=thkit(L,His,P)
im=His;
Q=length(P)+1;
w=zeros(1,L);
muk=zeros(1,L);
si=zeros(1,L);
mu=0;
sigb=0;
    for j=1:Q
        if(j==1)
            for i=1:P(j)
                w(j)=w(j)+im(i);
                muk(j)=muk(j)+im(i)*i;
            end
        end
        if(j==Q)
            for i=(P(j-1)+1):L
                w(j)=w(j)+im(i);
                muk(j)=muk(j)+im(i)*i;
            end
        end
        if(j>1 && j<Q)
            for i=(P(j-1)+1):P(j)
                w(j)=w(j)+im(i);
                muk(j)=muk(j)+im(i)*i;
            end
        end
    end
end
```

```
for i=1:Q
    muk(i)=muk(i)/w(i);
end

for j=1:Q
    if(j==1)
        for i=1:P(j)
            si(j)=si(j)+im(i)*(i-muk(j))*(i-muk(j));
        end
    end
    if(j==Q)
        for i=(P(j-1)+1):L
            si(j)=si(j)+im(i)*(i-muk(j))*(i-muk(j));
        end
    end
    if(j>1 && j<Q)
        for i=(P(j-1)+1):P(j)
            si(j)=si(j)+im(i)*(i-muk(j))*(i-muk(j));
        end
    end
end

for i=j:Q
    si(i)=si(i)/w(i);
end

for i=j:Q
    si(i)=sqrt(si(i));
end

for i=1:Q
    if(si(i)>0 && w(i)>=0)
        sigb=sigb+1+2*w(i)*(log(si(i)-log(w(i))));
    else
        sigb=10000;
    end
end
fv1=sigb;
tr=P;
end
```

Figure 7.6 Sample code of Func⁴ and Func⁵ for QINSGA-II.

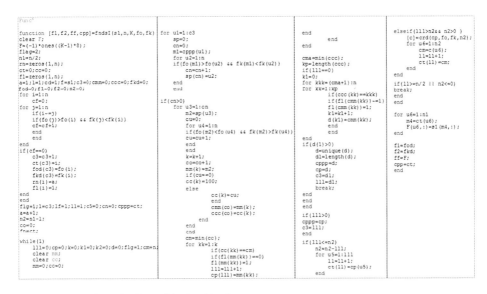

```
Func6
function [f1,f2,ff,cpp]=fndsI(s1,n,K,fo,fk)      for u1=1:c3                                         end                                              elseif(lll>n2&& n2>0 )
clear F;                                            sp=0;                                                end                                              [c]=crd(cp,fo,fk,n2);
F=(-1)*ones((K-1)*8);                               cn=0;                                            end                                                for u6=1:n2
flag=2;                                             m1=cppp(u1);                                      cma=min(ccc);                                       cm=c(u6);
n1=n/2;                                             for u2=1:n                                        kp=length(ccc);                                     ll=ll+1;
rn=zeros(1,n);                                      if(fo(m1)>fo(u2) && fk(m1)<fk(u2))               if(lll==0)                                          ct(ll)=cm;
ct=0;cc=0;                                          cn=cn+1;                                          kl=0;                                              end
f1=zeros(1,n);                                      sp(cn)=u2;                                        for kkk=(cma+1):n                                 end
a=1;l=1;cd=1;f=s1;c3=0;cmm=0;ccc=0;fkd=0;           end                                              for kk=1:kp                                        if(ll>=n/2 || n2<=0)
fod=0;f1=0;f2=0;n2=0;                               end                                              if(ccc(kk)==kkk)                                    break;
for i=1:n                                          if(cn>0)                                          if(f1(cmm(kk))~=1)                                 end
    cf=0;                                           for u3=1:cn                                      f1(cmm(kk))=1;                                    end
for j=1:n                                            m2=sp(u3);                                       k1=k1+1;                                          for u6=1:n1
    if(i~=j)                                         cu=0;                                            d(k1)=cmm(kk);                                     m4=ct(u6);
    if(fo(j)>fo(i) && fk(j)<fk(i))                  for u4=1:n                                       end                                                F(u6,:)=s1(m4,:);
    cf=cf+1;                                         if(fo(m2)<fo(u4) && fk(m2)>fk(u4))              end                                               end
    end                                             cu=cu+1;                                         end                                              f1=fod;
    end                                             end                                              if(d(1)>0)                                        f2=fkd;
end                                                 end                                              d=unique(d);                                      ff=F;
if(cf==0)                                           k=k+1;                                           dl=length(d);                                     cpp=ct;
    c3=c3+1;                                         co=co+1;                                         cppp=d;                                           end
    ct(c3)=i;                                        mm(k)=m2;                                        cp=d;
    fod(c3)=fo(i);                                   if(cu==0)                                        c3=dl;
    fkd(c3)=fk(i);                                   cc(k)=100;                                       lll=dl;
    rn(i)=a;                                         else                                            break;
    f1(i)=1;                                         cc(k)=cu;                                        end
end                                                 cmm(co)=mm(k);                                   end
end                                                 ccc(co)=cc(k);                                   end
flg=1;l=c3;lf=1;ll=1;c5=0;cn=0;cppp=ct;             end                                              if(lll>0)
a=a+1;                                              end                                              cppp=cp;
n2=n1-1;                                            end                                              c3=lll;
co=0;                                               cm=min(cc);                                      end
fn=ct;                                              for kk=1:k                                       if(lll<=n2)
while(1)                                            if(cc(kk)==cm)                                    n2=n2-lll;
    lll=0;cp=0;k=0;k1=0;k2=0;d=0;flg=1;cm=n;        if(f1(mm(kk))==0)                                for u5=1:lll
    clear mm;                                       f1(mm(kk))=1;                                     ll=ll+1;
    clear cc;                                       lll=lll+1;                                        ct(ll)=cp(u5);
    mm=0;cc=0;                                       cp(lll)=mm(kk);                                  end
```

Figure 7.7 Sample code of Func[6] for QINSGA-II.

```
Func7                                              for i=1:n1                                         end
function [nd,fu1,fu2]=nondom(nc,fo,fk,v,n)         while 1                                            if(j>1 && j<=(K-1))
if(v==1)                                             m1=randi([1,n1],1,1);                            pos=randi([[(K-1)*8-7,(K-1)*8],1,1);
c3=0;                                                m2=randi([1,n1],1,1);                            s1(m1,(K-1)*8-7:pos)=S(m1,(K-1)*8-7:pos);
fod=0;                                              if(m1~=m2)                                        s1(m1,pos+1:(K-1)*8)=S(m1,pos+1:(K-1)*8);
fkd=0;                                               break;                                          s1(m2,pos+1:(K-1)*8)=S(m1,pos+1:(K-1)*8);
end                                                end                                               s1(m2,(K-1)*8-7:pos)=S(m2,(K-1)*8-7:pos);
c3=nc;                                              end                                               end
f1=fo;                                             if(fv(m1)>fv(m2))                                  end
f2=fk;                                               s1(i,:)=s(m1,:);                                end
for i=1:n                                           else                                             S(m1,:)=s1(m1,:);
    cf=0;                                            s1(i,:)=s(m2,:);                                S(m2,:)=s1(m2,:);
for j=1:n                                           end                                              end
    if(i~=j)                                        end                                              C=S;
    if(f1(j)>f1(i) && f2(j)<f2(i))                  S=s1;                                            end
    cf=cf+1;                                        end
    end                                                                                             Func10
    end                                            Func9
end                                                                                                 function MU=mute(C,mu,n1,K)
if(cf==0)                                           function C=cross(S,cr,K,n1)                      for i=1:n1
    c3=c3+1;                                        s1=zeros(n1,(K-1)*8);                            for j=1:(K-1)*8
    fod(c3)=f1(i);                                  b=0;                                             if(rand<mu)
    fkd(c3)=f2(i);                                  for i=1:floor(n1/2)                              C(i,j)=randi([0,1],1,1);
end                                                 while 1                                          end
end                                                  m1=randi([1,n1],1,1);                           end
nd=c3;                                               m2=randi([1,n1],1,1);                           end
fu1=fod;                                            if(m1~=m2)                                        MU=C;
fu2=fkd;                                             break;                                          end
end                                                end
Func8                                              end
                                                   if(rand<cr)
function S=selection(s,n1,fa,cpp)                   for j=1:(K-1)
fv=zeros(1,n1);                                      if(j==1)
for j=1:n1                                           pos=randi([1,8],1,1);
    c=cpp(j);                                        s1(m1,1:pos)=S(m1,1:pos);
    fv(j)=fa(c);                                     s1(m1,pos+1:8)=S(m2,pos+1:8);
end                                                 s1(m2,pos+1:8)=S(m1,pos+1:8);
                                                    s1(m2,1:pos)=S(m2,1:pos);
```

Figure 7.8 Sample code of Func[7], Func[8], Func[9], and Func[10] for QINSGA-II.

```
Func¹¹

function [ta,fb,fv,ff1,ff2]=fdv(n,ss,L,His,K)
mx=1000;tt=1;f=0;fg1=0;fg2=0;
for i=1:n
d=ss(i,:);
P=partition(K,d);
[tro,fvo]=thck(L,His,K,P);
[trk,fvk]=thli(L,His,P);
[tr,fvl]=thkit(L,His,P);
f(i)=fvl;
if(mx>fvl)
mx=fvl;
fg1=fvo;
fg2=fvk;
tt=tr;
end
end
ff1=fg1;
ff2=fg2;
fv=f;
ta=tt;
fb=mx;
end

Func¹²

function ss=totpop(s,s2,n1)
for i=1:n1
    s(n1+i,:)=s2(i,:);
end
ss=s;
end
```

Figure 7.9 Sample code of Func[11] and Func[12] for QINSGA-II.

8

Conclusion

This chapter is the epilogue, and proposes future directions of research in this area. Several issues in connection with the proposed works have been addressed. All the algorithms are properly discussed and implemented. Application of these algorithms is demonstrated with a set of synthetic and real-life images of gray-scale and color versions. Experimental results have been analyzed properly in several aspects. In conclusion, the further scope of research is focused and described in this book in the light of the soft computing approach, based on the central theme of the book.

Image analysis extracts pertinent and important information from images. There exists a variety of automatic/semi-automatic techniques which include computer/machine vision, pattern recognition, image description, image understanding, etc. of image analysis. Image segmentation is the most popular, fundamental and significant step in a handful of image analysis techniques. Thresholding is commonly acknowledged as the most effective tool/technique in the context of image processing (image segmentation) and pattern recognition. In the last few decades, image analysis has been widely and successfully used to investigate images that are obtained from different imaging technologies, such as magnetic resonance imaging (MRI), radiology, ultrasound, etc. It is also widely used in cancer detection and diagnosis.

Quantum Computer basically works on several quantum physical features. It primarily focuses on developing computer technology using the principles of quantum theory, which tenders the nature of energy and substance and its behavior on the quantum level (atomic/subatomic level). Several classical and other approaches are used in image analysis. In the last few years, many researchers have developed different techniques that can efficiently be applied in various fields of image analysis.

The foremost contributions of this book are to focus on the development of quantum inspired meta-heuristic algorithms by suggesting modifications of classical meta-heuristics with due recourse to basic quantum features and thresholding techniques. The primary contributions are concentrated on the realization of greater efficacy and applicability of these works to the task of bi-level, multi-level, gray-scale and color image thresholding. Efforts have been made to resolve the inherent computational complexities of conventional algorithms regarding the thresholding of multi-level, gray-scale and color images. The shortcoming of conventional algorithms has been addressed by apposite modifications. The possible modification of the functional characteristics of the conventional algorithms was suggested in Chapter 4. In this chapter, genetic algorithm and particle swarm optimization-based quantum inspired algorithms were introduced. The proposed approaches targeted in the design of quantum inspired

Quantum Inspired Meta-heuristics for Image Analysis, First Edition.
Sandip Dey, Siddhartha Bhattacharyya, and Ujjwal Maulik.
© 2019 John Wiley & Sons Ltd. Published 2019 by John Wiley & Sons Ltd.

meta-heuristic algorithms incorporate the basic principles of a quantum mechanical system into the algorithmic structure. These algorithms have been developed to find the optimal threshold value with several thresholding techniques as standard objective functions used in this process. The different steps of these algorithms have been illustrated in detail. The effectiveness and usefulness of the proposed approaches were exhibited with application to the thresholding of bi-level images. The performance of the proposed algorithms has been strengthened by using the features of quantum computing. Continuing with the bi-level image thresholding, the aim of this book also is the development of several quantum inspired meta-heuristic algorithms for the thresholding of gray-scale images. Several quantum versions of the meta-heuristics based optimization algorithms have been developed to find the optimal threshold values in the multi-level and gray-scale domains and these algorithms are presented in Chapter 5. The application of these algorithms has been demonstrated on several synthetic/real-life gray-scale images. The effectiveness of the proposed approaches has been established with the assistance of a standard objective function for the underlying optimization problem of multi-level image thresholding. An investigation has been carried out to judge the performance of the participating algorithms in different directions. It reveals that the proposed quantum inspired particle swarm optimization outperforms the others. It can be noted that parameters with different combinations have been tuned for a number of times for the final assessment.

As an extension of multi-level and gray-scale image thresholding, this book also intends to develop several quantum inspired meta-heuristic algorithms for the thresholding of pure color images. A parallel extension of the algorithmic structures of the proposed works of Chapter 5 has been made. The parallel versions of several meta-heuristic-based quantum inspired algorithms have been presented in Chapter 6. Note that a pure color image is defined by three primary color components, namely, *Red, Green* and *Blue* information. These modified versions are capable of handling this component level image information in parallel. The proposed algorithms find the optimal threshold values of color images by utilizing the component level processing of these primary color components. The application of the proposed quantum inspired algorithms was exhibited on a set of real-life true color images. The performance of the proposed approaches has been proved using a number of standard objective functions. Different combinations of parameter specifications were used for experimental purposes. Another notable contribution of this book is to throw light on the process of upgrading the mode of optimization procedures for the gray-scale image thresholding. First, an NSGA-II-based quantum inspired algorithm has been developed in the multi-objective perspective. The proposed algorithm is designed for multi-level gray-scale image thresholding and is presented in Chapter 5. Later, other forms of meta-heuristic-based quantum inspired multi-objective algorithms have been suggested in this chapter. These algorithms are developed to find the optimal threshold values for both bi-level and multi-level gray-scale image thresholding. The proposed NSGA-II-based quantum inspired algorithm was compared with other notable multi-objective algorithms using a standard objective function. Application of this proposed approach is exhibited on several real-life gray-scale images and Berkeley images (Benchmark dataset). The superiority of this proposed approach is established quantitatively and visually in different aspects. For other approaches, a wide investigation has been made to demonstrate their application on real-life bi-level and multi-level gray scale images. The efficiency of these algorithms has also been established visually and quantitatively using a number of standard objective functions and several combinations of parameter settings.

In this book, several algorithms have been built for bi-level and multi-level image thresholding on the gray-scale and color domains. The prime motivation of the book is to develop efficient quantum inspired meta-heuristic algorithms of the aforesaid versions. The inherent drawbacks of computational paradigm for conventional algorithms have been addressed and resolved to a greater extent. Special attention has been taken to avoid the shortcomings of conventional algorithms with regard to space and time efficiency for bi-level, multi-level, gray-scale, and color image segmentation. For this purpose, the concept of developing quantum inspired algorithms has come in handy. Within the scope of the book, the aim is not confined to developing quantum algorithms in the process of image thresholding. An attempt has been made in the design of quantum version of conventional algorithms by incorporating the basic principles of quantum computing into the algorithmic structure.

Although extensive research work has been accomplished so far, scope always exists for future research. Therefore, a substantial study of the performance evaluation with other comparable strategies can be conducted in a systematic approach. Challenges can be acknowledged in the development of quantum inspired algorithms for automatic image segmentation in single-objective and multi-objective perspective. Needless to say, medical images (MRI, PET, CT) play a significant role in medical science nowadays. A comprehensive study can be conducted on the application of the proposed algorithms in this area of application. Automatic segmentation of images of this category would be a challenging task. The application of the proposed algorithms can be extended to processing of 3D images. It would be an interesting and promising task to apply W-state in the process of development of these algorithms for expected better performances.

Bibliography

1 Benchmark dataset. https://www.eecs.berkeley.edu/Research/Projects/CS/vision/bsds/BSDS300/html/dataset/images.html, 2016.

2 T. Abak, U. Baris, and B. Sankur. The performance of thresholding algorithms for optical character recognition. *Proceedings of International Conference on Document Analysis and Recognition (ICDAR97)*, pages 697–700, 1997.

3 A.S. Abutaleb. Automatic thresholding of gray-level pictures using two-entropy. *Computer Vision, Graphics, Image Processing*, 47:22–32, 1989.

4 F. Alfares and I.I. Esat. Quantum algorithms: how useful for engineering problems? In: *Proceedings of the Seventh World Conference on Integrated Design and Process Technology, Austin, Texas, USA*, pages 669–673, 2003.

5 E. Alpaydin. *Introduction to Machine Learning*. The MIT Press, Cambridge, MA, 2010.

6 E. Alpaydin and F. Gürgen. Introduction to pattern recognition and classification in medical and astrophysical images. In: C. T. Leondes (ed.), *Image Processing and Pattern Recognition*, Academic Press, San Diego, 1998.

7 P. Angeline. Using selection to improve particle swarm optimization. In: *Proceedings of the IEEE Congress on Evolutionary Computation, Anchorage, AK*, pages 84–89, 1998.

8 T.P.M. Araujo, N. Nedjah, and L. de. M. Mourelle. Quantum-inspired evolutionary state assignment for synchronous finite state machines. *Journal of Universal Computer Science*, 14(15):2532–2548, 2008.

9 A.Z. Arifin and A. Asano. Image segmentation by histogram thresholding using hierarchical cluster analysis. *Pattern Recognition Letters*, 27(13): 1515–1521, 2006.

10 S. Aroraa, J. Acharyab, A. Vermac, and P.K. Panigrahic. Multilevel thresholding for image segmentation through a fast statistical recursive algorithm. *Pattern Recognition Letters*, 29(2):119–125, 2008.

11 C. Audet and D.E. Dennis. Mesh adaptive DS algorithms for constrained optimization. *Siam Journal of Optimization*, 17(1):188–217, 2006.

12 S.K.C. Aytekin and M. Gabbouj. Quantum mechanics in computer vision: Automatic object extraction. In: *Proceedings of the 20th IEEE International Conference on Image Processing (ICIP), Melbourne, VIC*, pages 2489–2493, 2013. doi: 10.1109/ICIP.2013.6738513.

13 T. Bäck and F. Hoffmeister. Extended selection mechanisms in genetic algorithms. In: *Proceedings of the Fourth International Conference on Genetic Algorithms*, Morgan Kaufmann, San Mateo, CA, pages 92–99, 1991.

Quantum Inspired Meta-heuristics for Image Analysis, First Edition.
Sandip Dey, Siddhartha Bhattacharyya, and Ujjwal Maulik.
© 2019 John Wiley & Sons Ltd. Published 2019 by John Wiley & Sons Ltd.

14 T. Back, D. Fogel, and Z. Michalewicz. *Handbook of Evolutionary Computation.* IOP Publishing Ltd, Bristol, UK, 1997.

15 J. Bader and E. Zitzler. Hype: An algorithm for fast hypervolume-based many-objective optimization. *Evolutionary Computation,* 19(1):45–76, 2011.

16 S. Bag and G. Harit. Topographic feature extraction for Bengali and Hindi character images. *International Journal of Signal & Image Processing,* 2 (2), 2011.

17 J.E. Baker. Adaptive selection methods for genetic algorithms. In: *Proceedings of the International Conference on Genetic Algorithms and their Applications,* pages 101–111, 1985.

18 S. Bandyopadhyay and S.K. Pal. *Classification and Learning Using Genetic Algorithms: Application in Bioinformatics and Web Intelligence.* Springer, Berlin, 2007.

19 S. Bandyopadhyay and S. Saha. *Unsupervised Classification.* Springer, Berlin, 2013.

20 S. Bandyopadhyay, U. Maulik, and M.K. Pakhira. Clustering using simulated annealing with probabilistic redistribution. *International Journal of Pattern Recognition and Artificial Intelligence,* 15(2):269–285, 2001.

21 S. Bandyopadhyay, U. Maulik, and A. Mukhopadhyay. Multiobjective genetic clustering for pixel classification in remote sensing imagery. *IEEE Transactions on GeoScience and Remote Sensing,* 45(5):1506–1511, 2007.

22 M.R. Banham and A.K. Katsaggelos. IEEE signal processing magazine. *Neuron,* 14(2):24–41, 1997. doi: 10.1016/j.neuron.2012.03.002.

23 P. Benioff. Quantum mechanical models of Turing machines that dissipate no energy. *Physical Review Letters,* 48(23):1581–1585, 1982.

24 J.C. Bezdek. On the relationship between neural networks, pattern recognition and intelligence. *International Journal of Approximate Reasoning,* 6:85–107, 1992.

25 S. Bhandarkar and H. Zhang. Image segmentation using evolutionary computation. *IEEE Transactions on Evolutionary Computation,* 3(1): 1–20, 1999.

26 S. Bhattacharyya. Object Extraction In A Soft Computing Framework. Ph.D thesis, Jadavpur University, India, 2007.

27 S. Bhattacharyya. A brief survey of color image preprocessing and segmentation techniques. *Journal of Pattern Recognition Research,* 1: 120–129, 2011.

28 S. Bhattacharyya and S. Dey. An efficient quantum inspired genetic algorithm with chaotic map model based interference and fuzzy objective function for gray level image thresholding. *Proceedings of International Conference on Computational Intelligence and Communication Networks (CICN),* Gwalior, India, pages 121–125, 2011.

29 S. Bhattacharyya, U. Maulik, and S. Bandyopadhyay. Soft computing and its applications. In: Y. Dai, B. Chakraborty, and M. Shi (eds) *Kansei Engineering and Soft Computing: Theory and Practice,* IGI Global, Hershey, PA, USA, 2011.

30 F. Bin, Z. Wang, and J. Sun. Niche quantum-behaved particle swarm optimization with chaotic mutation operator. *Computer Applications and Software,* 26(1):50–52, 2009.

31 C. Blum and A. Roli. ACM computing surveys (CSUR). *Metaheuristic in Combinatorial Optimization: Overview and Conceptual Comparison,* 35 (3):268–308, 2003.

32 C.-W. Bong and M. Rajeswari. Multi-objective nature-inspired clustering and classification techniques for image segmentation. *Applied Soft Computing,* 11(4):3271–3282, 2011.

33 A. Boyarsky and P. Góra. A random map model for quantum interference. *Communications in Nonlinear Science and Numerical Simulation*, 15(8): 1974–1979, 2010.

34 A.D. Brink. Gray level thresholding of images using a correlation criterion. *Pattern Recognition*, 9(5):335–341, 1989.

35 A.D. Brink and N.E. Pendock. Minimum cross entropy threshold selection. *Pattern Recognition*, 29(1):179–188, 1996.

36 Y. Brink, L. Bruzzone, and F. Melgani. Image thresholding based on the EM algorithm and the generalized Gaussian distribution. *Pattern Recognition*, 40(2):619–634, 2007.

37 R. Brits, A. Engelbrecht, and F. van den Bergh. A niching particle swarm optimizer. In: *Proceedings of the 4th Asia-Pacific Conference on Simulated Evolutionary Learning, Singapore*, pages 692–696, 2002.

38 C.J.C. Burges. A tutorial on support vector machines for pattern recognition. *Data Mining and Knowledge Discovery*, 2(2):121–167, 1998.

39 L. Cao, P. Bao, and Z.K. Shi. The strongest schema learning GA and its application to multilevel thresholding. *Image and Vision Computing*, 26 (5):716–724, 2008.

40 R. Caves, S. Quegan, and R. White. Quantitative comparison of the performance of SAR segmentation algorithms. *The Journal of Chemical Physics*, 7(11):1534–1546, 1998.

41 F. Chan, F. Lam, and H. Zhu. Adaptive thresholding by variational method, *IEEE Transactions on Image Processing*, 7(3): 468–473.

42 S. Chatterjee and A. Bhattacherjee. Genetic algorithms for feature selection of image analysis-based quality monitoring model: an application to an iron mine. *Engineering Applications of Artificial Intelligence*, 24(5): 786–795, 2011.

43 H.C. Chen, W.J. Chien, and S.J. Wang. Contrast-based color image segmentation. *IEEE Signal Processing Letters*, 11(7):641–644, 2004.

44 H. Cheng, J. Shan, W. Ju, Y. Guo, and L. Zhang. Automated breast cancer detection and classification using ultrasound images: a survey. *Pattern Recognition*, 43(1):299–317, 2010.

45 P. Civicioglu. Backtracking search optimization algorithm for numerical optimization problems. *Applied Mathematics and Computation*, 219(15): 8121–8144, 2013.

46 C.A. Coello and N. CruzCortes. Solving multiobjective optimization problems using an artificial immune system. *Genetic Programming and Evolvable Machines*, 6(2):163–190, 2005.

47 C.C. Coello. Evolutionary multiobjective optimization: a historical view of the field. *IEEE Computational Intelligence Magazine*, 1(1):28–36, 2002.

48 W.J. Conover. *Practical Nonparametric Statistics*. edition, Wiley, Chichester, UK, 1999.

49 D.W. Corne, J.D. Knowles, and M.J. Oates. The Pareto envelope-based selection algorithm for multi-objective optimization. In: *Parallel Problem Solving from Nature (PPSN VI), LNCS*, Springer, Heidelberg, 1917: 839–848, 2000.

50 C. Cortes and V.N. Vapnik. Support vector networks. *Machine Learning*, 20(3):273–297, 1995.

51 E. Cuevas, D. Zaldivar, M. Pérez-Cisneros, and M. Ortega-Sánchez. Circle detection using discrete differential evolution optimization. *Pattern Analysis and Applications*, 14(1):93–107, 2011.

52 P. Czyzak and A. Jaszkiewicz. Pareto simulated annealing - a metaheuristic technique for multiple-objective combinatorial optimization. *Journal of Multi-Criteria Decision Analysis*, 7:34–47, 1998.

53 L. Davis. *Handbook of Genetic Algorithms*. Van Nostrand Reinhold, New York, 1991.

54 K. Deb. *Multi-objective optimization using evolutionary algorithms*. John Wiley & Sons Ltd., Chichester, UK, 2001.

55 K. Deb, A. Pratap, S. Agarwal, and T. Meyarivan. A fast elitist nondominated sorting genetic algorithm for multi-objective optimization: NSGA-II. *Proceedings of the Parallel Problem Solving from Nature (PPSN VI)*, pages 849–858, 2000.

56 K. Deb, A. Pratap, S. Agarwal, and T. Meyarivan. A fast and elitist multiobjective genetic algorithm: NSGA-II. *IEEE Transactions on Evolutionary Computation*, 6(2):182–197, 2002.

57 D. Deutsch. Quantum theory, the Church-Turing principle and the universal quantum computer. *Proceedings of the Royal Society of London. Series A, Mathematical and Physical Sciences*, 400(1818):97–117, 1985.

58 D. Deutsch and R. Jozsa. Rapid solution of problems by quantum computation. *Proceedings of the Royal Society of London. Series A, Mathematical and Physical Sciences*, 439(1907):553–558, 1992.

59 S.K. Dewangan. Human authentication using biometric recognition. *International Journal of Computer Science & Engineering Technology (IJCSET)*, 6(4), 2015.

60 D. Dey, S. Bandyopadhyay, and U. Maulik. Quantum behaved multi-objective PSO and ACO optimization for multi-level thresholding. In: *Proceedings of International Conference on Computational Intelligence and Communication Networks (ICCICN 2014), RCCIIT, Kolkata, India*, pages 242–246, 2014.

61 S. Dey, S. Bhattacharyya, and B.K. Maulik, U. Chaotic map model-based interference employed in quantum inspired genetic algorithm to determine the optimum gray level image thresholding. In: U. Tripathy and D.P. Acharjya (eds) *Global Trends in Intelligent Computing Research and Development*, IGI Global, Hershey, PA, 2013.

62 S. Dey, S. Bhattacharyya, and U. Maulik. Quantum inspired meta-heuristic algorithms for multi-level thresholding for true colour images. In: *Proceedings of the IEEE Indicon 2013, Mumbai, India*, pages 1–6, 2013.

63 S. Dey, I. Saha, S. Bhattacharyya, and U. Maulik. New quantum inspired meta-heuristic methods for multi-level thresholding. In: *Proceedings of the 2013 International Conference on Advances in Computing, Communications and Informatics (ICACCI)*, 2013.

64 S. Dey, S. Bhattacharyya, and U. Maulik. Quantum inspired genetic algorithm and particle swarm optimization using chaotic map model based interference for gray level image thresholding. *Swarm and Evolutionary Computation*, 15:38–57, 2014.

65 S. Dey, S. Bhattacharyya, and U. Maulik. New quantum inspired tabu search for multi-level colour image thresholding. In: *Proceedings of 8th International Conference on Computing for Sustainable Global Development (INDIACom-2014), BVICAM, New Delhi*, pages 311–316, 2014.

66 S. Dey, S. Bhattacharyya, and U. Maulik. Quantum behaved multi-objective pso and aco optimization for multi-level thresholding. *Proceedings of the International Conference on Computational Intelligence and Communication Networks (ICCICN 2014), RCCIIT, Kolkata, India*, pages 242–246, 2014.

67 S. Dey, I. Saha, S. Bhattacharyya, and U. Maulik. Multi-level thresholding using quantum inspired meta-heuristics. *Knowledge-Based Systems*, 67: 373–400, 2014.

68 S. Dey, S. Bhattacharyya, and S. Maulik. Optimum gray level image thresholding using a quantum inspired genetic algorithm. In: S. Bhattacharyya, P. Banerjee, D. Majumdar, and P. Dutta (eds) *Handbook of Advanced Research on Hybrid Intelligent Techniques and Applications*, IGI Global, Hershey, PA, 2016.

69 S. Dey, S. Bhattacharyya, and S. Maulik. Quantum inspired multi-objective SA for bi-level image thresholding. In: S. Bhattacharyya, U. Maulik, and P. Dutta (eds) *Quantum Inspired Computational Intelligence: Research and Applications*, Morgan-Kaufmann, Burlington, MA, 2016.

70 S. Dey, S. Bhattacharyya, and U. Maulik. New quantum inspired meta-heuristic techniques for multi-level colour image thresholding. *Applied Soft Computing*, 46:677–702, 2016.

71 S. Dey, S. Bhattacharyya, and U. Maulik. Quantum inspired multi-objective SA for bi-level image thresholding. In: U. Maulik, S. Bhattacharyya, and P. Dutta (eds) *Quantum Inspired Computational Intelligence: Research and Applications*, Morgan-Kaufmann, Burlington, MA, 2016.

72 S. Dey, S. Bhattacharyya, and U. Maulik. Efficient quantum inspired meta-heuristics for multi-level true colour image thresholding. *Applied Soft Computing*, 56:472–513, 2017.

73 S. Dey, S. Bhattacharyya, and U. Maulik. *Quantum Inspired Nondominated Sorting Based Multi-objective GA for Multi-level Image Thresholding*. World Scientific Publishing Co. Pte. Ltd., Singapore, 2018.

74 S. Dey, S. Bhattacharyya, and U. Maulik. Quantum inspired nondominated sorting based multi-objective GA for multi-level image thresholding. *Hybrid Metaheuristics: Research and Applications*, pages 141–170, 2018.

75 K. Doksum. Robust procedures for some linear models with one observation per cell. *Annals of Mathematical Statistics*, 38(3):878–883, 2010.

76 M. Dorigo, V. Maniezzo, and A. Colorni. The ant system: Optimization by a colony of cooperating agents. *IEEE Transactions on Systems, Man, Cybernetics-Part B*, 26(1):29–41, 1996.

77 L. Dos and S. Coelho. A quantum particle swarm optimizer with chaotic mutation operator. *Chaos, Solitons & Fractals*, 37(5):1409–1418, 2008.

78 G. Dougherty. Image analysis in medical imaging: recent advances in selected examples. *Biomedical Imaging and Intervantion Journal*, 6(3), 2010.

79 C.-J. Du and D.-W. Sun. Object classification methods. In D.-W. Sun (ed.), *Computer Vision Technology for Food Quality Evaluation*, Academic Press, Burlington, 2007.

80 R.O. Duda, P.E. Hart, and D.G. Stork. *Pattern classification*, 2nd edition. John Wiley and Sons, New York, 2001.

81 R.P.W. Duin and D.M.J. Tax. Statistical pattern recognition. In C. H. Chen and P.S.P. Wang (eds), *Handbook of Pattern Recognition and Computer Vision*. Singapore: World Scientific, Singapore, 2005.

82 M. Egmont-Petersen, D. De. Ridder, and H. Handels. Image processing with neural networks-a review. *Pattern Recognition*, 35(10):2279–2301, 2002.

83 A.E. Eiben and J.E. Smith. *Introduction to Evolutionary Computing*. Springer, Berlin, 2003.

84 M. Emmerich, N. Beume, and B. Naujoks. *An EMO Algorithm Using the Hypervolume Measure as Selection Criterion.* Springer-Verlag, Berlin, 2005.

85 L.J. Eshelman and J.D. Schaffer. Real-coded genetic algorithms and interval schemata, volume 2. In; L. Whitley (ed), *Foundation of Genetic Algorithms,* Morgan Kaufmann, San Mateo, CA, 2007.

86 U. Faigle and W. Kern. Some convergence results for probabilistic tabu search. *ORSA Journal on Computing,* 4(1):32–37, 1992.

87 P.F. Felzenszwalb and D.P. Huttenlocher. Efficient graph-based image segmentation. *International Journal of Computer Vision,* 59(2):167–181, 2004.

88 S. Fengjie, W. He, and F. Jieqing. 2d Otsu segmentation algorithm based on simulated annealing genetic algorithm for iced-cable images. In: *Proceedings of International Forum on Information Technology and Applications (IFITA 2009),* Chengdu, China, 2:600–602, 2009.

89 R. Feynman. Simulating physics with computers. *International Journal of Theoretical Physics,* 21(6):467–488, 1982.

90 C.B. Filho, C.A. Mello, J. Andrade, M. Lima, W.D. Santos, A. Oliveira, and D. F. (Fritzsche). Image thresholding of historical documents based on genetic algorithms. *Tools in Artificial Intelligence,* pages 93–100, 2008.

91 L.J. Fogel. On the organization of intellect. Ph.D thesis, University of California (UCLA): Los Angeles, CA, USA, 1964.

92 C.M. Fonseca and P.J. Fleming. Genetic algorithms for multiobjective optimization: Formulation, discussion and generalization. In: S. Forrest (ed.), *Proceedings of the Fifth International Conference on Genetic Algorithms,* San Mateo, CA: Morgan Kauffman, pages 416–423, 1993.

93 B. Fox. Integrating and accelerating tabu search, simulated annealing and genetic algorithms. *Annals of Operations Research,* 41(2):47–67, 1993.

94 M. Friedman. The use of ranks to avoid the assumption of normality implicit in the analysis of variance. *Journal of the American Statistical Association,* 31(200):675–701, 1937.

95 M. Friedman. A comparison of alternative tests of significance for the problem of m rankings. *Annals of Mathematical Statistics,* 11(1):86–92, 1940.

96 M. Friedman and A. Kandel. *Introduction to Pattern Recognition: Statistical, Structural, Neural and Fuzzy Logic Approaches.* Imperial College Press, London, 2001.

97 K.S. Fu and J.K. Mu. A survey of image segmentation. *Pattern Recognition,* 13(1):3–16, 1981.

98 M. García, A. Fernández, J. Luengo, and F. Herrera. Advanced nonparametric tests for multiple comparisons in the design of experiments in computational intelligence and data mining: Experimental analysis of power. *Information Sciences,* 180(10):2044–2064, 2010.

99 G.E.J. Garrido, S.S. Furuie, and A.C.F. Orgambide. Refinement of left ventricle segmentation in MRI based on simulated annealing. In: *Proceedings of the 20th Annual International Conference of the IEEE Engineering in Medicine and Biology Society, Hong Kong SAR, China,* 2: 598–601, 1998.

100 S. Geman and D. Geman. Stochastic relaxation, Gibbs distributions and the Bayesian restoration of images. *IEEE Transactions on Pattern Analysis and Machine Intelligence,* 6(6):721–741, 1984.

101 L. El Ghaoui, F. Oustry, and M. AitRami. A cone complementary linearization algorithm for static output-feedback and related problems. *IEEE Transactions on Automatic Control*, 42(8):1171–1176, 1997.

102 A. Ghosh and S.K. Pal. Neural network, self-organization and object extraction,. *Pattern Recognition Letters*, 13(5):387–397, 1992.

103 A. Ghosh and A. Sen. *Soft Computing Approach to Pattern Recognition and Image Processing*. World Scientific, Singapore, 2002.

104 A. Ghosh, N.R. Pal, and S.K. Pal. Self-organization for object extraction using a multilayer neural network and fuzziness measures. *IEEE Transactions on Fuzzy Systems*, 1(1): 54–68, 1993.

105 F. Glover. Tabu search, part I. *ORSA Journal on Computing*, 1(3): 190–206, 1989.

106 F. Glover. Tabu search, part II. *ORSA Journal on Computing*, 2(1):4–32, 1990.

107 F. Glover. Tabu search: A tutorial. *Interfaces*, 20(4):74–94, 1990.

108 F. Glover and A. Kochenberger. *Handbook on Metaheuristics*. Kluwer Academic Publishers, New York, 2003.

109 F. Glover and M. Laguna. *Modern Heuristic Techniques for Combinatorial Problems (Tabu Search)*. John Wiley & Sons, Inc., New York, NY, USA, 1993.

110 D.E. Goldberg. *Genetic Algorithms in Search Optimization and Machine Learning*. Addison-Wesley Longman Publishing Co., Inc., Boston, MA, USA, 1989.

111 D.E. Goldberg and K. Deb. A comparative analysis of selection schemes used in genetic algorithms. In: G.J.E. Rawlins (Ed.), *Foundations of Genetic Algorithms*, Morgan Kaufmann, Los Altos, 1991.

112 R.C. Gonzalez and R.E. Woods. *Digital Image Processing*. Prentice-Hall, Englewood Cliffs, NJ, 2002.

113 M. Gravel, W.L. Price, and C. Gagné. Scheduling continuous casting of aluminium using a multiple objective ant colony optimization metaheuristic. *European Journal of Operational Research*, 142(1): 218–229, 2002.

114 I. Grigorenko and M. Garcia. Ground-state wave functions of two-particle systems determined using quantum genetic algorithms. *Physica A, Statistical Mechanics and its Applications*, 291(1–4):439–441, 2001.

115 I. Grigorenko and M. Garcia. Calculation of the partition function using quantum genetic algorithms. *Physica A, Statistical Mechanics and its Applications*, 313(3–4):463–470, 2002.

116 F. Glover. Future paths for integer programming and links to artificial intelligence. *Computers and Operations Research*, 13(5):533–549, 1986.

117 L. K. Grover. Quantum computers can search rapidly by using almost any transformation. *Physical Review Letters*, 80(19):4329–4332, 1998.

118 I. Guyon and A. Elisseeff. An introduction to variable and feature selection. *Journal of Machine Learning Research*, 3(3):1157–1182, 2003.

119 K. Hammouche, M. Diaf, and P. Siarry. A multilevel automatic thresholding method based on a genetic algorithm for a fast image segmentation. *Computer Vision and Image Understanding*, 109(2): 163–175, 2008.

120 K. Hammouche, M. Diaf, and P. Siarry. A comparative study of various meta-heuristic techniques applied to the multilevel thresholding problem. *Engineering Applications of Artificial Intelligence*, 23(5):678–688, 2010.

121 K.H. Han and J.H. Kim. A multilevel automatic thresholding method based on a genetic algorithm for a fast image segmentation. *IEEE Transaction on Evolutionary Computation*, 6(6):580–593, 2002.

122 K.H. Han and J.H. Kim. Quantum-inspired evolutionary algorithms with a new termination criterion, hepsilon gate, and two-phase scheme. *IEEE Transaction on Evolutionary Computation*, 8(2):156–169, 2004.

123 M. Hapke, A. Jaszkiewicz, and R. Slowinski. Pareto simulated annealing for fuzzy multiobjective combinatorial optimization. *Journal of Heuristics*, 6: 329–345, 2000.

124 H. Hirsh. A quantum leap for AI. *IEEE Intelligent System*, 14(4):9–16, 1999.

125 T. Hogg. Highly structured searches with quantum computers. *Physical Review Letters*, 80(11):2473–2476, 1998.

126 T. Hogg and D.A. Portnov. Quantum optimization. *Information Sciences*, 128(3):181–197, 2000.

127 J.H. Holland. *Adaptation in Natural and Artificial Systems.* University of Michigan Press, Ann Arbor, 1975.

128 J. Horn, N. Nafpliotis, and D.E. Nafpliotis. A niched Pareto genetic algorithm for multiobjective optimization. *Proceedings of the First IEEE Conference on Evolutionary Computation, IEEE World Congress on Computational Intelligence, Piscataway, New Jersey, IEEE Service Center*, pages 82–87, 1994.

129 D.Y. Huang and C.H Wang. Optimal multi-level thresholding using a two-stage otsu optimization approach. *Pattern Recognition Letters*, 30(3): 275–284, 2009.

130 L.K. Huang and M.J.J. Wang. Image thresholding by minimizing the measures of fuzziness. *Pattern Recognition*, 28(1):41–51, 1995.

131 N. Ikonomakis, K.N. Plataniotis, and A.N. Venetsanopoulos. Color image segmentation for multimedia applications. *Journal of Intelligent and Robotic Systems*, 28(1–2):5–20, 2000.

132 A.K. Jain. *Fundamentals of Digital Image Processing.* Englewood Cliffs, NJ: Prentice-Hall, 1991.

133 A.K. Jain, M.N. Murty, and P.J. Flynn. Data clustering: A review. *ACM Computing Surveys*, 31(3):264–323, 1999.

134 L. Jiao, J. Liu, and W. Zhong. An organizational coevolutionary algorithm for classification. *IEEE Transaction on Evolutionary Computation*, 10(1): 67–80, 2006.

135 G. Johannsen and J. Bille. A threshold selection method using information measures. In: *Proceedings 6th International Conference on Pattern Recognition*, Munich, Germany, pages 140–143, 1982.

136 A.K. De Jong. *Evolutionary computation: a unified approach.* MIT Press, Cambridge, MA, 2006.

137 S. Kak. Quantum computing and AI. *IEEE Intelligent System*, pages 9–11, 1999.

138 M. Kalakech, L. Macaire, P. Biela, and D. Hamad. Constraint scores for semisupervised feature selection: a comparative study. *Pattern Recognition Letters*, 32(5):656–665, 2011.

139 A. Kandel. *Mathematical Techniques with Applications.* Addison-Wesley, New York, 1986.

140 J.N. Kapur, P.K. Sahoo, and A.K.C. Wong. A new method for gray-level picture thresholding using the entropy of the histogram. *Computer Vision, Graphics, and Image Processing*, 29(3):273–285, 1985.

141 N. Kasamov. *Foundations of Neural Networks, Fuzzy Systems and Knowledge Engineering*. Addison-Wesley, Cambridge, MA, 1996.

142 T. Kavzoglu and P.M. Mather. The use of backpropagating artificial neural networks in land cover classification. *International Journal of Remote Sensing*, 24(23):4907–4938, 2003.

143 J.M. Keller, P. Gader, H. Tahani, J.-H. Chiang, and M. Mohamed. Advances in fuzzy integration for pattern recognition. *Fuzzy Sets and Systems*, 65(2–3):273–283, 1994.

144 K. Kennedy and R. Eberhart. Particle swarm optimization. In: *Proceedings of the IEEE International Conferenceon Neural Networks (ICNN95), Perth, Australia*, 4:1942–1948, 1995.

145 F.J. Kenney, and E.S. Keeping. *Mathematics of Statistics*, Pt. 2, 2nd ed. Princeton, NJ: Van Nostrand, 1951.

146 K.H. Kim and J.H. Kim. Genetic quantum algorithm and its application to combinatorial optimization problem. In: *Proceedings of 2000 Congress on Evolutionary Computation, La Jolla, CA*, 2:1354–1360, 2000.

147 S. Kirkpatrick, C.D. Gelatt, and M.P. Vecchi. Optimization by simulated annealing. *Science*, 220(4598):671–680, 1983.

148 J. Kittler and J. Illingworth. Minimum error thresholding. *Pattern Recognition*, 19(1):41–47, 1986.

149 J.D. Knowles and D.W. Corne. Approximating the nondominated front using the Pareto archived evolution strategy. *Evolutionary Computation*, 8(2):149–172, 2000.

150 B. Kosko. *Neural Networks and Fuzzy Systems: A Dynamical Systems Approach to Machine Intelligence*. Prentice-Hall, New Jersey, 1992.

151 S. Kullback. *Information Theory and Statistics*. Dover, New York, 1968.

152 F. Kurugollu, B. Sankar, and A.E. Harmanc. Color image segmentation using histogram multithresholding and fusion. *Image and Vision Computing*, 19:915–928, 2001.

153 S.H. Kwon. Pattern recognition letters. *Mathematics of Computation*, 25 (9):1045–1050, 2004.

154 J.C Lagarias, J.A. Reeds, M.H. Wright, and M.E. Wright. Convergence properties of the Nelder-Mead simplex method in low dimensions. *SIAM Journal of Optimizationl*, 9(1):112–147, 1998.

155 J. Lan, M.Y. Hu, E. Patuwo, and G.P. Zhang. An investigation of neural network classifiers with unequal misclassification costs and group sizes. *Decision Support Systems*, 48(4):581–591, 2010.

156 P. Langley. The changing science of machine learning. *Machine Learning*, 82(3):275–279, 2011.

157 B. Li and Z.Q. Zhuang. Genetic algorithm based on the quantum probability representation. *Intelligent Data Engineering and Automated Learning IDEAL 2002*, 2412:500–505, 2002.

158 C.H. Li and C.K. Lee. Minimum cross-entropy thresholding. *Pattern Recognition*, 26(4):617–625, 1993.

159 Z. Li, C. Liu, G. Liu, X. Yang, and Y. Cheng. Statistical thresholding method for infrared images. *Pattern Analysis & Applications*, 14(2):109–126, 2011.

160 J.J. Liang, A.K. Qin, P.N. Suganthan, and S. Baskar. Comprehensive learning particle swarm optimizer for global optimization of multimodal functions. *IEEE Transactions on Evolutionary Computation*, 10(3): 281–295, 2006.

161 P.S. Liao, T.S. Chen, and P.C. Chung. A fast algorithm for multi-level thresholding. *Journal of Information Science and Engineering*, 17:713–723, 2001.

162 J. Lie, M. Lysaker, and X. Tai. A variant of the level set method and applications to image segmentation. *Mathematics of Computation*, 75: 1155–1174, 2006.

163 E. López-Ornelas. High resolution images: Segmenting, extracting information and GIS integration. *World Academy of Science, Engineering and Technology*, 3(6):150–155, 2009.

164 M. Lukac and M. Perkowski. Evolving quantum circuits using genetic algorithm. In: *Proceedings of the NASA/DoD Conference on Evolvable Hardware, IEEE*, pages 177–185, 2002.

165 W. Luo, W. Wang, and H. Liao. Image segmentation on colonies images by a combined algorithm of simulated annealing and genetic algorithm. In: *Proceedings of Fourth International Conference on Image and Graphics (ICIG 2007), Chengdu, China*, pages 342–346, 2007.

166 M.M. Tabb and N. Ahuja. Multiscale image segmentation by integrated edge and region detection. *IEEE Transactions on Image Processing*, 6(5): 642–655, 1997.

167 M. Maitra and A. Chatterjee. A hybrid cooperative comprehensive learning based pso algorithm for image segmentation using multilevel thresholding. *Expert Systems with Applications*, 34(2):1341–1350, 2008.

168 J. Malik, S. Belongie, T. Leung, and J. Shi. Contour and texture analysis for image segmentation. *International Journal of Computer Vision*, 43(1): 7–27, 2001.

169 U. Maulik and I. Saha. Modified differential evolution based fuzzy clustering for pixel classification in remote sensing imagery. *Pattern Recognition*, 42(9):2135–2149, 2009.

170 U. Maulik and I. Saha. Automatic fuzzy clustering using modified differential evolution for image classification. *IEEE Transactions on Geoscience and Remote Sensing*, 48(9):3503–3510, 2009.

171 U. Maulik, S. Bandyopadhyay, and I. Saha. Integrating clustering and supervised learning for categorical data analysis. *IEEE Transactions on Systems, Man and Cybernetics Part-A*, 40(4):664–675, 2010.

172 U. Maulik, S. Bandyopadhyay, and A. Mukhopadhyay. *Multiobjective Genetic Algorithms for Clustering: Applications in Data Mining and Bioinformatics*. Springer-Verlag, Berlin, 2011.

173 D. McMohan. *Quantum Computing Explained*. John Wiley & Sons, Inc., Hoboken, New Jersey, 2008.

174 N. Metropolis, A. Rosenbluth, M. Rosenbluth, A. Teller, and E. Teller. Equations of state calculations by fast computing machines. *The Journal of Chemical Physics*, 21(6):1087–1092, 1953.

175 Z. Michalewicz. *Genetic Algorithms + Data Structures = Evolution Programs*. Springer-Verlag, New York, 1992.

176 Z.-H. Michalopoulou, D. Alexandrou, and C. de Moustier. Application of neural and statistical classifiers to the problem of seafloor characterization. *IEEE Journal of Oceanic Engineering*, 20(3):190–197, 1995.

177 K. Miettine. *Nonlinear Multiobjective Optimization*. Kluwer Academic Publishers, Boston: MA, 1998.

178 T. Mitchell. *Machine Learning*. McGraw-Hill, Maidenhead, 1997.

179 S. Mitra and S.K. Pal. Fuzzy sets in pattern recognition and machine intelligence. *Fuzzy Sets and Systems*, 156(3):381–386, 2005.

180 N.E. Mitrakis, C.A. Topaloglou, T.K. Alexandridis, J.B. Theocharis, and G.C. Zalidis. Decision fusion of GA self-organizing neuro-fuzzy multilayered classifiers for land cover classification using textural and spectral features. *IEEE Transactions on Geoscience and Remote Sensing*, 46(7):2137–2151, 2008.

181 M.P. Moore and A. Narayanan. *Quantum-Inspired Computing*. Old Library, University of Exeter, Exeter: Department of Computer Science, 1995.

182 A. Mukhopadhyay and U. Maulik. A multiobjective approach to MR brain image segmentation. *Applied Soft Computing*, 11(1):872.–880, 2011.

183 T. Murata and H. Ishibuchi. MOGA: multi-objective genetic algorithms. In: *Proceedings of the 1995 IEEE international conference on evolutionary computation, (Perth, WA, Australia), 29 November–1 December*, 1995.

184 A. Nakib, H. Oulhadj, and P. Siarry. Non-supervised image segmentation based on multiobjective optimization. *Pattern Recognition Lettersv*, 29 (2):161–172, 2008.

185 A. Nakiba and P.H. Oulhadja, and Siarry. Image histogram thresholding based on multiobjective optimization. *Signal Processing*, 87(11): 2516–2534, 2007.

186 A. Nakiba and P.H. Oulhadja, and Siarry. Image thresholding based on Pareto multi-objective optimization. *Engineering Applications of Artificial Intelligence*, 23(3):313–320, 2010.

187 D.K. Nam and D.K. Park. Multiobjective simulated annealing: a comparative study to evolutionary algorithms. *International Journal of Fuzzy Systems*, 2(2):87–97, 2000.

188 A. Narayanan and T. Manneer. Quantum artificial neural network architectures and components. *Information Sciences*, 128(3–4):231–255, 2000.

189 A. Narayanan and M. Moore. Quantum inspired genetic algorithm. In: *Proceedings of the IEEE Conference on Evolutionary Computation (ICEC 96)*, Nayoya University, Japan, pages 61–66, 1996.

190 A.M. Nielsen, and L.I. Chuang. *Quantum Computation and Quantum Information*. Cambridge University Press, Cambridge, 2000.

191 D. Oliva, E. Cuevas, G. Pajares, D. Zaldivar, and V. Osuna. A multilevel thresholding algorithm using electromagnetism optimization. *Neurocomputing*, 139:357–381, 2014.

192 N. Otsu. A threshold selection method from gray level histograms. *IEEE Transactions on Systems, Man, and Cybernetics*, 9(1):62–66, 1979.

193 N.R. Pa and D. Bhandari. Image thresholding: some new techniques. *Signal Processing*, 33:139–158, 1993.

194 N.R. Pal and S.K. Pal. A review on image segmentation techniques. *Pattern Recognition*, 26(9):1277–1294, 1993.

195 G. Pampara, A. Engelbrecht, and N. Franken. Binary differential evolution. In: *Proceedings of the IEEE Congress on Evolutionary Computation (CEC 2006), Vancouver, BC*, pages 1873–1879, 2006.

196 R. Parpinelli, H. Lopes, and A. Freitas. Data mining with an ant colony optimization algorithm. *IEEE Transactions on Evolutionary Computing*, 6(4):321–332, 2002.

197 D. Parrott and X. Li. Locating and tracking multiple dynamic optima by a particle swarm model using speciation. *IEEE Transactions on Evolutionary Computation*, 10(4):440–458, 2006.

198 N. Patel and S.K. Dewangan. An overview of face recognition schemes. *International Conference of Advance Research and Innovation (ICARI-2015), Institution of Engineers (India), Delhi State Centre, Engineers Bhawan*, New Delhi, India, 2015.

199 S. Paterlini and T. Krink. Differential evolution and particle swarm optimisation in partitional clustering. *Computational Statistics & Data Analysis*, 50(5):1220–2006, 2006.

200 S. Patil and S. Dewangan. Neural network based offline handwritten signature verification system using Hu's moment invariant analysis. *International Journal of Computer Science & Engineering Technology (IJCSET)*, 1(1), 2011.

201 J. Pearl. *Probabilistic Reasoning in Intelligent Systems: Networks of Plausible Inference*. Morgan Kaufmann Publishers Inc., San Francisco, CA, USA, 1988.

202 J. Pearl. *Causality: Models, Reasoning and Inference*. Cambridge University Press, New York, USA, 2000.

203 J.C. Pichel, D.E. Singh, and F.F. Rivera. Image segmentation based on merging of sub-optimal segmentations. *Pattern Recognition Letters*, 27 (10):1105–1116, 2006.

204 W.K. Pratt. *Digital Image Processing*. Wiley, New York, 1991.

205 P. Pudil, J. Novoviov, and P. Somol. Feature selection toolbox software package. *Pattern Recognition Letters*, 35(12):487–492, 2002.

206 T. Pun. A new method for gray-level picture threshold using the entropy of the histogram. *Signal Processing*, 2(3):223–237, 1980.

207 N. Ramesh, J.H. Yoo, and I.K. Sethi. Thresholding based on histogram approximation. *IEE Proceedings - Vision, Image and Signal Processing*, 142(4):271–279, 1985.

208 A. Ratnaweera, S. Halgamuge, and H. Watson. Self-organizing hierarchical particle swarm optimizer with time-varying acceleration coefficients. *IEEE Transactions on Evolutionary Computation*, 8(3):240–255, 2004.

209 I. Rechenberg. Evolutions Strategie: Optimierung technischer Systeme nach Prinzipien der biologischen Evolution. Ph.D thesis, Stuttgart, Germany: Frommann-Holzboogl, 1973.

210 C.R. Reeves. Using genetic algorithms with small populations. In: *Proceedings of the Fifth International Confrcncc on Genctic Algorithms, Morgan Kaufman, San Mateo, CA, USA*, pages 92–99, 1993.

211 X. Ren. An optimal image thresholding using genetic algorithm. In: *Proceedings of International Forum on Computer Science-Technology and Applications (IFCSTA09)*, pages 169–172, 2009.

212 J.A. Richards and J. Xiuping. *Remote Sensing Digital Analysis*. Springer-Verlag, Berlin, enlarged edition, 1999.

213 A. Rosenfeld and P. de la Torre. Histogram concavity analysis as an aid in threshold selection. *IEEE Transactions on Systems Man & Cybernetics*, 13(2):231–235, 1983.

214 E.M. Rosheim. *Robot Evolution: The Development of Anthrobotics*. Wiley, Chichester, UK, 1994.

215 J.T. Ross. *Fuzzy Logic with Engineering Applications*. John Wiley & Sons Ltd, Chichester, UK, 2004.

216 J.F.B. Rylander, T. Soule, and J. Alves-Foss. Quantum evolutionary programming. In: *Proceedings of the Genetic and Evolutionary Computation Conference (GECCO-2001), Morgan Kaufmann*, pages 1005–1011, 2001.

217 H. Ryu and Y. Miyanaga. A study of image segmentation based on a robust data clustering method. *Electronics and Communications in Japan (Part III: Fundamental Electronic Science)*, 87(7):27–35, 2004.

218 I. Saha, U. Maulik, and D. Plewczynski. A new multi-objective technique for differential fuzzy clustering. *Journal of Intelligent and Robotic Systems*, 11(2):2765–2776, 2011.

219 M. Sahin, U. Atav, and M. Tomak. Quantum genetic algorithm method in self-consistent electronic structure calculations of a quantum dot with many electrons. *International Journal of Modern Physics C*, 16(9): 1379–1393, 2005.

220 P. Sahoo and J.C. Wilkins, and Yeager. Threshold selection using renyis entropy. *Pattern Recognition*, 30(1):71–84, 1997.

221 A.A.H. Salavati and S. Mozafari. Provide a hybrid method to improve the performance of multilevel thresholding for image segmentation using GA and SA algorithms. In: *Proceedings of 7th Conference onInformation and Knowledge Technology (IKT)*, pages 1–6, 2015.

222 P.D Sathya and R. Kayalvizhi. Modified bacterial foraging algorithm based multilevel thresholding for image segmentation. *Engineering Applications of Artificial Intelligence*, 24(4):595–615, 2011.

223 Y. Sayes, I. Inza, and P. Larranaga. A review of feature selection techniques in bioinformatics. *Bioinformatics*, 29(19):2507–2517, 2007.

224 J.D. Schaffer, R.A. Caruana, L.J. Eshelman, and R. Das. A study of control parameters affecting online performance of genetic algorithms for function optimization. In: *Proceedings of the Third International Conference on Genetic Algorithms, Morgan Kaufman*, pages 51–60, 1989.

225 R. J. Schalkoff. *Digital Image Processing and Computer Vision Sol*. John Wiley & Sons, Ltd, Singapore, 1992.

226 Jr. J. Schmid. The relationship between the coefficient of correlation and the angle included between regression lines. *The Journal of Educational Research*, 41(4):311–313, 1947.

227 A.R. Schowengerdt. *Remote Sensing: Models and Methods for Image Processing* 3rd edition. Academic Press, Orlando, FL, 2007.

228 G. Sedmak. Image processing for astronomy. *Meeting on Advanced Image Processing and Planetological Application, Vulcano, Italy, Sept. 16-18, 1985 Società Astronomica Italiana, Memorie (ISSN 0037-8720)*, 57(2), 1986.

229 J. Serra. *Image Analysis and Mathematical Morphology*. Academic Press, Inc., Orlando, FL, USA, 1982.

230 M. Sezgin and B. Sankar. A new dichotomization technique to multi-level thresholding devoted to inspection applications. *Pattern Recognition Letter*, 21(2):151–161, 2000.

231 M. Sezgin and B. Sankar. Survey over image thresholding techniques and quantitative performance evaluation. *Journal of Electronic Imaging*, 13 (1):146–165, 2004.

232 K.M. Shaaban and N.M. Omar. Region-based deformable net for automatic color image segmentation. *Image and Vision Computing*, 27(2): 1504–1514, 2009.

233 J.D. Shaffer. Some experiments in machine learning using vector evaluated genetic algorithms. Ph. D. thesis, Vanderbilt University, Nashville, TN, 1984.

234 A.G. Shanbag. Utilization of information measure as a means of image thresholding. *CVGIP: Graphical Models and Image Processing*, 56(5): 414–419, 1994.

235 P.W. Shor. Polynomial-time algorithms for prime factorization and discrete log-
arithms on a quantum computer. *SIAM Journal of Computing*, 26(5):1484–1509,
1997.

236 J.M.M. da Silva, R.D. Lins, and V.C. da Rocha. Binarizing and filtering historical documents with back-to-front interference. In: *Proceedings of the 2006 ACM
symposium on Applied Computing*, pages 853–858, 2006.

237 K. Smith, R. Everson, and J. Fieldsend. Dominance measures for multi-objective
simulated annealing. In: *Proceedings of 2004 IEEE Congr. Evol. Comput.* (Zitzler, E.
and Deb, K. and Thiele, L. and Coello, C. A. C. and Corne, D. eds.), pages 23–30,
2004.

238 T. Sousa, A. Silva, and A. Neves. Particle swarm based data mining algorithms for
classification tasks. *Parallel Computing*, 30:767–783, 2004.

239 L. Spector, H. Barnum, and H. Bernstein. *Advances in Genetic Programming*, volume 3. MIT Press, Englewood Cliffs, 1999.

240 N. Srinivas and K. Deb. Multiobjective optimization using nondominated sorting in
genetic algorithms. *Evolutionary Computation*, 2(3):221–248, 1994.

241 T. Stewart. Extrema selection: accelerated evolution on neural networks. In: *Proceedings of the 2001 IEEE Congress on Evolutionary Computation*, 1:25–29, 2001.

242 R. Storn and K. Price. Differential evolution: a simple and efficient heuristic for
global optimization over continuous spaces. *Journal of Global Optimization*,
11(4):341–359, 1997.

243 B. Suman. Study of self-stopping PDMOSA and performance measure in multiobjective optimization. *Computers & Chemical Engineering*, 29(5): 1131–1147,
2005.

244 B. Suman and P. Kumar. A survey of simulated annealing as a tool for single
and multiobjective optimization. *Journal of the Operations Research Society*,
57(10):1143–1160, 2006.

245 Y. Sun and G. He. Segmentation of high-resolution remote sensing image based on
marker-based watershed algorithm. In: *Proceedings of Fifth International Conference
on Fuzzy Systems and Knowledge Discovery*, 4: 271–276, 2008.

246 A. Suppapitnarm, K.A. Seffen, G.T. Parks, and P. Clarkson. A simulated annealing
algorithm for multiobjective optimization. *Engineering Optimization*, 33:59–85,
2000.

247 H. Talbi, A. Draa, and M. Batouche. A new quantum inspired genetic algorithm
for solving the traveling salesman problem. In: *Proceedings of IEEE International
Conference on Industrial Technology (ICIT04)*, 3: 1192–1197, 2004.

248 K.S. Tan and N.A.M. Isa. Color image segmentation using histogram thresholding:
fuzzy c-means hybrid approach. *Pattern Recognition*, 44(1): 1–15, 2011.

249 M. Tanaka and T. Tanino. Global optimization by the genetic algorithm in a
multi-objective decision support system. In: *Proceedings of the 10th International
Conference on Multiple Criteria Decision Making*, 2: 261–270, 1992.

250 L. Tang, K. Wang, and Y. Li. An image segmentation algorithm based on the
simulated annealing and improved snake model. In: *Proceedings of International
Conference on Mechatronics and Automation (ICMA 2007)*, Harbin, Heilongjiang,
China, pages 3876–3881, 2007.

251 W. Tao, J. Tian, and J. Liu. Image segmentation by three-level thresholding based on maximum fuzzy entropy and genetic algorithm. *Pattern Recognition Letters*, 24(16):3069–3078, 2003.

252 W.B. Tao, H. Jin, and L.M. Liu. Object segmentation using ant colony optimization algorithm and fuzzy entropy. *Pattern Recognition Letters*, 28(7):788–796, 2008.

253 A.R. Teixera, A.M. Tom, and E.W. Lang. Unsupervised feature extraction via kernel subspace techniques. *Neurocomputing*, 74(5):820–830, 2011.

254 D.R. Thedens, D.J. Skorton, and S.R. Fleagle. Methods of graph searching for border detection in image sequences with applications to cardiac magnetic resonance imaging. *IEEE Transactions on Medical Imaging*, 14 (1):42–55, 1995.

255 S. Theodoridis and K. Koutroumbas. *Pattern Recognition*, 4th edition. Academic Press, Burlington, 2009.

256 R. Tiwari, A. Shukla, C. Prakash, D. Sharma, R. Kumar, and S. Sharma. Face recognition using morphological method. In: *Proceedings of IEEE International Advance Computing Conference, Patiala, India*, 2009.

257 H.R. Tizhoosh. Image thresholding using type II fuzzy sets. *Pattern Recognition*, 38(12):2363–2372, 2005.

258 O.J. Tobias and R. Seara. Image segmentation by histogram thresholding using fuzzy sets. *IEEE Transactions on Image Processing*, 11(12): 1457–1465, 2002.

259 V.J. Torczon. Multidirectional search: A direct search algorithm for parallel machines. Ph.D thesis, Rice University, 1989.

260 A. Tremeau and N. Borel. A region growing and merging algorithm to color segmentation. *Pattern Recognition*, 30(7):1191–1203, 1997.

261 W. Tsai. Moment-preserving thresholding: a new approach. *Computer Vision, Graphics, and Image Processing*, 29(3):377–393, 1985.

262 G. Turinici, C.L. Bris, and H. Rabitz. Efficient algorithms for the laboratory discovery of optimal quantum controls. *Physical Review E: Statistical, Nonlinear, and Soft Matter Physics*, 40(016704), PMID: 15324201, 2004.

263 E.L. Ulungu, J. Teghaem, P. Fortemps, and D. Tuyttens. MOSA method: a tool for solving multiobjective combinatorial decision problems. *Journal of Multi-Criteria Decision Analysis*, 8:221–236, 1999.

264 N Vandenbroucke, L. Macaire, and J.G. Postaire. Color image segmentation by pixel classification in an adapted hybrid color space. application to soccer image analysis. *Computer Vision and Image Understanding*, 90(2):190–216, 2003.

265 D.V. Veldhuizen and G. Lamont. Multiobjective evolutionary algorithm test suites. In: *Proceedings of ACM Symposium on Applied Computing*, pages 351–357, 1999.

266 V. Vendral, M.B. Plenio, and M.A Rippin. Quantum entanglement. *Physical Review Letters*, 78(12):2275–2279, 1997.

267 D. Venturak. Quantum computational intelligence: Answers and questions. *IEEE Intelligent System*, pages 14–16, 1999.

268 S. Wang and R. Haralick. Automatic multithreshold selection. *Computer Vision, Graphics, and Image Processing*, 25:46–67, 1984.

269 Y. Rui and T.S. Huang, Image retrieval: Current techniques, promising directions and open issues. *Journal of Visual Communication and Image Representation*, 10(1): 39–62, 1999.

270 Y. Wang, Z. Cai, and Q. Zhang. Differential evolution with composite trial vector generation strategies and control parameters. *IEEE Transactions on Evolutionary Computation*, 15(1):55–56, 2011.

271 L. Wei, Y. Yang, and R.M. Nishikawa. Microcalcification classification assisted by content based image retrieval for breast cancer diagnosis. *IEEE International Conference on Image Processing*, 5:1–4, 2007.

272 T.C. Weinacht and P.H. Bucksbaum. Using feedback for coherent control of quantum systems. *Journal of Optics B: Quantum and Semiclassical Optics*, 4:R35–R52, 2002.

273 D. Whitley. The genitor algorithm and selection pressure: why rank-based allocation of reproductive trials is best. In: *Proceedings of the Third International Conference on Genetic Algorithms*, Morgan Kaufmann Publishers Inc., San Francisco, CA, USA, 37(3):116–121, 1989.

274 L.U. Wu, M.A. Songde, and L.U. Hanqing. An effective entropic thresholding for ultrasonic imaging. *Proceedings of the Fourteenth International Conference on Pattern Recognition(ICPR98), Brisbane, Qld., Australia*, 2:1552–1554, 1998.

275 F. Xue. Multi-objective differential evolution: Theory and applications. PhD thesis, Rensselaer Polytechnic Institute, Troy, New York, 2004.

276 W. Yang, C. Sun, H.S. Du, and J. Yang. Feature extraction using laplacian maximum margin criterion. *Neural Processing Letters*, 33(1):99–110, 2011.

277 J.C. Yen, F.Y. Chang, and S. Chang. A new criterion for automatic multilevel thresholding. *IEEE Transactions on Image Processing*, 4(3): 370–378, 1995.

278 P.Y. Yin. Multilevel minimum cross entropy threshold selection based on particle swarm optimization. *Applied Mathematics and Computation*, 184 (2):503–513, 2007.

279 P.Y. Yin and L.H. Chen. A fast iterative scheme for multilevel thresholding methods. *Signal Processing*, 60(3):305–313, 1997.

280 C. Yufei, Z. Weidong, and W. Zhicheng. Level set segmentation algorithm based on image entropy and simulated annealing. In: *Proceedings of 1st International Conference on Bioinformatics and Biomedical Engineering (ICBBE 2007), China*, pages 999–1003, 2007.

281 Z. Zabinsky. *Stochastic Adaptive Search for Global Optimization*. Springer, Berlin, 2003.

282 L.A. Zadeh. Fuzzy logic, neural networks and soft computing. *Communications of the ACM*, 37(3):77–84, 1994.

283 L.A. Zadeh. Fuzzy sets. *Information and Control*, 8(3):338–353, 1965.

284 E. Zahara, S.K.S. Fan, and D.M. Tsai. Optimal multi-thresholding using a hybrid optimization approach. *Pattern Recognition Letters*, 26(8): 1082–1095, 2005.

285 G.X. Zhang, W.D.J.N. Li, and L.Z. Hu. A novel quantum genetic algorithm and its application. *ACTA Electronica Sinica*, 32(3):476–479, 2004.

286 M. Zhang, L. Zhang, and H.D. Cheng. A neutrosophic approach to image segmentation based on watershed method. *Signal Processing*, 90(5): 1510–1517, 2010.

287 W.J.G. Zhang and N. Li. An improved quantum genetic algorithm and its application. *Lecture Notes in Artificial Intelligence*, 2639:449–452, 2003.

288 Y. Zhang and L. Wu. Optimal multi-level thresholding based on maximum tsallis entropy via an artificial bee colony approach. *Entropy*, 13(4): 841–859, 2011.

289 A. Zhigljavsky. *Stochastic Global Optimization*. Springer, Berlin, 2008.

290 E. Zitzler and L. Thiele. An evolutionary algorithm for multiobjective optimization: The strength Pareto approach. TIK-Report No. 43, Zurich, Switzerland, 1998.

291 E. Zitzler and L. Thiele. Multiobjective evolutionary algorithms: A comparative case study and the strength Pareto approach. *IEEE Transactions on Evolutionary Computation*, 3(4):257–271, 1999.

292 E. Zitzler and L. Thiele. Performance assessment of multiobjective optimizers: An analysis and review. *IEEE Transactions on Evolutionary Computation*, 7(2):117–132, 2003.

293 E. Zitzler, M. Laumanns, and L. Thiele. SPEA3: Improving the strength Pareto evolutionary algorithm. *IEEE Transactions on Evolutionary Computation*, 3(4):257–271, 2002.

Index

Quantum Inspired Meta-heuristics for Image Analysis, First Edition.
Sandip Dey, Siddhartha Bhattacharyya, and Ujjwal Maulik.
© 2019 John Wiley & Sons Ltd. Published 2019 by John Wiley & Sons Ltd.